ENTIRE FUNCTIONS

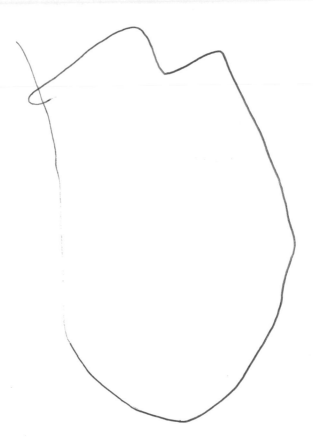

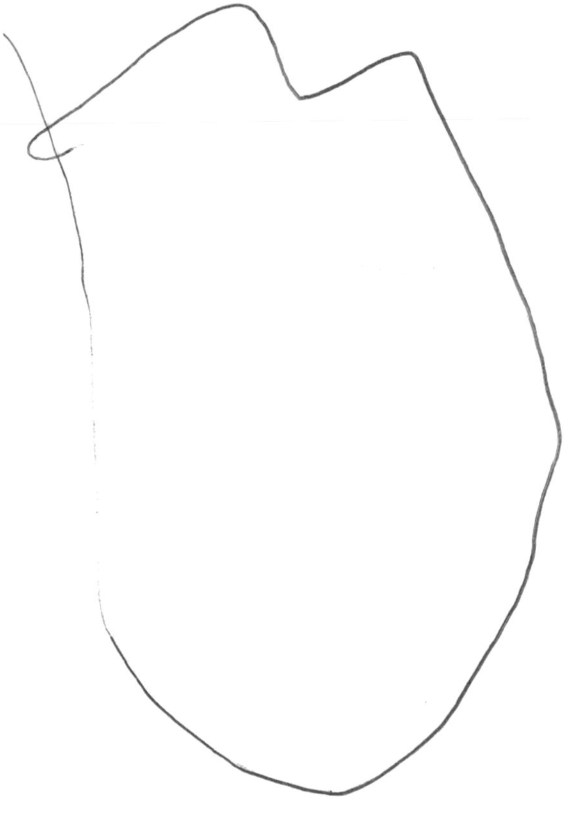

This is Volume 5 in
MATHEMATICS IN SCIENCE AND ENGINEERING
A series of monographs and textbooks
Edited by RICHARD BELLMAN, *University of Southern California*

The complete listing of books in this series is available from the Publisher upon request.

Entire Functions

RALPH PHILIP BOAS, Jr.
Northwestern University, Evanston, Illinois

1954
ACADEMIC PRESS INC., PUBLISHERS
NEW YORK, N. Y.

TO MARY

ACADEMIC PRESS, INC.
111 Fifth Avenue, New York, New York 10003

United Kingdom Edition published by
ACADEMIC PRESS, INC. (LONDON) LTD.
Berkeley Square House, London W.1

LIBRARY OF CONGRESS CATALOG CARD NUMBER: 54-11061

Third Printing, 1973

PRINTED IN THE UNITED STATES OF AMERICA

PREFACE

My chief aim has been to give an account of the extensive modern theory of functions of exponential type; the natural domain for these functions is often a half plane or an angle rather than the whole plane. Thus this book is not a comprehensive treatise on entire functions, and is not concerned exclusively with entire functions. However, a short and reasonably accurate title seemed preferable to a longer and more descriptive one. Even the limited subject of entire functions is too vast to be dealt with in a single volume with any approach to completeness, and I have preferred to omit altogether those topics which I could not cover fairly thoroughly. Thus there is no mention at all even of such important matters as Picard's theorem and the whole circle of ideas connected with it. Functions of exponential type have many applications in other fields; to cover the applications adequately would demand a book in itself, but I have discussed selected examples from a variety of fields to indicate how some of the applications arise.

I begin, in Chapter 2, with an account of some of the fundamental but elementary results on entire functions of finite order in general; much of this material is to be found in almost any comprehensive book on the theory of functions. Otherwise I have selected mostly material which is not covered in detail in existing books.

A reader familiar with the theory may notice a number of simplifications of proofs and slight improvements of theorems, as well as an attempt to use unified methods wherever possible. However, I do not claim for my own all material which is not explicitly credited to somebody else, since I have not attempted to locate the original sources for many things which are well known to workers in the field.

I assume that the reader is familiar with the basic theory of functions of a complex variable, as presented in any modern text intended for mathematicians; I assume that he knows, or is willing to look up, such things as Jensen's and Carleman's theorems, the ideas associated with the names of Phragmén and Lindelöf (although I state the theorems for reference), and the elements of the theory of harmonic and subharmonic functions in the plane. I also assume a certain command of the technique of "hard" analysis: I take for granted manipulations with lim sup and lim inf, with O and o, with Hölder's, Minkowski's and Jensen's inequalities, and with Lebesgue and Stieltjes integrals. There no longer seems to be any justification for depriving one's self of the convenience of the Lebesgue integral even though almost everything in the theory of entire functions can be done without it. At a few points I require the L^2 theory of Fourier transforms.

I have not tried to quote all papers connected with a given topic nor have I tried systematically to assign theorems to their discoverers. Names like "Pólya's theorem" are intended merely as catch phrases to identify results that are frequently referred to. The bibliography contains only those works which are referred to in the text. Bibliographies up to the middle 1920's are given by Valiron [3], [5], and for older work I usually refer to Valiron rather than to the primary sources. Would-be investigators in the field of entire functions are warned that the non-inclusion of a given topic or specific problem in this book, even when it may seem germane to topics which are included, is no guarantee that it is not already discussed in the literature. The field has already suffered more than most from repeated rediscovery of results and apparent reluctance of investigators to read each other's writings.

The reader will find only one or two new pieces of terminology, and no abbreviations (other than standard symbols). The temptation to introduce ad hoc abbreviations and portmanteau words is almost overwhelming when one works for a long time on a subject, but I believe that the saving of pencil for the author and of type for the printer is far offset by the inconvenience for the reader, especially in a book which is intended to be consulted by the non-specialist in search of possibly applicable results as well as to be studied systematically by a student who desires to become acquainted with the field.

Theorems which are stated but not proved in the book are identified by stars; some of these are suitable for exercises while others are too complicated to have their proofs included.

The reference numbers attached to formulas, theorems, etc., are intended to be read as integers in the scale of 100, with the dots indicating the space between "digits." Superscript letters refer to the notes at the ends of the chapters.

I am indebted to the John Simon Guggenheim Memorial Foundation and to Northwestern University for financial support during the academic year 1951–52, when most of the book was written.

R. P. Boas, Jr.

Evanston, Illinois
June, 1954

CONTENTS

ERRATA

ENTIRE FUNCTIONS

Ralph Philip Boas, Jr.

p. 2, 1.2.3: Replace $\log\left|\dfrac{z - z_k}{z - \bar{z}_k}\right|$ by $\log\left|\dfrac{z - z_k}{z - \bar{z}_k} \cdot \dfrac{R^2 - z_k z}{R^2 - \bar{z}_k z}\right|$. Make corresponding changes on pp. 93, 121.

p. 4 Theorem 1.4.3, add the hypothesis: $|f(z)| = O(\exp(r^{1+\epsilon}))$.

p. 6, line 2: interchange δ and Δ; line 10: the density is the limit of the measure divided by r. Omit Lemma 1.5.3 and remark after 10.7.4 (p. 200) which was based on it.

p. 24 delete the sentence in lines 4–5; in the third line of 2.9, and in the last part of p. 28, read ρ for p.

p. 33 line 6 from below, for "of divergence class" read: "of divergence call if of the same order as $g(z)$."

pp. 86, 87: The proof of 6.3.14 requires revision.

p. 90 6.4.8; p. 97, 6.6.6; and in Chapter 7 generally, add the hypothesis that the zeros have no finite limit point.

p. 99, 6.7.10; p. 196, 10.6.2: the condition on θ should read "three directions θ which do not all belong to the same half plane."

p. 119, line 8: an additional calculation is needed to show that $\sup x^{-1} \log|B(z)| = 0$.

p. 155 9.3.7, add: and $-\int_1^R \sigma(t)\,dt < O(R^2)$; replace "$h(\pi/2) = \pi$" by "the point πi lies on the indicator diagram."

p. 163, 9.6.11: add the hypothesis that $H(z)$ is of exponential type.

p. 163 line 3, for "are the zeros," read: "are among the zeros"; in the line before 9.6.13, read: "$H(z)$ is of order 1."

p. 168 line 3 before 9.9, for 9.8.4 read: "(9.8.7) for F."

pp. 178 ff: Throughout chapter 10, add the hypothesis that $f(z)$ is of exponential type.

p. 200 delete lines 14–10 from below.

p. 241 starting from line 11, for a, read: c.

CHAPTER 1

INTRODUCTION

1.1. Terminology. In this chapter I shall define some terms and introduce some notations which will be used consistently; and collect, in most cases without proof, some auxiliary theorems which will be used more or less frequently but do not altogether belong to the subject matter of the book.

A function of the complex variable z will be called regular in a region if it is analytic and single-valued there. An entire function[a] is one which is regular for all finite z. We consistently write $z = x + iy = re^{i\theta}$, and understand that x, y, r, θ have these meanings unless something is said to the contrary; similarly, writing z_1 implies that $z_1 = x_1 + iy_1$, etc. We write $\bar{z} = x - iy$, but $\bar{f}(z)$ means the function such that $\bar{f}(\bar{z}) = \overline{f(z)}$. We call r the modulus and $\theta = \arg z$ the argument of z. We also write $x = \Re(z)$, $y = \Im(z)$. "The upper half plane" means the half plane $y > 0$ or $y \geq 0$ according to context; similarly "the right-hand half plane" is $x > 0$ or $x \geq 0$, and "the unit circle" is $|z| < 1$ or $|z| \leq 1$. A contour is a rectifiable Jordan arc or curve.

If $f(z)$ is regular in $|z| < R$ (or entire), $M(r)$ denotes the maximum modulus of $f(z)$ for $|z| = r < R$ (or $< \infty$), and $m(r)$ denotes the minimum modulus for $|z| = r$. We write $M_f(r)$, etc., when it is necessary to call attention to the particular function that is being considered. By $n(r)$, or $n_f(r)$, etc., we denote the number of zeros of $f(z)$ in $|z| \leq r$, and by $N(r)$ the integral $\int_0^r t^{-1}n(t)\,dt$, provided that $n(0) = 0$. The zeros themselves are frequently denoted by $z_k = r_k e^{i\theta_k}$, and are supposed to be arranged in order of increasing modulus, with multiple zeros counted according to their multiplicities.

A number appearing without explanation in a formula is defined by the formula. We occasionally use the same symbol A to mean different numbers in different parts of the same formula or chain of formulas; any ambiguity can be avoided, if desired, by the reader's numbering such A's serially as they occur. Greatest lower bounds and least upper bounds are denoted by inf and sup, respectively. When $t > 0$, $\log^+ t$ means sup $(\log t, 0)$ and $\log^- t$ means inf $(\log t, 0)$; similarly for other functions. By $[x]$ we mean the greatest integer not exceeding x, except when it is clear from the context that the brackets are used merely as parentheses.

The statements $f(x) = O\{g(x)\}$, $f(x) = o\{g(x)\}$, $f(x) \sim g(x)$ mean, as usual, that $f(x)/g(x)$ is bounded, approaches zero, or approaches 1, respectively; and $f(x) \leq O\{g(x)\}$, $f(x) \leq o\{g(x)\}$ mean that $f(x)/g(x)$ is bounded above by a constant or by $o(1)$.

The notation $\int_a^{\to\infty} f(x)\,dx$ means $\lim_{R\to\infty} \int_a^R f(x)\,dx$. The principal value (at c) of $\int_a^b f(x)\,dx$, $a < c < b$, means $\lim_{\epsilon\to 0} \left\{ \int_a^{c-\epsilon} + \int_{c+\epsilon}^b \right\} f(x)\,dx$. The class L^p consists of the measurable functions $f(x)$ for which $|f(x)|^p$ is integrable (over whatever set is under consideration).

1.2. Jensen's, Carleman's and Nevanlinna's formulas. These are formulas connecting the zeros of $f(z)$ with its behavior on the boundary of a circle or a half plane.

1.2.1. *Jensen's Theorem.*[a] *If $f(z)$ is regular in $r < R$, and $f(0) \neq 0$, then for $r < R$ we have*

$$N(r) = (2\pi)^{-1} \int_0^{2\pi} \log | f(re^{i\theta}) | \, d\theta - \log | f(0) |.$$

1.2.2. *Carleman's Theorem.*[b] *If $f(z)$ is regular for $y \geq 0$ and $f(0) = 1$, and if z_k are the zeros of $f(z)$ in the upper half plane,*

$$\sum_{r_k \leq R} \left(\frac{1}{r_k} - \frac{r_k}{R^2} \right) \sin \theta_k = \frac{1}{\pi R} \int_0^{\pi} \log | f(Re^{i\theta}) | \sin \theta \, d\theta + \frac{1}{2\pi} \int_0^R \left(\frac{1}{x^2} - \frac{1}{R^2} \right) \log | f(x) f(-x) | \, dx + \tfrac{1}{2} \Im\{ f'(0) \}.$$

1.2.3. *Nevanlinna's Theorem (Poisson's Formula for a Semicircle).*[c] *If $f(z)$ is regular for $y \geq 0$ and z_k are its zeros,*

$$\log | f(re^{i\theta}) | = \sum_{|z_k| < R} \log \left| \frac{z - z_k}{z - \bar{z}_k} \right| + \frac{y}{\pi} \int_{-R}^{R} P_1(r, z, t) \log | f(t) | \, dt + \frac{2Ry}{\pi} \int_0^{\pi} P_2(R, z, \phi) \log | f(Re^{i\phi}) | \, d\phi,$$

where

$$P_1 = \frac{1}{t^2 - 2tx + r^2} - \frac{R^2}{R^4 - 2tR^2x + r^2t^2},$$

$$P_2 = \frac{(R^2 - r^2) \sin \phi}{| R^2 e^{2i\phi} - 2Rxe^{i\phi} + r^2 |^2}.$$

We may obtain 1.2.2 formally from 1.2.3 by letting $z \to 0$.

1.3. Carathéodory's inequality.[a] This gives an upper bound for the modulus of an analytic function in a circle when we know an upper bound for its real part (not the absolute value of its real part) in a larger circle.

1.3.1. *Theorem. If $f(z)$ is regular in $|z| \leq R$ and $\Re\{f(z)\} \leq Q(r)$ in*

$|z| \leq r$ *then for* $0 < r < R$

$$|f(z)| \leq |f(0)| + \frac{2r}{R-r}\{Q(R) - \Re f(0)\}.$$

We shall need the result of applying this to the logarithm of a zero-free function.

1.3.2. *Theorem. Let* $f(z)$ *be regular in* $|z| \leq R$ *and have no zeros there. If* $f(0) = 1$,

$$\log m(r) \geq -\frac{2r}{R-r}\log M(R), \qquad 0 < r < R. \tag{1.3.3}$$

Carathéodory's inequality gives

$$|\log f(z)| \leq \frac{2r}{R-r}\log M(R);$$

taking the z of modulus r for which $|f(z)| = m(r)$, we have (1.3.3).

Carathéodory's inequality also furnishes a convenient means of dealing with the analogue of Liouville's theorem in which the real part of the function is considered instead of the modulus.

1.3.4. *Theorem. If* $f(z)$ *is entire and* $\Re\{f(z)\} \leq A(\epsilon)r^{\rho+\epsilon}$ *on arbitrarily large circles for each positive* ϵ, $f(z)$ *is a polynomial of degree at most* ρ.

Supposing for convenience that $f(0) = 0$, we have by 1.3.1, for $r = R/2$, where R is one of the values of r for which the hypothesis is satisfied,

$$|f(z)| \leq 2^{\rho+1+\epsilon} A(\epsilon)\, r^{\rho+\epsilon},$$

and so by Liouville's theorem $f(z)$ reduces to a polynomial of degree at most ρ.

1.4. Phragmén-Lindelöf theorems.[a] These are generalizations of the maximum principle in which we infer the boundedness of a function inside an unbounded region from the hypotheses that the function is bounded on the boundary and not of too rapid growth inside.

1.4.1. *Theorem. Let* $f(z)$ *be regular in the half plane* $x > 0$, *continuous in the closed half plane*, $|f(iy)| \leq M$ *and* $f(z) = O(e^{r^\beta})$, $\beta < 1$, *uniformly in* θ, *for a sequence* $r = r_n \to \infty$. *Then* $|f(z)| \leq M$ *for* $x \geq 0$.

Consider $F(z) = f(z)\exp(-\epsilon z^\gamma)$, where $\beta < \gamma < 1$ and $\epsilon > 0$. Then $|F(z)| \leq |f(z)| \exp(-\epsilon r^\gamma \cos \gamma\theta)$. Since $\gamma < 1$, $\cos\gamma\theta > 0$ in the half plane and so $|F(iy)| \leq |f(iy)| \leq M$. On the arc $|\theta| \leq \frac{1}{2}\pi$ of $|z| = r_n$, $|F(z)| \leq |f(z)| \exp(-\epsilon r_n^\gamma \cos \pi\gamma/2)$, and the right-hand side approaches zero as $r_n \to \infty$ because $\gamma > \beta$. Hence if r_n is large enough, $|F(z)| \leq M$ on $|z| = r_n$, $|\theta| \leq \frac{1}{2}\pi$, and so $|F(z)| \leq M$ in the semicircular region which this arc cuts from the half plane, and hence in the entire half plane since r_n can be arbitrarily large. That is, $|f(z)| \leq M\exp(\epsilon r^\gamma)$ and we may let $\epsilon \to 0$ (for each fixed z) to obtain the conclusion of 1.4.1.

Of course the theorem is true for any half plane, since an arbitrary half plane may be translated and rotated to become $x > 0$. In addition, the theorem may be transformed by conformal mapping into one for any angle. We give the general statement since we shall need to use it.

1.4.2. *Theorem. Let* $f(z)$ *be regular in the angle* $D: |\arg z| < \frac{1}{2}\pi/\alpha$, *continuous in the closed angle,* $|f(z)| \le M$ *on the boundary,* $f(z) = O(e^{r^\beta})$, $\beta < \alpha$, *uniformly in* θ, *for* $r = r_n \to \infty$. *Then* $|f(z)| \le M$ *throughout* D.

The conclusion of 1.4.1 still holds under a more general hypothesis.

1.4.3. *Theorem. Let* $f(z)$ *be regular in* $x > 0$, *continuous in* $x \ge 0$, $|f(iy)| \le M$ *and* $f(z) = O(e^{ar})$, $r \to \infty$, *for* $z = re^{i\alpha}$, *where* α *is a number between* $-\pi/2$ *and* $\pi/2$. *Then* $|f(z)| \le Me^{(a \sec \alpha) r \cos \theta}$ for $x \ge 0$. *In particular,* $|f(z)| \le M$ *if the hypotheses are satisfied with arbitrarily small positive* a; *and* $f(z) \equiv 0$ *if they are satisfied with arbitrarily small* a, *i.e. if* $f(re^{i\alpha}) = O(e^{-r\omega(r)})$, $\omega(r) \to \infty$.

Now we consider $F(z) = e^{-bz \sec \alpha} f(z)$, $b > a$, which is bounded by M on the positive imaginary axis and by some number N on the positive real axis (since $F(x) \to 0$ as $x \to \infty$). By 1.4.2, $|F(z)| \le \max(M, N)$ in the first quadrant, and similarly $|F(z)| \le \max(M, N)$ in the fourth quadrant. Since $|F(z)| = N$ at some finite point we must have $N \le M$, since otherwise $|F(z)|$ would take its maximum value in the right-hand half plane at an interior point and so would be a constant, which could only be M. Since now $|f(z)e^{-bz \sec \alpha}| \le M$ for all $b > a$, we can let $b \to a$ to obtain the result.

1.4.4. *Theorem. If* $f(z) \to a$ *as* $z \to \infty$ *along two rays, and* $f(z)$ *is regular and bounded in the angle between them,* $f(z) \to a$ *uniformly in the whole angle.*

We may suppose without loss of generality that $a = 0$ and that the angle is $|\arg z| \le \phi < \frac{1}{2}\pi$. *Let* $F(z) = z(z + \lambda)^{-1}f(z)$, $\lambda > 0$. Let $\epsilon > 0$ and take r_1 so large that $|f(re^{\pm i\phi})| < \epsilon$ for $r \ge r_1$. Then, with r_1 fixed, take λ so large that $|F(z)| < \epsilon$ for $r \le r_1$ and $|\arg z| \le \phi$. By 1.4.2, $|F(z)| < \epsilon$ throughout $|\arg z| \le \phi$, and

$$|f(z)| \le (1 + \lambda/r)\,|F(z)| < 2\epsilon, \qquad r > \lambda.$$

This establishes the result.

By conformal mapping we can obtain a similar result for a strip.

1.4.5. *Corollary. If* $f(z) \to a$ *as* $x \to \infty$ *for* $y = y_1$ *and* $y = y_2$, *and* $f(z)$ *is regular and bounded for* $x \ge b$, $y_1 \le y \le y_2$, *then* $f(z) \to a$ *uniformly for* $y_1 \le y \le y_2$.

1.4.6. *Beurling's Theorem. If* $f(z)$ *is regular in* $x > 0$ *and continuous in* $x \ge 0$, *if* $|f(iy)| \le \phi(|y|)$, *where* $\phi(r)$ *is continuous and*

$$\limsup r^{-\rho} \log \phi(r) = 0, \qquad 0 < \rho < 1,$$

and $|f(z)| = O(e^{\epsilon r})$ *for a sequence* $r_n \to \infty$, *for each* $\epsilon > 0$, *then either* $f(z)$

is bounded for $x \geq 0$ *or*

(1.4.7) $$\log M(r) < \log \phi(r) \cdot \sec \tfrac{1}{2} \pi\rho$$

for a sequence $r = R_n \to \infty$.

If $\phi(r)$ were bounded we should have $f(z)$ bounded, by 1.4.3; hence we may assume $\phi(r)$ unbounded. Consider $F(z) = f(z) \exp(-\epsilon z^\rho - a)$, where z^ρ is the branch which is positive for positive z. For $z = iy$ we have

$$\log |F(z)| \leq \log \phi(|y|) - \epsilon \cos \tfrac{1}{2} \pi\rho - a.$$

Now take $a = a(\epsilon)$ so that

$$\log \phi(r) \leq r^\rho \epsilon \cos \tfrac{1}{2} \pi\rho + a(\epsilon)$$

for all r, with equality for some $r = r(\epsilon)$. Then we have $\log |F(z)| \leq 0$ for $z = iy$, and hence $|F(z)| \leq 1$ for $x \geq 0$ by 1.4.3. Then for $r = r(\epsilon)$ we have $\log M(r) \leq \epsilon r^\rho + a(\epsilon)$, $\log \phi(r) = \epsilon \cos \tfrac{1}{2} \pi\rho \cdot r^\rho + a(\epsilon)$, and so

$$\log \phi(r) \geq \cos \tfrac{1}{2} \pi\rho \cdot \log M(r) + a(\epsilon)(1 - \cos \tfrac{1}{2} \pi\rho).$$

Since $r(\epsilon) \to \infty$ as $\epsilon \to 0$ (because $\phi(r)$ is unbounded), the conclusion follows.

A result very similar to 1.4.4 is

1.4.8. *Montel's Theorem.*[b] *If $f(z)$ is regular and bounded in the angle between two rays, and $f(z) \to a$ as $z \to \infty$ on one ray in the interior of the angle, then $f(z) \to a$ uniformly in any interior angle.*

The equivalent result for a strip is

1.4.9. *Theorem.*[b] *If $f(z)$ is regular and bounded for $x \geq b$, $y_1 \leq y \leq y_2$, and $f(z) \to a$ as $x \to \infty$ for $y = y_3$, $y_1 < y_3 < y_2$, then $f(z) \to a$ uniformly for $y_1 + \delta \leq y \leq y_2 - \delta$.*

1.5. Density of sequences and sets. If $\{\lambda_n\}$ is a nondecreasing sequence of positive numbers it is said to have density D if $\lim_{n\to\infty} n/\lambda_n = D$ (D may be 0, finite, or infinite). If the numbers λ_n may have either sign, $\{\lambda_n\}$ has density D if $\{|\lambda_n|\}$ has density $2D$. We shall have frequent use for the following elementary lemma.

1.5.1. *Lemma. If $\{\lambda_n\}_1^\infty$ is a nondecreasing sequence of positive numbers and $n(r)$ denotes the number of λ_n not exceeding r, the statements $n/\lambda_n \to D$ and $n(r)/r \to D$ are equivalent.*

On one hand, $n(\lambda_n)/\lambda_n = n/\lambda_n$, so if $n(r)/r \to D$, $n/\lambda_n \to D$. On the other hand, if $n/\lambda_n \to D$, suppose that λ_m is the first λ greater than λ_n; then $\lambda_{m-1} = \lambda_n$, so $(m-1)/\lambda_n \to D$. For $\lambda_n < x < \lambda_m$, $n(x) = m - 1$, so $n(x)/x < m/\lambda_n \to D$. Thus $\limsup n(x)/x \leq D$. Similarly $\liminf n(x)/x \geq D$.

Even if the sequence $\{\lambda_n\}$, $\lambda_n > 0$, fails to have a density, it always has (finite or infinite) upper and lower densities, $\limsup n/\lambda_n$, $\liminf n/\lambda_n$.

It also has maximum and minimum densities,[a]

$$\Delta = \lim_{\xi \to 1-} \limsup_{r \to \infty} \frac{n(r) - n(r\xi)}{r - r\xi}, \qquad \delta = \lim_{\xi \to 1-} \liminf_{r \to \infty} \frac{n(r) - n(r\xi)}{r - r\xi}.$$

These have the property that a sequence of minimum density δ contains a sequence of density δ, but not any sequence having a larger density, while a sequence of maximum density Δ is contained in a sequence of density Δ, but not in any sequence of smaller density. For example, the sequence consisting of the integers in the intervals $(3^k, 3^k + 3^{k-1})$ has maximum density 1 and minimum density 0.

In a somewhat similar way, if E is a measurable set on $(0, \infty)$ we define its density (or linear density) as $\lim_{r\to\infty}$ meas $\{E \cap (0, r)\}$, if the limit exists, and its upper density by replacing lim by lim sup. r

The logarithmic length of E is $\int_E x^{-1}\, dx$.

1.5.2. *Lemma. If E has finite logarithmic length, it has linear density* 0. For, if E_r is the part of E in $(0, r)$, we have

$$r^{-1} \int_{E_r} dx \leq \int_{E_r} x^{-1}\, dx.$$

1.5.3. *Lemma. If E is a set of disjoint intervals* (a_n, b_n) *and has finite logarithmic length, and* $\{\lambda_n\}$ *is an increasing sequence of positive numbers such that* $\lambda_{n+1} - \lambda_n \geq \delta > 0$, *then if* $\sum 1/\lambda_n$ *diverges there are infinitely many* λ_n *which are not in E.*

For, $\sum \log(b_n/a_n) = \sum \log\{1 + (b_n - a_n)/a_n\}$ converges, so $b_n/a_n \to 1$ and $\sum (b_n - a_n)/a_n$ converges. There are at most $A(b_n - a_n)$ of the λ's in (a_n, b_n), where A is a constant, and so $\sum 1/\lambda_k$, taken over the λ's which are in some (a_n, b_n), is dominated by the convergent series

$$\sum A(b_n - a_n)/a_n.$$

1.6. Stirling's formula. This is

$$n! = n^n e^{-n} (2\pi n)^{1/2} e^{\delta/(12n)},$$

where $0 < \delta < 1$. We usually need only weaker forms.

1.7. Mittag-Leffler summability.[a] Let $\{s_n\}$ be a sequence of complex numbers with $|s_n|^{1/n} = O(1)$, and let $E(z) = \sum_{n=0}^{\infty} d_n z^n$ be an entire function, not a polynomial, with $d_n \geq 0$. The series $\sum_{n=0}^{\infty} d_n s_n w^n$ converges for $0 \leq w < \infty$ to $H(w)$, say. If $\lim_{w\to\infty} H(w)/E(w) = s$, we say that $\{s_n\}$ is E-summable to s. The E-method is totally regular, i.e. a sequence which converges to a finite limit or $+\infty$ is E-summable to the same limit or to $+\infty$, respectively. When $E(z) = \sum z^n/\{\log(n + 2)\}^n$, E-summability is called Mittag-Leffler summability. If $s_n(z) = \sum_{k=0}^{n} a_k z^k$, and $f(z) =$

$\sum_{k=0}^{\infty} a_k z^k$, the sequence $\{s_n(z)\}$ is Mittag-Leffler summable to $f(z)$ in the Mittag-Leffler star of $f(z)$, which is the z-plane omitting the parts of the rays $\arg z = \theta$ extending from the singular points of $f(z)$ to ∞.

1.8. Laplace and Fourier transforms. We require only some special results from the general theory of Laplace transforms. Let $\phi(t)$ be integrable in every finite interval $(0, R)$; if $f(z) = \int_0^{\to\infty} e^{-zt}\phi(t)\,dt$ converges for some z, it converges for every z of larger real part, and represents a regular function of z in the half plane in which it converges; the derivatives of $f(z)$ are obtained by differentiating under the integral sign.

If $f(z)$ belongs to L^2 on $(-\infty, \infty)$, its Fourier transform

$$F(x) = \int_{-\infty}^{\infty} e^{ixt} f(t)\,dt$$

exists for almost all x, with $\int_{-\infty}^{\infty}$ defined as $\text{l.i.m.}_{T\to\infty} \int_{-T}^{T}$ (limit in the mean, limit in the L^2 metric). Furthermore, $F(x)$ belongs to L^2 and, almost everywhere, $f(t) = (2\pi)^{-1} \int_{-\infty}^{\infty} e^{-ixt}F(x)\,dx$, with the same interpretation of the integral. We have

$$2\pi \int_{-\infty}^{\infty} |f(t)|^2\,dt = \int_{-\infty}^{\infty} |F(x)|^2\,dx.$$

If $f(t)$ and $g(t)$ both belong to L^2 and L, so does their convolution,

$$\phi(t) = \int_{-\infty}^{\infty} f(u)g(t-u)\,du,$$

and $\Phi(x) = F(x)G(x)$.

Notes for Chapter 1

1.1a. An entire function is an integral function in British usage.

1.2a. See e.g. Titchmarsh [3], p. 125.

1.2b. See e.g. Titchmarsh [3], p. 130.

1.2c. F. and R. Nevanlinna [1], R. Nevanlinna [1], Levinson [4], p. 245.

1.3a. Landau [1], vol. 1, p. 299, vol. 2, p. 894; [2], Satz 225; Titchmarsh [3], p. 174. The version in the first edition of Titchmarsh [3] is incorrect. For further developments and references cf. Rajagopal [1].

1.4a. For everything in this section except 1.4.6 and 1.4.8 see Phragmén and Lindelöf [1] or Titchmarsh [3]; Titchmarsh states slightly less general results than we need, but his proofs suffice. Beurling's theorem is quoted by Kjellberg [1].

1.4b. Titchmarsh [3], p. 170.

1.5a. For maximum and minimum density see Pólya [2]. For some other kinds of density see Buck [2], Tsuji [1], Mandelbrojt [2].

1.7a. See e.g. Dienes [1], p. 311.

CHAPTER 2

GENERAL PROPERTIES OF ENTIRE FUNCTIONS OF FINITE ORDER

2.1. Measures of rate of growth. We shall be concerned only with entire functions whose rate of growth is not too large, and more precisely with those which are of finite order according to the following definition.

2.1.1. *Definition. The entire function $f(z)$ is of order ρ if*

$$\limsup_{r\to\infty} \frac{\log\log M(r)}{\log r} = \rho \qquad (0 \le \rho \le \infty). \tag{2.1.2}$$

A constant has order 0, by convention.

If ρ is finite, as we shall always suppose, and not zero, we define another number, the type of $f(z)$, which gives a more precise description of the rate of growth of $f(z)$.

2.1.3. *Definition. The entire function $f(z)$ of positive order ρ is of type τ if*

$$\limsup_{r\to\infty} r^{-\rho} \log M(r) = \tau \qquad (0 \le \tau \le \infty). \tag{2.1.4}$$

Thus $f(z)$ is of finite order ρ if, and only if, for every positive ϵ but for no negative ϵ,

$$M(r) = O(e^{r^{\rho+\epsilon}}), \qquad r \to \infty. \tag{2.1.5}$$

According as $\tau = \infty$, $0 < \tau < \infty$, or $\tau = 0$, $f(z)$ is said to be of maximum (or infinite), mean (or normal), or minimum (or zero) type of order ρ. It is of finite type τ if and only if, for every positive ϵ but for no negative ϵ,

$$M(r) = O(e^{(\tau+\epsilon)r^{\rho}}), \qquad r \to \infty. \tag{2.1.6}$$

It is often convenient to have a term to describe a function which is of order not exceeding ρ and of type not exceeding τ if of order ρ; there is no standard term for this, but we shall say in this book that such an $f(z)$ is "of growth (ρ, τ)." A function of growth $(1, \tau)$, $\tau < \infty$, is called a function of exponential type,[a] or of exponential type τ if it is necessary to specify the value of τ. The reader should note that functions of exponential type τ include all functions of order 1 and type less than or equal to τ, as well as all functions of order less than 1.

There is occasional use for the lower order,[b]

$$\lambda = \liminf_{r\to\infty} \frac{\log\log M(r)}{\log r}. \tag{2.1.7}$$

When still more precise specification of the rate of growth of $f(z)$ is desired, one can use the proximate order, a function $V(r)$ which is monotone nondecreasing, piecewise differentiable, and satisfies

$$\lim_{r\to\infty} V(r) = \rho, \qquad \lim_{r\to\infty} V'(r)r \log r = 0, \qquad \lim_{r\to\infty} r^{-V(r)} \log M(r) = 1.$$

A proximate order always exists,[c] but we shall not use proximate orders in this book.

The rate of growth of an entire function in different directions can be specified by the Phragmén-Lindelöf indicator function,

$$h(\theta) = \limsup_{r\to\infty} r^{-\rho} \log | f(re^{i\theta}) |. \tag{2.1.8}$$

Sometimes it is desirable to extend the preceding definitions to functions which are not entire, but are regular in an angle with vertex at the origin. They can be carried over unchanged if $M(r)$ is taken to mean the maximum of $| f(z) |$ for $| z | \leq r$ and z in the angle.

We take for granted such frequently used and easily proved results as that the order of a sum or product of two entire functions is at most the larger of the respective orders.

2.2. Order and type in terms of the coefficients. Let

$$f(z) = \sum_{n=0}^{\infty} a_n z^n, \qquad a_n \to 0, \tag{2.2.1}$$

be the Maclaurin series of an entire function (if a_n does not approach zero, $f(z)$ is not entire). Since the sequence $\{a_n\}$ determines the function completely, it should in principle be possible to discover all the properties of the function by examining the coefficients. For the order and type this can be done without difficulty.

2.2.2. *Theorem. The entire function $f(z)$* (2.2.1) *is of finite order if and only if*

$$\mu = \limsup_{n\to\infty} \frac{n \log n}{\log (1/| a_n |)} \tag{2.2.3}$$

is finite; and then the order ρ of $f(z)$ is equal to μ.

In calculating μ, the quotient on the right of (2.2.3) is to be taken as 0 if $a_n = 0$.

We prove first that $\rho \geq \mu$, where if $\mu = \infty$ this statement is to be interpreted as meaning that $\rho = \infty$ or else $f(z)$ is not entire. We need the elementary fact that

$$| a_n | = | f^{(n)}(0)/n! | \leq (2\pi)^{-1} \int_{|z|=r} | z |^{-n-1} | f(z) | \, | dz | \leq r^{-n} M(r). \tag{2.2.4}$$

If $\mu = 0$, $\rho \geq \mu$ since ρ is not negative.

Suppose now that $0 < \mu \leq \infty$, and let $0 < \epsilon < \mu$. Then from (2.2.3) we have, for an infinity of n,

$$n \log n \geq R \log (1/ | a_n |), \tag{2.2.5}$$

where $R = \mu - \epsilon$ (ϵ "small") if $\mu < \infty$, $R = \epsilon$ (ϵ "large") if $\mu = \infty$; hence

$$\log | a_n | \geq -nR^{-1} \log n,$$

and by (2.2.4),

$$\log M(r) \geq \log | a_n | + n \log r \geq n(\log r - R^{-1} \log n).$$

To simplify the appearance of the right-hand side, take $r = (en)^{1/R}$, and then

$$\log M(r) \geq n/R = r^R/(eR),$$

$$\frac{\log \log M(r)}{\log r} \geq R - \frac{\log (eR)}{\log r} \tag{2.2.6}$$

for the r's corresponding to the n's in (2.2.5). Since R is independent of r, we have

$$\rho = \limsup_{r\to\infty} \frac{\log \log M(r)}{\log r} \geq R = \begin{cases} \mu - \epsilon, & \mu < \infty; \\ \epsilon \ , & \mu = \infty. \end{cases}$$

Since ϵ is arbitrary, it follows that $\rho \geq \mu$.

Next we show that $\rho \leq \mu$. If $\mu = \infty$ there is nothing to prove, so we suppose that $\mu < \infty$. Let $\epsilon > 0$; by (2.2.3) we have, for all large n,

$$0 \leq \frac{n \log n}{\log (1/| a_n |)} \leq \mu + \epsilon,$$

so

$$| a_n | \leq n^{-n/(\mu+\epsilon)}. \tag{2.2.7}$$

First of all, (2.2.7) implies that $f(z)$ is entire. Since adding a polynomial to an entire function does not change its order, we may suppose that (2.2.7) is satisfied for all n (interpreting its right-hand side as 1 for $n = 0$). Then we have

$$M(r) \leq \sum_{n=0}^{\infty} | a_n | r^n \leq \sum_{n=0}^{\infty} n^{-n/(\mu+\epsilon)} r^n = S_1 + S_2 ,$$

where S_1 contains the terms for which $n < (2r)^{\mu+\epsilon}$. (The reason for splitting the series at approximately this point is that the maximum of $n^{-n/(\mu+\epsilon)} r^n$ occurs when $n = e^{-1} r^{\mu+\epsilon}$.) We estimate S_1 by taking the largest value for

r^n; then

$$S_1 \le r^{(2r)^{\mu+\epsilon}} \sum n^{-n/(\mu+\epsilon)} = O(e^{(2r)^{\mu+\epsilon}} \log r) = O(e^{r^{\mu+2\epsilon}}), \tag{2.2.8}$$

since the series in (2.2.8) is convergent if extended to ∞ and is independent of r.

On the other hand, in S_2 we have $rn^{-1/(\mu+\epsilon)} \le \frac{1}{2}$ and so $S_2 < 1$. Hence $\rho \le \mu + 2\epsilon$; letting $\epsilon \to 0$, we obtain $\rho \le \mu$.

Next we obtain a formula for the type of a function of finite order. Suppose that $0 < \rho < \infty$ and define

$$\nu = \limsup_{n\to\infty} n \mid a_n \mid^{\rho/n}. \tag{2.2.9}$$

2.2.10. *Theorem. If* $0 < \nu < \infty$, *the function* (2.2.1) *is of order* ρ *and type* τ *if and only if* $\nu = e\tau\rho$. *If* $\nu = 0$ *or* ∞, $f(z)$ *is, respectively, of growth* $(\rho, 0)$ *or of growth not less than* (ρ, ∞); *and conversely.*

Using Stirling's formula (§1.6), we see that an equivalent statement is

$$\limsup_{n\to\infty} (n/e)^{1-\rho} \mid f^{(n)}(z) \mid^{\rho/n} = \tau\rho, \tag{2.2.11}$$

where z is any (fixed) complex number; when $\rho = 1$, this becomes

$$\limsup_{n\to\infty} \mid f^{(n)}(z) \mid^{1/n} = \tau, \tag{2.2.12}$$

the finiteness τ of being necessary and sufficient for $f(z)$ to be of exponential type.[a]

We prove first that if $\nu < \infty$, $f(z)$ is of order ρ at most, while if $\nu > 0$, $f(z)$ is of order ρ at least. Let $\epsilon > 0$; for large n we have $n \mid a_n \mid^{1/n} \le \nu + \epsilon$, and hence

$$\frac{n \log n}{\log (1/\mid a_n \mid)} \le \frac{\rho}{1 - \log \{(\nu + \epsilon)/n\}}.$$

By 2.2.2, the order of $f(z)$ is at most ρ. Similarly, if $\nu > 0$ the order of $f(z)$ is at least ρ.

Suppose now that $0 \le \nu < \infty$; we prove that $\tau \le \nu/(e\rho)$. Let $\epsilon > 0$. For large n we have

$$\mid a_n \mid \le \{(\nu + \epsilon)/n\}^{n/\rho}; \tag{2.2.13}$$

since we may add a polynomial to $f(z)$ without affecting its type, we may suppose that (2.2.13) holds for all n, interpreting its right-hand side as 1 for $n = 0$. Then

$$\mid f(z) \mid \le \sum_{n=0}^{\infty} \mid a_n \mid r^n \le \sum_{n=0}^{\infty} \{r^\rho(\nu + \epsilon)/n\}^{n/\rho}.$$

The general term of the right-hand side does not exceed its maximum, which is $\exp\{(\nu+\epsilon)r^\rho/(e\rho)\}$, attained for $n = (\nu+\epsilon)r^\rho/e$. If S_1 denotes the part of the series for which $n \le (\nu + 2\epsilon)r^\rho$, S_1 does not exceed the number of terms multiplied by the maximum term, so

$$S_1 \le (\nu + 2\epsilon)r^\rho \exp\{(\nu+\epsilon)r^\rho/(e\rho)\} = O[\exp\{(\nu+\epsilon)\, r^\rho/(e\rho)\}].$$

In S_2, the remainder of the series, $r^\rho < n/(\nu + 2\epsilon)$, and so

$$S_2 < \sum_{n=0}^{\infty} \{(\nu+\epsilon)/(\nu+2\epsilon)\}^{n/\rho} = O(1).$$

Thus $\tau \le \nu/(e\rho)$, $0 \le \nu < \infty$.

Suppose on the other hand that $0 < \nu \le \infty$. To show that $\tau \ge \nu/(e\rho)$, we observe that, for an infinity of n, $|a_n| \ge \{(\nu-\epsilon)/n\}^{n/\rho}$, $0 < \epsilon < \nu$, where $\nu - \epsilon$ is to be interpreted as an arbitrarily large number if $\nu = \infty$. By (2.2.4), if we take r so that $r^\rho = ne/(\nu - \epsilon)$ for these values of n, we have

$$M(r) \ge |a_n| r^n \ge \{r^\rho(\nu-\epsilon)/n\}^{n/\rho} = e^{n/\rho} = \exp\{(\nu-\epsilon)r^\rho/(e\rho)\},$$

for a sequence of values of r tending to ∞, so that $\tau \ge \nu/(e\rho)$.

2.3. Other connections between growth and coefficients. The lower order (2.1.7) cannot readily be found from the coefficients unless they satisfy conditions which prevent them from varying too rapidly; however, there are partial analogues of the theorems of §2.2.

2.3.1.* *Theorem. The lower order* λ *satisfies*[a]

$$\lambda \ge \liminf_{n\to\infty} \frac{n \log n}{\log\{1/|a_n|\}} \ge \liminf_{n\to\infty} \frac{\log n}{\log|a_n/a_{n+1}|}.$$

2.3.2.* *Theorem. If* $0 < \rho < \infty$ *then*[b]

$$e\rho \liminf_{r\to\infty} \frac{\log M(r)}{r^\rho} \ge \liminf_{n\to\infty} n\, |a_n|^{\rho/n}.$$

See also 2.12.15.

A quite different group of results connect the order with the maximum term in the power series. In §2.2 we made use of the fact that $M(r)$ is at least as large as the largest of the absolute values of the individual terms; one may hope that the maximum term will actually give a fair approximation to $M(r)$. We define

(2.3.3) $$\mu(r) = \max_{0\le n<\infty} |a_n| r^n,$$

the maximum term, and $\nu(r)$ as the value of n for which the maximum is attained (the largest such n if there is more than one). We quote some of the relationships which connect $\mu(r)$ and $\nu(r)$ with each other and with $M(r)$ for entire functions of positive finite order.[c] Suppose for simplicity that $|f(0)| = 1$. Then:

2.3.4.* $$\log \mu(r) = \int_0^r t^{-1}\nu(t)\,dt.$$

2.3.5.* $$\liminf_{r\to\infty} \frac{\nu(r)}{\log \mu(r)} \le \liminf_{r\to\infty} \frac{\log \nu(r)}{\log r} = \liminf_{r\to\infty} \frac{\log\log \mu(r)}{\log r} = \lambda$$

$$< \rho = \limsup_{r\to\infty} \frac{\log\log \mu(r)}{\log r} = \liminf_{r\to\infty} \frac{\log \nu(r)}{\log r} \le \limsup_{r\to\infty} \frac{\nu(r)}{\log \mu(r)}.$$

The inequalities not involving λ and ρ are direct consequences of (2.3.4) without reference to the meaning of $\mu(r)$ and $\nu(r)$.

2.3.6*. $$\lim_{r\to\infty} \frac{\log M(r)}{\log \mu(r)} = 1.$$

2.4. The order and type of a derivative. It follows from 2.2.2 and 2.2.10 that

2.4.1. *Theorem. $f(z)$ and $f'(z)$ are of the same order and type.*

This can be proved more directly as follows. Let $M_1(r)$ be related to $f'(z)$ as $M(r)$ is to $f(z)$. Then

$$f(z) = \int_0^z f'(w)\,dw + f(0),$$

with integration along a line segment. Hence

$$M(r) \le rM_1(r) + |f(0)|.$$

On the other hand, if $|z| = r$,

$$f'(z) = (2\pi i)^{-1} \int_{|z|=2r} (w - z)^{-2} f(w)\,dw,$$

$$M_1(r) \le M(r+1)/r.$$

Thus

$$\{M(r) - |f(0)|\}/r \le M_1(r) \le M(r+1)/r$$

and the result follows from the definition of order and type.

A more precise description of the relationships between $M_1(r)$ and $M(r)$ is given by the following results.[a]

2.4.2*. $$\lambda = \liminf_{r\to\infty} \frac{\log \{rM_1(r)/M(r)\}}{\log r} \le \limsup_{r\to\infty} \frac{\log \{rM_1(r)/M(r)\}}{\log r} = \rho.$$

2.4.3*. $\liminf_{r\to\infty} M_1(r)/M(r) \le \liminf_{r\to\infty} \nu(r)/r \le \limsup_{r\to\infty} \nu(r)/r$

$$\le \limsup_{r\to\infty} M_1(r)/M(r).$$

2.5. Rate of growth and distribution of zeros. It is evident from Jensen's theorem (1.2.1) that the more zeros an entire function has, the faster it must increase. In a general way, Jensen's theorem seems to tell the whole truth unless some special restriction is imposed on the growth of the function in various directions or on the position of the zeros. This section contains some simple theorems under very general hypotheses; a considerable part of this book is devoted to more difficult and more useful theorems which demand special hypotheses and furnish stronger conclusions.

For the purposes of this section the removal of a finite number of zeros or multiplication of the function by a non-zero constant is irrelevant, so to simplify the formulas we shall suppose that $f(0) = 1$. We denote by $0 < r_1 \le r_2 \le \cdots$ the absolute values of the zeros (if any) of $f(z)$, by $n(t)$ the number (counted according to multiplicity) of zeros in $|z| \le t$, and put

$$N(r) = \int_0^r t^{-1} n(t)\, dt. \tag{2.5.1}$$

2.5.2. *Definition. The convergence exponent of the zeros of* $f(z)$ *(for short, the convergence exponent of* $f(z)$*) is the infimum of positive numbers* α *for which*

$$\sum_{n=1}^{\infty} r_n^{-\alpha} \tag{2.5.3}$$

converges; it will be denoted by ρ_1.

2.5.4. *Definition. The smallest positive integer* α *for which* (2.5.3) *converges will be denoted by* $p + 1$, *and* p *will be called the genus of the set of zeros of* $f(z)$. (*This is not necessarily the same as the genus of* $f(z)$, *which will be defined in* 2.7.3.)

2.5.5. *Lemma. The series* (2.5.3) *and the integral*

$$\int_0^{\infty} t^{-\alpha-1} n(t)\, dt \tag{2.5.6}$$

converges or diverge together if $\alpha > 0$.

In fact, a partial sum of the series is

$$\int_0^T t^{-\alpha}\, dn(t) = T^{-\alpha} n(T) + \alpha \int_0^T t^{-\alpha-1} n(t)\, dt. \tag{2.5.7}$$

If the left-hand side is bounded as $T \to \infty$, the integral on the right does not exceed that on the left and hence (2.5.6) converges. On the other hand, if (2.5.6) converges, since $n(t)$ increases we have

$$\alpha^{-1}(1 - 2^{-\alpha})T^{-\alpha}n(T) = n(T)\int_T^{2T} t^{-\alpha-1}\,dt \le \int_T^{2T} t^{-\alpha-1}n(t)\,dt$$

$$\le \int_0^{2T} t^{-\alpha-1}n(t)\,dt,$$

and so

$$n(T) = O(T^{\alpha}).$$

Thus the right-hand side of (2.5.7) is bounded and therefore so is the left-hand side, i.e. (2.5.3) converges.

An alternative definition of ρ_1 is given by

2.5.8. *Theorem. If $f(z)$ has at least one zero,*

$$\rho_1 = \limsup_{r\to\infty} \frac{\log n(r)}{\log r}.$$

For, if we denote the limit superior by σ, we have $n(r) = O(r^{\sigma+\epsilon})$, $\epsilon > 0$, and so (2.5.6) convergent for $\alpha > \sigma + \epsilon$, hence $\rho_1 \le \sigma$. On the other hand, if $\sigma > 0$ there is a sequence $r_k \uparrow \infty$ such that $n(r_k) > r_k^{\sigma-\epsilon}$, and hence, since $n(r)$ increases, if $s > 2^{1/\alpha}r_k$ we have

$$\int_{r_k}^{s} t^{-1-\alpha}n(t)\,dt \ge r_k^{\sigma-\epsilon}\int_{r_k}^{s} t^{-1-\alpha}\,dt \ge \tfrac{1}{2}\,\alpha^{-1}r_k^{\sigma-\alpha-\epsilon},$$

so that the left-hand side is arbitrarily large if $\alpha < \sigma - \epsilon$; this means that (2.5.6) must diverge for $\alpha < \sigma$, and so $\rho_1 \ge \sigma$.

We are now going to infer from Jensen's theorem (1.2.1) some explicit connections between $n(r)$ and $\log M(r)$. We have, by Jensen's theorem,

$$(2.5.9)\quad N(r) = \int_0^r t^{-1}n(t)\,dt = (2\pi)^{-1}\int_0^{2\pi} \log|f(re^{i\theta})|\,d\theta \le \log M(r).$$

Suppose first that $f(z)$ is of order ρ. Then by (2.1.5),

$$(2.5.10)\qquad \log M(r) \le A(\epsilon)r^{\rho+\epsilon}, \qquad \epsilon > 0,\ r > r(\epsilon),$$

and since $n(t)$ increases,

$$(2.5.11)\quad n(r)\log 2 = n(r)\int_r^{2r} t^{-1}\,dt \le \int_r^{2r} t^{-1}n(t)\,dt$$

$$\le \log M(2r) \le A(\epsilon)(2r)^{\rho+\epsilon}, \qquad r > r(\epsilon).$$

Hence we have

2.5.12. *Theorem. If $f(z)$ is of order ρ, $n(r) = O(r^{\rho+\epsilon})$ for every positive ϵ.*

For another upper estimate of $n(r)$, see 2.5.21.

In general, we cannot make any lower estimate for $n(r)$, since (in the most extreme case) $f(z)$ may have large order but no zeros at all (for example, $f(z) = e^{z^n}$). When ρ is not an integer we shall see that there are always zeros, and indeed approximately the "right" number: see §2.9. This is an example of the difference between integral and nonintegral orders, which dominates much of the theory. For the present we consider only results which do not involve this distinction.

The ϵ cannot be dropped from 2.5.12[a], but it can for functions of finite type.

2.5.13. *Theorem. If $f(z)$ is of positive order ρ and finite type τ then*

$$L = \limsup_{r\to\infty} r^{-\rho} n(r) \le e\rho\tau, \tag{2.5.14}$$

$$l = \liminf_{r\to\infty} r^{-\rho} n(r) \le \rho\tau. \tag{2.5.15}$$

The gap between (2.5.14) and (2.5.15) cannot be narrowed.[b] However, it is not possible to have equality simultaneously in (2.5.14) and (2.5.15) if $\tau > 0$:

2.5.16.* *Theorem.*[c] *Under the hypotheses of* 2.5.13, $Le^{l/L} \le e\rho\tau$, *and in particular* $L + l \le e\rho\tau$.

Thus if there is equality in (2.5.14), $l = 0$; while if there is equality in (2.5.15), $L = l$.

To prove 2.5.13, we start from the fact that by (2.1.6) we have

$$\log M(r) \le (\tau + \epsilon) r^{\rho}, \qquad \epsilon > 0,\ r > r(\epsilon),$$

and so

$$N(r)\, r^{-\rho} \le \tau + \epsilon, \qquad r > r(\epsilon). \tag{2.5.17}$$

If $n(t) \ge \sigma t^{\rho}$ for $t \ge t_0$, (2.5.17) gives

$$r^{-\rho} \int_0^{t_0} t^{-1} n(t)\, dt + \sigma r^{-\rho} \int_{t_0}^{r} t^{\rho-1}\, dt \le \tau + \epsilon, \qquad r > t_0,$$

and (2.5.15) follows.

If $\beta > 1$, (2.5.17) also gives

$$n(r) \log \beta \le \int_r^{\beta r} t^{-1} n(t)\, dt \le \int_0^{\beta r} t^{-1} n(t)\, dt \le \beta^{\rho} (\tau + \epsilon) r^{\rho},$$

$$n(r) r^{-\rho} \le \frac{(\tau + \epsilon)\beta^{\rho}}{\log \beta}.$$

The right-hand side is smallest when $\beta = e^{1/\rho}$, and (2.5.14) follows.

An immediate corollary of 2.5.12 and 2.5.5 is

2.5.18. *Theorem. If $f(z)$ is of order ρ and has exponent of convergence ρ_1, then $\rho_1 \le \rho$.*

Thus we have the following relations among the order ρ, the exponent of convergence ρ_1, and the genus p of the set of zeros:[d]

$$\begin{cases} p = [\rho_1] \text{ if } \rho_1 \text{ is not an integer;} \\ \rho_1 - 1 \le p \le \rho_1 \le \rho \text{ in all cases.} \end{cases} \tag{2.5.19}$$

It is often desirable to distinguish between the cases where (2.5.3) converges or diverges for $\alpha = \rho_1$; we introduce the following definition.

2.5.20. *Definition. The entire function $f(z)$ of positive order ρ is said to be of convergence or divergence class according as $\sum_{n=1}^{\infty} r_n^{-\rho}$ converges or diverges.*

Connections between the properties of being of convergence or divergence class and other properties of the function will be given in §2.11.

A somewhat stronger result than (2.5.12) is

2.5.21. *Theorem.*[e] *If $f(z)$ has at least one zero (but $f(z) \not\equiv 0$),*

$$\liminf_{r\to\infty} \frac{n(r)}{\log M(r)} \le \liminf_{r\to\infty} \frac{\log n(r)}{\log r} = \lambda_1 \le \rho_1 . \tag{2.5.22}$$

The number λ_1 is sometimes called the lower order of the zeros, by analogy with (2.5.8). We have $\lambda_1 \le \lambda$ because, by (2.5.11),

$$\log n(r) + \log\log 2 \le \log\log M(2r),$$

$$\frac{\log n(r)}{\log r} \le \frac{\log\log M(2r)}{\log (2r)} (1 + o(1)).$$

Since $N(r) \le \log M(r)$ by (2.5.9), (2.5.22) will be established if we prove that

$$\liminf_{r\to\infty} \frac{n(r)}{N(r)} \le \lambda_1 . \tag{2.5.23}$$

Suppose that (2.5.23) is not true; then for a positive ϵ and all sufficiently large R we have

$$n(r) > (\lambda_1 + 2\epsilon)\, N(r), \qquad r \ge R.$$

Then if $S > R$ we have

$$\begin{aligned}(\lambda_1 + 2\epsilon) \int_R^S x^{-\lambda_1-\epsilon-1} N(x)\, dx &\le \int_R^S x^{-\lambda_1-\epsilon-1} n(x)\, dx \\ &= \int_R^S x^{-\lambda_1-\epsilon}\, dN(x) \\ &\le N(S)S^{-\lambda_1-\epsilon} + (\lambda_1 + \epsilon) \int_R^S N(x) x^{-\lambda_1-\epsilon-1}\, dx.\end{aligned}$$

Hence

$$(2.5.24) \quad \epsilon \int_R^S N(x)x^{-\lambda_1-\epsilon-1}\,dx \le N(S)S^{-\lambda_1-\epsilon} \le (\lambda_1 + 2\epsilon)^{-1}n(S)S^{-\lambda_1-\epsilon}.$$

By the definition of λ_1 in (2.5.22), there is a sequence of arbitrarily large values of S for which $n(S)S^{-\lambda_1-\epsilon} \to 0$. However, the integrand in (2.5.24) is positive and R is independent of S, so this makes (2.5.24) impossible. Hence (2.5.23) must be true.

2.6. Lemmas on infinite products. To proceed further with the connections between the growth of a function and the location of its zeros we need the analogue for transcendental entire functions of the factorization of a polynomial into linear factors. Complications arise both from the fact that an entire function is determined by its zeros only up to multiplication by a nonvanishing entire function, and from the fact that a straightforward infinite product exhibiting linear factors corresponding to the zeros will usually not converge. According to a familiar theorem of Weierstrass, any entire function can be expressed as an infinite product, but this theorem is so general that it is of little use. For functions of finite order there is a much more specific factorization due to Hadamard, and it is on this that much of the more detailed part of the theory is based.

In this section we collect some preliminary material on infinite products of the kind which we shall need.[a]

We introduce the Weierstrass primary factors,

$$(2.6.1) \quad E(u, 0) = 1 - u,$$
$$E(u, p) = (1 - u) \exp \{u + \tfrac{1}{2} u^2 + \cdots + p^{-1} u^p\}, \qquad p > 0,$$

and note that

$$(2.6.2) \qquad \log E(u, p) = -\sum_{k=p+1}^{\infty} k^{-1}u^k, \qquad |u| < 1,$$

$$(2.6.3) \qquad |\log E(u,p)| \le \sum_{k=p+1}^{\infty} |u|^k \le |u|^{p+1}/(1 - \epsilon), \qquad |u| < \epsilon < 1,$$

where the branch of the logarithm is that which is zero for $u = 0$.

Let $\{z_n\}_1^\infty$ be a sequence of complex numbers, numbered in order of increasing modulus, with $z_1 \neq 0$, and with convergence exponent ρ_1 (2.5.2) and genus p (2.5.4). Then the infinite product

$$(2.6.4) \qquad P(z) = \prod_{n=1}^{\infty} E(z/z_n, p)$$

is called a canonical product of genus p. By using (2.6.3) we see that it converges uniformly in every finite region and so represents an entire function. More precisely, we have

2.6.5. *Theorem. A canonical product $P(z)$ of genus p is an entire function of order equal to the convergence exponent of its zeros.*

Let ρ be the order of $P(z)$; we know that $\rho_1 \leq \rho$ (2.5.18), and we have to show that $\rho_1 \geq \rho$. Let $r_n = |z_n|$. It is convenient to put the essential part of the proof in the form of two lemmas which will also be useful later.

2.6.6. *Lemma.*

$$\sum_{n=1}^{N} \log |E(z/z_n, p)| \leq \begin{cases} Ar^p \sum_{n=1}^{N} r_n^{-p}, & p > 0; \\ \sum_{n=1}^{N} \log (1 + r/r_n), & p = 0. \end{cases}$$

2.6.7. *Lemma.*

$$\sum_{n=N+1}^{\infty} \log |E(z/z_n, p)| \leq Ar^{p+1} \sum_{n=N+1}^{\infty} r_n^{-p-1}.$$

Here A denotes a number depending only on p. By writing the sums as Stieltjes integrals and integrating by parts, and combining the results with an appropriate N, we have

$$\sum_{n=1}^{\infty} \log |E(z/z_n, p)| \leq K \left\{ r^p \int_0^r t^{-p-1} n(t)\, dt + r^{p+1} \int_r^{\infty} t^{-p-2} n(t)\, dt \right\}, \tag{2.6.8}$$

where K is another number depending only on p. It is sometimes convenient to have an estimate involving only a single integral; the inequality

$$\sum_{n=1}^{\infty} \log |E(z/z_n, p)| \leq Kr^{p+1} \int_0^{\infty} \frac{n(t)}{t^{p+1}(r+t)}\, dt \tag{2.6.9}$$

(with a different K) is equivalent to (2.6.8).

To prove 2.6.6 we break the sum into two parts, S_1 and S_2, in which $r_n \leq 2r$ and $r_n > 2r$, respectively. In S_2 we apply (2.6.3) with $\epsilon = \frac{1}{2}$ and obtain

$$\log |E(z/z_n, p)| \leq 2 |z/z_n|^{p+1} \leq |z/z_n|^p, \tag{2.6.10}$$

hence

$$S_2 \leq \sum_{2r < r_n \leq r_N} (r/r_n)^p = r^p \sum_{2r < r_n \leq r_N} r_n^{-p}. \tag{2.6.11}$$

For $p > 0$, in S_1 we have $r/r_n \geq \frac{1}{2}$, and so

$$(r/r_n)^k \leq 2^{p-k} (r/r_n)^p, \qquad 0 \leq k \leq p. \tag{2.6.12}$$

Thus

$$\begin{aligned}\log|E(z/z_n, p)| &\le \log|1 - z/z_n| + |z/z_n| + \cdots + p^{-1}|z/z_n|^p \\ &\le \log|1 - z/z_n| + 2^p|z/z_n|^p \\ &\le 2^{p+1}|z/z_n|^p,\end{aligned} \tag{2.6.13}$$

since

$$\log|1 + w| \le \log(1 + |w|) \le |w|. \tag{2.6.14}$$

Hence

$$S_1 \le 2^{p+1}r^p \sum_{r_n \le 2r} r_n^{-p}, \qquad p > 0. \tag{2.6.15}$$

Combining this with (2.6.11), we have 2.6.6. For $p = 0$, we have 2.6.6 directly from the first part of (2.6.14).

For 2.6.7 we proceed similarly, breaking the sum into parts S_1 and S_2 with $r_n \le 2r$ and $r_n > 2r$. In S_2 we use the first instead of the second inequality in (2.6.10), so that

$$S_2 \le 2 \sum_{r_n > 2r} (r/r_n)^{p+1} = 2r^{p+1} \sum_{r_n > 2r} r_n^{-p-1}. \tag{2.6.16}$$

For S_1, we have $2r/r_n \ge 1$, and so (2.6.13) leads to

$$\log|E(z/z_n, p)| \le 2^{p+2}(r/r_n)^{p+1},$$

hence

$$S_1 \le 2^{p+2}r^{p+1} \sum_{r_n \le 2r} r_n^{-p-1}; \tag{2.6.17}$$

(2.6.16) and (2.6.17) combine to give 2.6.7.

To prove 2.6.5 write

$$\log|P(z)| = \Bigl(\sum_{r_n \le 2r} + \sum_{r_n > 2r}\Bigr) \log|E(z/z_n, p)| = S_1 + S_2.$$

If we apply 2.6.6 to S_1, for $\epsilon > 0$ we find

$$S_1 \le Ar^p \sum_{r_n \le 2r} r_n^{-p} \le Ar^p(2r)^{\rho_1 - p + \epsilon} \sum_{r_n \le 2r} r_n^{-\rho_1 - \epsilon} = O(r^{\rho_1 + \epsilon}), \qquad p > 0;$$

$$\begin{aligned}S_1 &\le \log(1 + r/r_1) \sum_{r_n \le 2r} (r/r_n)^{\rho_1 + \epsilon}(r_n/r)^{\rho_1 + \epsilon} \\ &\le (2r)^{\rho_1 + \epsilon} \log(1 + r/r_1) \sum_{r_n \le 2r} r_n^{-\rho_1 - \epsilon} \\ &\le O(r^{\rho_1 + 2\epsilon}), \qquad p = 0,\end{aligned}$$

since $\sum r_n^{-\rho_1 - \epsilon}$ is bounded (because ρ_1 is the exponent of convergence of $\{z_n\}$).

For S_2 we have either $p = \rho_1 - 1$ or $p > \rho_1 - 1$. In the first case, 2.6.7 gives us

$$S_2 \leq Ar^{p+1} \sum_{r_n > 2r} r_n^{-p-1} = Ar^{\rho_1} \sum_{r_n > 2r} r_n^{-p-1} = O(r^{\rho_1}),$$

since $\rho_1 = p + 1$ and $\sum r_n^{-p-1}$ converges by the definition of p. In the second case, $\rho_1 + \epsilon < p + 1$ for sufficiently small positive ϵ, and then, again by 2.6.7,

$$S_2 \leq Ar^{p+1} \sum_{r_n > 2r} r_n^{-p-1} < Ar^{p+1}(2r)^{-p-1+\rho_1+\epsilon} \sum_{r_n > 2r} r_n^{-\rho_1-\epsilon} = O(r^{\rho_1+\epsilon}).$$

Hence $S_1 + S_2 = O(r^{\rho_1+\epsilon})$ in either case, so $\rho \leq \rho_1$ and 2.6.5 follows.

Theorem 2.6.5 determines the order of a canonical product. To determine its type is a more complicated matter, and we shall postpone discussion of the type until we are in a position to consider entire functions which are not canonical products as well as those which are (see §2.10).

It is convenient to prove here a lemma which gives a lower bound for a canonical product outside neighborhoods of its zeros. (More precise results in this direction will be discussed in Chapter 3.)

2.6.18. *Lemma. Let $P(z)$ be a canonical product of order ρ. If σ and ϵ are positive numbers, then for all sufficiently large r ("sufficiently large" depending on σ and ϵ), we have*

$$\log | P(z) | > -r^{\rho+\epsilon} \tag{2.6.19}$$

provided z is outside the circles of center z_n and radius $r_n^{-\sigma}$, $| r_n | > 1$.

We have

$$\log | P(z) | = \sum_{r_n \leq 2r} \log | 1 - z/z_n |$$
$$+ \sum_{r_n \leq 2r} \log | \exp \{z/z_n + \cdots + p^{-1}(z/z_n)^p\} | + \sum_{r_n \leq 2r} \log | E(z/z_n, p) |.$$

In the second sum we use the middle line of (2.6.13), and in the third sum we use the first inequality of (2.6.10). This gives

$$\log | P(z) | \geq \begin{cases} \sum_{r_n \leq 2r} \log | 1 - z/z_n | - 2 \sum_{r_n > 2r} r/r_n, & p = 0; \\ \sum_{r_n \leq 2r} \log | 1 - z/z_n | - 2^p \sum_{r_n \leq 2r} (r/r_n)^p \\ \qquad - 2 \sum_{r_n > 2r} (r/r_n)^{p+1}, & p > 0, \end{cases}$$
$$= \sum_{r_n \leq 2r} \log | 1 - z/z_n | - O(r^{\rho+\epsilon}),$$

just as in the proof of 2.6.5.

If now $|z - z_n| > r_n^{-\sigma}$ and $r_n \leq 2r$,

$$|1 - z/z_n| > r_n^{-1-\sigma} > (2r)^{-1-\sigma}.$$

Hence

$$\sum_{1<r_n\leq 2r} \log|1 - z/z_n| > - \sum_{1<r_n\leq 2r} (1+\sigma)\log(2r)$$

$$\geq -(1+\sigma)n(2r)\log(2r)$$

$$\geq O(r^{\rho+2\epsilon})$$

by 2.5.12 applied to $P(z)$; and

$$\sum_{r_n\leq 1} \log|1 - z/z_n| > 0$$

if $r > 2$. Hence (2.6.19) follows.

2.7. Hadamard's factorization theorem. We can now establish the standard factorization of an entire function of finite order.

2.7.1. *Hadamard's Factorization Theorem.*[a] *If $f(z)$ is an entire function of order ρ with an m-fold zero at the origin, we have*

(2.7.2) $$f(z) = z^m e^{Q(z)} P(z),$$

where $Q(z)$ is a polynomial of degree $q \leq \rho$ and $P(z)$ is the canonical product (of genus p) formed with the zeros (other than $z = 0$) of $f(z)$.

2.7.3. *Definition. The genus of $f(z)$ is* $\max(p, q)$.

We know that $P(z)$ is an entire function of order $\rho_1 \leq \rho$. Then $z^{-m}f(z)/P(z)$ is entire and has no zeros, and so can be written as $e^{Q(z)}$, where $Q(z)$ is entire; it remains to show that $Q(z)$ is a polynomial of degree at most ρ. If we choose $\sigma > \rho$ in 2.6.18, the sum of the radii of the excluded circles is finite and so

$$\Re(Q(z)) = \log|e^{Q(z)}| \leq r^{\rho+\epsilon}, \qquad \epsilon > 0,$$

on arbitrarily large circles. By (1.3.4) this is possible only if $Q(z)$ is a polynomial of degree ρ at most.

As an application of 2.7.1 we deduce a result on the minimum modulus $m(r)$. The behavior of $m(r)$ is of course much more complicated than that of $M(r)$; it vanishes when $f(z)$ has a zero of modulus r, but it has a tendency to be, sometimes, not much smaller than $1/M(r)$. The following result is fairly crude, but still accurate enough to be useful.

2.7.4. *Theorem. If $f(z)$ is of order ρ, then*

(2.7.5) $$m(r) \neq o(e^{-r^{\rho+\epsilon}})$$

for any positive ϵ.

For, in (2.7.2), $| P(z) | > e^{-r^{\rho+\epsilon}}$ for each positive ϵ and some z of arbitrarily large modulus, by 2.6.18. Also $| e^{Q(z)} | > e^{-Br^{\rho}}$ for some B and all large r, and (2.7.5) follows.

In the next chapter we shall give some results on the minimum modulus of functions of order at most 1 which are much more precise than 2.7.4. It was formerly conjectured that, for every positive ϵ, $m(r) \neq o\{M(r)\}^{-1-\epsilon}$ at least for entire functions of finite order, but this has recently been disproved by Hayman.[b]

However, the following result is true for all entire functions.

2.7.6.* *Theorem.*[c] *If* $f(z)$ *is entire and not constant, for each fixed* θ *and every positive* ϵ, $| f(re^{i\theta}) | \neq o\{M(r)\}^{-1-\epsilon}$.

2.8. Laguerre's theorem on separation of zeros.[a]

2.8.1. *Theorem. If* $f(z)$ *is an entire function, not a constant, which is real for real* z *and has only real zeros, and is of genus* 0 *or* 1, *then the zeros of* $f'(z)$ *are also real and are separated by the zeros of* $f(z)$.

We have, according as $f(z)$ is of genus 0 or 1,

$$f(z) = cz^k \prod_{n=1}^{\infty} (1 - z/z_n),$$

$$f(z) = cz^k e^{az} \prod_{n=1}^{\infty} (1 - z/z_n)e^{z/z_n},$$

where c, a, z_n are real. Then

$$f'(z)/f(z) = k/z + \sum_{n=1}^{\infty} (z - z_n)^{-1},$$

or

$$f'(z)/f(z) = k/z + a + \sum_{n=1}^{\infty} \left(\frac{1}{z - z_n} + \frac{1}{z_n}\right),$$

and in either case the imaginary part of $f'(z)/f(z)$ is

$$-iy \left\{\frac{k}{x^2 + y^2} + \sum_{n=1}^{\infty} \frac{1}{(x - z_n)^2 + y^2}\right\},$$

which cannot vanish unless $y = 0$, so that $f'(z)$ has only real zeros. Moreover,

$$\{f'(z)/f(z)\}' = -kz^{-2} - \sum_{n=1}^{\infty} (z - z_n)^2,$$

which is real and negative for real z, so that $f'(z)/f(z)$ decreases where it is continuous, i.e. between the zeros of $f(z)$, and so vanishes precisely once between each pair of zeros.

If $f(z)$ is of order less than 2 in 2.8.1 (it cannot be of order greater than 2), $f'(z)$ is of the same order and the theorem can be applied to $f'(z)$. (Actually $f'(z)$ is, in this case, of the same genus as $f(z)$, but this requires a result from the next chapter: see 3.3.9. If $f(z)$ is of genus 1 but order 2 it is possible that $f'(z)$ is of genus 2.) By using this fact repeatedly we can deduce an inequality satisfied by the coefficients in the power series of an entire function satisfying the hypotheses of 2.8.1.

2.8.2. *Theorem. If* $f(z)$ *is a transcendental entire function of order less than* 2, *real for real* z, *and with only real zeros, and* $f(z) = \sum_{n=0}^{\infty} c_n z^n$, *then*

$$(n + 1)c_{n+1}c_{n-1} < nc_n^2. \tag{2.8.3}$$

This means in particular that for sufficiently large indices n there are never two consecutive zero coefficients, and also that the coefficients on opposite sides of a gap have opposite signs.

We saw in the proof of 2.8.1 that

$$\{f'(x)/f(x)\}' = \frac{f(x)f''(x) - \{f'(x)\}^2}{\{f(x)\}^2} < 0;$$

hence $f(x)f''(x) < \{f'(x)\}^2$; applying this to $f^{(n-1)}(x)$ we have

$$f^{(n-1)}(x)f^{(n+1)}(x) < \{f^{(n)}(x)\}^2,$$

and taking $x = 0$ we have (2.8.3).

2.9. The zeros of functions of nonintegral order. A number of conclusions can be drawn directly from 2.7.1 when ρ is not an integer, since then $q < p$ and the behavior of $f(z)$ is dominated by that of the canonical product $P(z)$.

2.9.1. *Theorem. If* ρ *is not an integer,* $\rho = \rho_1$.

In fact, $\rho_1 \le \rho$. If $\rho_1 < \rho$, then $P(z)$ is of order ρ_1 and $e^{Q(z)}$ is of order $q < \rho$ (because q is an integer, $q \le \rho$, and ρ is not an integer). This would make $f(z)$ of order less than ρ.

As a corollary, we have

2.9.2. *Theorem. An entire function of nonintegral order has an infinite set of zeros.*

This result can be considerably sharpened. According to 2.5.21, we have

$$\liminf_{r \to \infty} n(r)/\log M(r) \le \rho;$$

thus $n(r)$ is sometimes almost as small as a multiple of $\log M(r)$; while by 2.5.12, $n(r)$ is never much larger than r^{ρ}. The following theorems are in the opposite direction and improve 2.9.2 by saying how large $n(r)$ must sometimes be if $f(z)$ is of nonintegral order ρ.

2.9.3. *Theorem. If* ρ *is not an integer, then for any positive* ϵ, $n(r) \ne o(r^{\rho-\epsilon})$.

2.9.4. *Theorem.*[a] *If* ρ *is not an integer,* $n(r) \ne o\{\log M(r)\}$.

2.9.5. *Theorem.*[b] *If ρ is not an integer, $f(z)$ is of zero type if and only if $n(r) = o(r^\rho)$, and of finite type if and only if $n(r) = O(r^\rho)$.*

The proof of 2.9.3 is simple. Since ρ is not an integer, $\rho = \rho_1$ and so

$$\int_0^\infty t^{-\rho-1+\epsilon} n(t)\,dt \tag{2.9.6}$$

diverges for every positive ϵ (2.5.5). If we had $n(t) = O(t^{\rho-2\epsilon})$ we should have, as $T \to \infty$,

$$\int_0^T t^{-\rho-1+\epsilon} n(t)\,dt = O\left\{\int_1^T t^{-1-\epsilon}\,dt\right\} = O(1),$$

and (2.9.6) would converge.

Theorem 2.9.4 is more difficult. We have to distinguish two cases, according as $f(z)$ is of convergence class (2.5.20) or not. In either case, $p < \rho < p + 1$ and $\int^\infty t^{-m}\, n(t)\,dt$ diverges for $m < \rho + 1$ and converges for $m > \rho + 1$. By (2.6.8) there is a finite K such that

$$\log M(r) \le K\left\{r^p \int_1^r t^{-p-1} n(t)\,dt + r^{p+1}\int_r^\infty t^{-p-2} n(t)\,dt\right\} \equiv K\psi(r). \tag{2.9.7}$$

It is therefore sufficient to show that it is impossible, for arbitrarily small positive ϵ, to have

$$n(r) \le \epsilon\psi(r), \qquad r > r(\epsilon). \tag{2.9.8}$$

For convenience suppose $n(r) = 0$ for $r \le 1$. Take m so that $\rho + 1 \le m < p + 2$ and so that $\int^\infty t^{-m}\, n(t)\,dt$ converges (it converges for $m > \rho + 1$ in both cases). If we multiply (2.9.8) by r^{-m}, integrate over (R, ∞) $(R > r(\epsilon))$, and change the order of integration in the resulting iterated integrals, we find

$$\begin{aligned}\int_R^\infty t^{-m} n(t)\,dt &\le \epsilon \int_1^R u^{-p-1} n(u)\,du \int_R^\infty t^{p-m}\,dt \\ &\quad + \epsilon\int_R^\infty u^{-p-1} n(u)\,du \int_u^\infty t^{p-m}\,dt + \epsilon\int_R^\infty u^{-p-2} n(u)\,du \int_R^u t^{p-m+1}\,dt \\ &\le \frac{\epsilon R^{p-m+1}}{m-p-1}\int_1^R n(u) u^{-p-1}\,du + \frac{\epsilon}{m-p-1}\int_R^\infty n(u) u^{-m}\,du \\ &\qquad + \frac{\epsilon}{p-m+2}\int_R^\infty n(u) u^{-m}\,du.\end{aligned}$$

Hence if $\epsilon < \frac{1}{2}(\rho - p)(p - \rho + 1)$, we find, on collecting terms,

$$\frac{1}{2}\int_R^\infty t^{-m} n(t)\, dt \le \frac{\epsilon R^{p-m+1}}{m - p - 1}\int_1^R u^{-p-1} n(u)\, du. \tag{2.9.9}$$

In case $f(z)$ is of divergence class, let $m \to \rho + 1$; the left-hand side becomes infinite while the right-hand side approaches a finite limit, so (2.9.8) leads to a contradiction.

In case $f(z)$ is of convergence class, we may take $m = \rho + 1$ to begin with; since $n(t)$ increases, (2.9.9) implies

$$\tfrac{1}{2}\, n(R)\rho^{-1}R^{-\rho} \le \frac{\epsilon R^{p-\rho}}{\rho - p}\int_1^R u^{-p-1} n(u)\, du,$$

and since this holds, for large enough R, for every positive ϵ, we have

$$n(r) = o\left\{r^p \int_1^r t^{-p-1} n(t)\, dt\right\}.$$

Since $\int^\infty t^{-p-m} n(t)\, dt$ diverges for $1 < m < \rho + 1 - p$, for such m we have, as $R \to \infty$,

$$\begin{aligned}\int_1^R n(r) r^{-p-m}\, dr &= o\left\{\int_1^R r^{-m}\, dr \int_1^r t^{-p-1} n(t)\, dt\right\}\\ &= o\left\{\int_1^R t^{-p-1} n(t)\, dt \int_t^R r^{-m}\, dr\right\}\\ &= o\left\{\int_1^R n(t) t^{-p-m}\, dt\right\},\end{aligned}$$

a contradiction. Theorem 2.9.4 is thus proved in both cases.

To prove 2.9.5, it is enough, by 2.5.13, to prove that when ρ is not an integer, $f(z)$ is of finite type if $n(r) = O(r^\rho)$ and of zero type if $n(r) = o(r^\rho)$.

In the Hadamard factorization of $f(z)$ we have $p < \rho < p + 1$, $q < \rho$, and so $\log|e^{Q(z)}| \le O(r^q) = o(r^\rho)$. Hence we have to deal only with the canonical product $P(z)$. By (2.6.9) we have, if $n(r) = O(r^\rho)$, a constant C for which

$$\begin{aligned}\log|P(z)| &\le C r^{p+1}\int_0^\infty \frac{dt}{t^{p-\rho+1}(r + t)}\\ &\le C\left(r^p \int_0^r \frac{dt}{t^{p-\rho+1}} + r^{p+1}\int_r^\infty \frac{dt}{t^{p-\rho+2}}\right)\\ &\le O(r^\rho),\end{aligned}$$

since $0 < p - \rho + 1 < 1$.

If $n(r) = o(r^\rho)$, we have, similarly, with a small positive ϵ and a sufficiently large R,

$$\log | P(z) | \leq Cr^p \int_0^R \frac{dt}{t^{p-\rho+1}} + \epsilon r^p \int_R^r \frac{dt}{t^{p-\rho+1}} + \epsilon r^{p+1} \int_r^\infty \frac{dt}{t^{p-\rho+2}},$$

and so $\log | P(z) | \leq o(r^\rho)$.

2.10. The zeros of functions of integral order. It is clear that no theorem as simple as 2.9.5 can hold for entire functions of integral order, since the order can be larger than the number of zeros would indicate. Thus e^z has no zeros; $1/\Gamma(z)$ has $n(r) = O(r)$ but is of infinite type. These examples indicate the two possibilities: the nonvanishing factor $e^{Q(z)}$ may dominate $f(z)$, or the canonical product $P(z)$ may be of larger type than the number of zeros suggests ($n(r) = O(r)$ and $P(z)$ of infinite type, or $n(r) = o(r)$ and $P(z)$ of positive type; cf. Chapter 8). We now prove the correct analogue of Theorem 2.9.5 for functions of integral order.

2.10.1. *Lindelöf's Theorem.*[a] *If ρ is a positive integer, the entire function $f(z)$ of order ρ is of finite type if and only if both $n(r) = O(r^\rho)$ and the sums*

(2.10.2) $$S(r) = \sum_{|z_n| \leq r} z_n^{-\rho}$$

are bounded.

2.10.3. *Theorem. If ρ is a positive integer, the entire function $f(z)$ of order ρ is of zero type if and only if either (a) $n(r) = o(r^\rho)$, $p = \rho$, and*

(2.10.4) $$\sum_{n=1}^{\infty} z_n^{-\rho} = -\rho\alpha_0,$$

where α_0 (possibly 0) is the coefficient of z^ρ in $Q(z)$ in the Hadamard factorization 2.7.1; or (b) $p = \rho - 1$ and $\alpha_0 = 0$.

In particular, an entire function of integral order is of zero type if its genus is less than its order.

We require the following consequence of Jensen's theorem (1.2.1).

2.10.5. *Lemma.*[b] *If $f(z)$ is an entire function, $f(0) = 1$, σ is a positive number, and $2\pi r l(\sigma, r)$ denotes the length of that part of the circumference $| z | = r$ on which $| f(re^{i\theta}) | < M(r)^{-\sigma}$, then for large r*

(2.10.6) $$l(\sigma, r) \leq (1 + \sigma)^{-1}.$$

In fact, on the specified part of $| z | = r$ we have

(2.10.7) $$(2\pi)^{-1} \int \log | f(re^{i\theta}) | \, d\theta \leq -\sigma l(\sigma, r) \log M(r);$$

while on the rest of the circle $\log |f(re^{i\theta}) | \leq \log M(r)$ and the corresponding contribution to the left-hand side of (2.10.7) is at most

$$\{1 - l(\sigma, r)\} \log M(r).$$

On the other hand, by Jensen's theorem the left-hand side is not negative and so

$$\{1 - l(\sigma, r)\} \log M(r) - \sigma l(\sigma, r) \log M(r) \geq 0;$$

(2.10.6) follows.

We now prove 2.10.1, which is a little simpler than 2.10.3. If $p = \rho - 1$, $\sum |z_n|^{-\rho}$ converges and so $S(r)$ is bounded. If $f(z)$ is of finite type, by (2.5.13) we have $n(r) = O(r^\rho)$. Hence it is enough to show that (a) if $n(r) = O(r^\rho)$ and $p = \rho$ the boundedness of $S(r)$ is necessary and sufficient for $f(z)$ to be of finite type; (b) if $p = \rho - 1$, and $S(r)$ is bounded, $f(z)$ is of finite type.

We have $E(u, \rho) = \exp(\rho^{-1}u^\rho)E(u, \rho - 1)$, and so from the Hadamard factorization

$$D \equiv \log |f(z)| - \Re\left\{z^\rho \left(\alpha_0 + \rho^{-1} \sum_{n=1}^{N} z_n^{-\rho}\right)\right\}$$

$$(2.10.8) \qquad = \Re(\alpha_1 z^{\rho-1} + \cdots + \alpha_\rho + m \log z) + \sum_{n=1}^{N} \log |E(z/z_n, \rho - 1)| + \sum_{n=N+1}^{\infty} \log |E(z/z_n, \rho)|,$$

where the integer N is to be specified later. Applying Lemmas 2.6.6 and 2.6.7, we have, with $|z| = r$,

$$D \leq O(r^{\rho-1}) + O\left(r^{\rho-1} \sum_{n \leq N} r_n^{-\rho+1}\right) + O\left(r^{\rho+1} \sum_{n>N} r_n^{-\rho-1}\right), \quad p > 1;$$

$$D \leq O(\log r) + O\left(\sum_{n \leq N} \log(1 + r/r_n)\right) + O\left(r^2 \sum_{n>N} r_n^{-2}\right), \quad p = 1.$$

Since $n(r) = O(r^\rho)$, we have $n < Br_n$ and so, if we choose $N = [r]$, we have

$$D \leq O(r^{\rho-1}) + O\left(r^{\rho-1} \sum_{n \leq N} n^{-(\rho-1)/\rho}\right) + O\left(r^{\rho+1} \sum_{n>N} n^{-(\rho+1)/\rho}\right)$$

$$(2.10.9) \qquad \leq O(r^{\rho-1}) + O(r^\rho) + O(r^\rho) = O(r^\rho), \qquad p > 1;$$

$$D \leq O(\log r) + O\left(\sum_{n \leq N} \log(1 + r/n)\right) + O\left(r^2 \sum_{n>N} n^{-2}\right)$$

$$\leq O(r), \qquad p = 1.$$

If $S(r)$ is bounded, it follows at once from (2.10.8) and (2.10.9) that $\log |f(z)| \leq O(r^\rho)$, i.e., that $f(z)$ is of finite type.

The inference in the other direction is harder, since (2.10.8) and (2.10.9) lead only to

$$(2.10.10)\qquad -\Re\left\{z^{\rho}\left[\alpha_0+\rho^{-1}\sum_{n=1}^{N} z_n^{-\rho}\right]\right\}\le -\log|f(z)|+O(r^{\rho}).$$

However, if $f(z)$ is of finite type we have $\log M(r) \le Ar^{\rho}$ with some constant A, and hence, for $\sigma = 3$, Lemma 2.10.5 shows that

$$\log|f(z)| < -3Ar^{\rho} \le -3\log M(r)$$

at most on one-fourth of the circumference $|z| = r$. Thus $|\log f(z)|| = O(r^{\rho})$ at least on three-fourths of $|z| = r$, so that by (2.10.10)

$$(2.10.11)\qquad -\Re\left\{z^{\rho}\left[\alpha_0+\rho^{-1}\sum_{n=1}^{N} z_n^{-\rho}\right]\right\}\le O(r^{\rho}),$$

on at least three-fourths of $|z| = r$. Suppose now that $S(r)$ is unbounded; then there is an increasing sequence of integers N such that

$$\alpha_0+\rho^{-1}\sum_{n=1}^{N} z_n^{-\rho} = R_N e^{i\phi_N},\qquad R_N\to\infty.$$

With $|z| = r_N$ and $r_N^{\rho} = N$,

$$\Re(z^{\rho}R_N\exp(i\phi_N)) = r_N^{\rho}R_N\cos(\theta+\phi_N) < -\tfrac{1}{2}r_N^{\rho}R_N$$

on the part of the circumference $|z| = r$ where $\cos(\theta+\phi)\le -\frac{1}{2}$, i.e. on at least one-third of the circumference, and so

$$(2.10.12)\qquad -\Re\left\{z^{\rho}\left[\alpha_0+\rho^{-1}\sum_{n=1}^{N} z_n^{-\rho}\right]\right\} > \tfrac{1}{2}\, r_N^{\rho}R_N$$

on at least one-third of $|z| = r_N$. Since $R_N\to\infty$, this contradicts the fact that (2.10.11) holds on three-fourths of every circle. This disposes of case (a).

Case (b) is an immediate consequence of the following lemma.

2.10.13. *Lemma. If* $\sum r_n^{-\rho}$ *converges, the canonical product*

$$(2.10.14)\qquad P(z) = \prod_{n=1}^{\infty} E(z/z_n, \rho-1)$$

is of growth $(\rho, 0)$.

By Lemma 2.5.5, $\int^{\infty} t^{-\rho-1}n(t)\,dt$ converges, so

$$n(r)r^{-\rho}\le \rho n(r)\int_r^{\infty} t^{-\rho-1}\,dt \le \rho\int_r^{\infty} t^{-\rho-1}n(t)\,dt = o(1),\qquad r\to\infty.$$

That is, $n(r) = o(r^\rho)$, so $n = o(r_n{}^\rho)$. Then by (2.6.9),

$$\log | P(z) | \le Kr^\rho \int_0^\infty \frac{n(t)}{t^\rho (r+t)}\, dt$$

$$\le Kr^{\rho-1} \int_0^r t^{-\rho} n(t)\, dt + Kr^\rho \int_r^\infty t^{-\rho-1} n(t)\, dt$$

$$\le Kr^{\rho-1} \int_0^r o(1)\, dt + Kr^\rho o(1)$$

$$\le o(r^\rho).$$

Next we prove 2.10.3. If $f(z)$ is of zero type, we have $n(r) = o(r^\rho)$, and so it is enough to prove that (a) if $n(r) = o(r^\rho)$, $p = \rho$, (2.10.4) holds if and only if $f(z)$ is of zero type; (b) if $p = \rho - 1$, $f(z)$ is of zero type if and only if $\alpha_0 = 0$.

The proof in case (a) follows the same lines as the proof of 2.10.1. We start from (2.10.8); since now we have $n < \epsilon r_n{}^\rho$ for an arbitrarily small ϵ and sufficiently large n, we have $D \le o(r^\rho)$ instead of (2.10.9), and so (2.10.4) implies $\log | f(z) | \le o(r^\rho)$, i.e. that $f(z)$ is of zero type. On the other hand, if $f(z)$ is of zero type, we find as before that

$$-\Re\left\{z^\rho \left[\alpha_0 + \rho^{-1} \sum_{n=1}^{N} z_n^{-\rho}\right]\right\} \le o(r^\rho)$$

on at least three-fourths of $| z | = r$, while if (2.10.4) is false, (2.10.12) still holds on at least one-third of $| z | = r_N$, where $r_N \to \infty$ and $\liminf R_N > 0$, and these two facts are in contradiction.

For case (b) of 2.10.3 we use 2.10.13. If $p = \rho - 1$, we have just seen that $n(r) = o(r^\rho)$, and if $\alpha_0 = 0$, $f(z)$ is the product of two functions of growth $(\rho, 0)$ and so of zero type if (as assumed) it is of order ρ. It remains to prove that if $p = \rho - 1$ and $\alpha_0 \ne 0$, in which case $e^{Q(z)}$ is of order ρ and positive type while $P(z)$ is at most of order ρ and zero type, $f(z)$ is not of zero type. By applying Lemma 2.10.5 to $P(z)$ we see that $| \log | P(z) || = o(r^\rho)$ at least on three-fourths of every sufficiently large circle. On the other hand, $\log | e^{Q(z)} |$ is dominated by $\Re(\alpha_0 z^\rho)$ for large r, and if $\alpha_0 = \beta e^{i\phi}$, $\Re(\alpha_0 z^\rho) = \beta r^\rho \cos(\phi + \rho\theta) > \frac{1}{2} \beta r^\rho$ on at least one-third of every circle; so $\log | f(z) | > \frac{1}{4} \beta r^\rho$ at some points of every large circle, and so $f(z)$ is not of zero type.

It is no longer true for functions of integral order that $n(r) \ne o\{\log M(r)\}$; but in the case where the behavior of $f(z)$ is dominated by that of its canonical product, there is an analogue of (2.9.4).

2.10.15.* *Theorem.*[c] *If $f(z)$, not a constant, is of integral order and of the same genus as its canonical product, then*

(2.10.16) $$\limsup_{r\to\infty} \frac{n(r)\phi(r)}{\log M(r)} = \infty$$

for any positive $\phi(r)$ *such that*

(2.10.17) $$\int^{\infty} x^{-1}\{\phi(x)\}^{-1}\,dx$$

converges.

The relation (2.10.16) may fail to hold when (2.10.17) diverges.[d]

It is convenient to add here another result which is qualitatively of the same kind as 2.10.5.

2.10.18. *Theorem.*[e] *If* $f(z)$ *is an entire function of positive order* ρ, *then*

(2.10.19) $$\int_0^{2\pi} |\log|f(re^{i\theta})||\,d\theta = O(r^{\rho+\epsilon}), \qquad \epsilon > 0;$$

if $f(z)$ *is of finite type,*

(2.10.20) $$\int_0^{2\pi} |\log|f(re^{i\theta})||\,d\theta = O(r^{\rho});$$

and if $f(z)$ *is of zero type*

(2.10.21) $$\int_0^{2\pi} |\log|f(re^{i\theta})||\,d\theta = o(r^{\rho}).$$

As a consequence of (for example) (2.10.20) we have the result that if $f(z)$ is of finite type of order ρ and $\psi(\theta)$ is a bounded measurable function, then

(2.10.22) $$\int_0^{2\pi} \psi(\theta)\log|f(re^{i\theta})|\,d\theta = O(r^{\rho}),$$

a fact which will be needed later.

We may suppose that $f(0) = 1$. Then from Jensen's theorem we have

$$0 \le \int_0^{2\pi} \log|f(re^{i\theta})|\,d\theta,$$

and hence

$$-\int_0^{2\pi} \log^-|f(re^{i\theta})|\,d\theta \le \int_0^{2\pi} \log^+|f(re^{i\theta})|\,d\theta \le 2\pi\log M(r),$$

so that

$$\int_0^{2\pi} |\log|f(re^{i\theta})||\,d\theta \le 4\pi\log M(r),$$

and the conclusions of 2.10.18 follow.

2.11. Further relations between growth and zeros. In 2.5.20, entire functions of order ρ were separated into two classes according to whether $\sum r_n^{-\rho}$ converges or diverges. In accordance with the general rule that a greater density of zeros is associated with more rapid growth, we should expect a tendency for $M(r)$ to be larger for a function of divergence class. This expectation is justified by the following theorem.

2.11.1. *Theorem.*[a] *If $f(z)$ is of order ρ and divergence class, the integral*

$$(2.11.2) \qquad \int^{\infty} r^{-1-\rho} \log M(r)\, dr$$

diverges; the converse is true when ρ is not an integer.

We also have

2.11.3. *Theorem. If $f(z)$ is of order ρ and positive type, the integral* (2.11.2) *diverges.*

Even more than this is true:

2.11.4. *Theorem.*[a] *If $f(z)$ is of positive integral order ρ and genus ρ then* (2.11.2) *diverges.*

Thus (2.11.2) can converge, for a function $f(z)$ of integral order ρ, only if $f(z)$ is not only of zero type but of genus $\rho - 1$.

If $f(z)$ is of positive type, there exist a positive number c and a sequence $\{r_n\}$, $r_{n+1} > 2r_n$, such that $\log M(r_n) > cr_n$. Hence $\log M(r) > cr_n^{\rho}$ for $r_n \le r \le 2r_n$, since $M(r)$ increases, and so

$$\int_{r_1}^{\infty} r^{-1-\rho} \log M(r)\, dr \ge \sum_{n=1}^{\infty} cr_n^{\rho} \int_{r_n}^{2r_n} r^{-1-\rho}\, dr = c\rho^{-1}(1 - 2^{-\rho}) \sum 1 = \infty.$$

This proves 2.11.3.

To establish 2.11.4 we have only to consider the case where $f(z)$ is of order and genus ρ but of zero type. If (2.11.2) converges, by 2.11.1, $\sum r_n^{-\rho}$ converges, so that the canonical product of $f(z)$ is of genus $\rho - 1$; then by 2.10.3 it follows that the genus of $f(z)$ is also $\rho - 1$, contrary to hypothesis.

We now prove 2.11.1. For the first part, we recall that by 2.5.5, $\sum r_n^{-\rho}$ converges or diverges with $\int_0^{\infty} x^{-\rho-1}\, n(x)\, dx$ (we assume without loss of generality that $f(0) \ne 0$). Now

$$(2.11.5) \qquad \int_0^r x^{-\rho-1} n(x) dx = \int_0^r x^{-\rho} dN(x) = r^{-\rho} N(r) + \rho \int_0^r x^{-\rho-1} N(x)\, dx$$

$$\le r^{-\rho} \log M(r) + \rho \int_0^r x^{-\rho-1} \log M(x)\, dx.$$

If (2.11.2) converges,

$$\rho \int_r^\infty x^{-\rho-1} \log M(x)\, dx \geq r^{-\rho} \log M(r),$$

and so the right-hand side of (2.11.5) is bounded; hence $\sum r_n^{-\rho}$ converges. That is, (2.11.2) must diverge if $f(z)$ is of divergence class.

For the second part of 2.11.1, suppose that ρ is not an integer; by (2.6.8) we have

$$\log M(r) - o(r^\rho) \leq K \left\{ r^p \int_0^r t^{-p-1} n(t)\, dt + r^{p+1} \int_r^\infty t^{-p-2} n(t)\, dt \right\}$$

$$= K(J_1(r) + J_2(r)),$$

say, where $p < \rho < p + 1$. We have to prove that the convergence of $\int^\infty t^{-\rho-1} n(t)\, dt$ implies the convergence of $\int^\infty r^{-\rho-1} J_k(r)\, dr$, $k = 1, 2$. Now

$$\int_R^\infty r^{-\rho-1} J_2(r)\, dr = \int_R^\infty r^{p-\rho}\, dr \int_r^\infty t^{-p-2} n(t)\, dt$$

$$= \int_R^\infty t^{-p-2} n(t) dt \int_R^t r^{p-\rho}\, dr$$

$$\leq (p - \rho + 1)^{-1} \int_R^\infty t^{-\rho-1} n(t)\, dt < \infty,$$

and

$$\int_0^R r^{-\rho-1} J_1(r)\, dr = \int_0^R r^{p-\rho-1}\, dr \int_0^r t^{-p-1} n(t)\, dt$$

$$= \int_0^R t^{-p-1} n(t)\, dt \int_t^R r^{p-\rho-1}\, dr$$

$$\leq (\rho - p)^{-1} \int_0^R t^{-\rho-1} n(t)\, dt < (\rho - p)^{-1} \int_0^\infty t^{-\rho-1} n(t)\, dt.$$

Theorem 2.11.1 leads to a number of conclusions about entire functions of nonintegral order, since (2.11.2) involves only $\log M(r)$ and not the zeros.[b] Thus if $g(z)$ is such that (2.11.2) diverges for $g(z)$ and $M_f(r) \geq \{M_g(r)\}^k$, $k > 0$, then $f(z)$ is also of divergence class. If $f(z)$ is of order ρ and $g(z)$ is of lower order, then $f + g$ is of the same class (convergence or divergence) as f.

It is also possible to connect the class of an entire function with its coefficients.[c] Let $f(z) = \sum c_n z^n$, and let the points $(n, -\log |c_n|)$ be plotted in a rectangular coordinate system. They determine a polygon (Newton polygon), above or on which all the points lie, and which is the graph of the largest convex function with this property. If we put G_n for the ordinate of this polygon at abscissa n, and define $R_n = \exp(G_n - G_{n-1})$, it turns out that

2.11.6.* *Theorem.*[d] *If $f(z)$ is of nonintegral order ρ, it is of convergence class if and only if*

$$\sum_{n=1}^{\infty} R_n^{-\rho} < \infty. \tag{2.11.7}$$

In one special case which often occurs in applications, the R_n can be computed more directly.

2.11.8.* *Theorem.*[e] *If $f(z)$ is of order at most* 1 *and its zeros are all real and negative, then $R_n = |c_{n-1}/c_n|$.*

The proof depends on 2.8.2.

Another result in the same general direction is as follows.

2.11.9.* *Theorem.*[f] *If $\sum_{n=q}^{\infty} |c_{n+1}/c_n|^\rho < \infty$, for some q, then $f(z)$ is at most of order ρ; if $f(z)$ is of order ρ it is of convergence class; if ρ is an integer the integral* (2.11.2) *also converges, and $f(z)$ is of genus $\rho - 1$.*

2.12. Functions of genus 0. The genus of an entire function of order ρ is $[\rho]$ when ρ is not an integer; but the genus of an entire function of positive integral order ρ can be either ρ or $\rho - 1$. Unless the function is initially defined by its Hadamard factorization, it is not always easy to find out what its genus is. In this section we give some theorems on the simplest case, that in which $f(z)$ is of order 1.

As an illustration of what can happen, we note the following result.

2.12.1.* *Theorem.*[a] *There is an entire function $f(z)$ of order* 1 *and genus* 0 *such that $f(z) - a$ is of genus* 1 *if $a \neq 0$, and $f(z) + f(-z)$ is also of genus* 1.

An example of such a function is

$$f(z) = \prod_{n=2}^{\infty} \left(1 + \frac{z}{n(\log n)^\alpha}\right), \qquad 1 < \alpha \leq 2.$$

This makes it clear that the genus of an entire function is not necessarily preserved under addition either of a function of the same genus or of one of lower genus. It also shows that the genus cannot be determined from the coefficients or from the maximum modulus in any very obvious way, and so we can expect simple theorems only under supplementary hypotheses. However, the situation described in 2.12.1 is in a sense exceptional, since we have the following result.

2.12.2.* *Theorem.*[b] *If $f(z)$ is an entire function of integral order, all the functions $f(z) - a$, except for at most one value of a, have the same genus.*

We have already seen (2.11.4) that a sufficient condition for the entire function $f(z)$ of order 1 to be of genus 0 is that

$$\int^{\infty} r^{-2} \log M(r)\, dr \text{ converges.} \tag{2.12.3}$$

This condition is not necessary because of 2.12.1. We now impose the restriction that $f(z)$ is an even function; and then the relationships among genus, maximum modulus and coefficients are much simplified—essentially because an even entire function of order 1 is reducible to an entire function of the fractional order ½. For an even entire function, 2.11.4 has a valid converse.

2.12.4. *Theorem.*[c] *If $f(z)$ is an even entire function of genus 0, then $\int^{\infty} r^{-2} \log M(r)\, dr$ converges.*

Therefore we may state

2.12.5. *Theorem. The even entire function $f(z)$ is of genus 0 if and only if $\int^{\infty} r^{-2} \log M(r)\, dr$ converges.*

In proving 2.12.4 we may assume that $f(0) = 1$ and that $f(z)$ is not a constant; then we have

$$f(z) = \prod_{n=1}^{\infty} (1 - z/z_n), \tag{2.12.6}$$

with

$$\sum 1/|z_n| < \infty. \tag{2.12.7}$$

Since (2.12.7) makes the product in (2.12.6) converge absolutely, we may rearrange it in the form

$$f(z) = \prod_{n=1}^{\infty} (1 - z^2/\beta_n^2), \qquad \Re(\beta_n) \geq 0. \tag{2.12.8}$$

Now we have

$$|f(z)| \leq \prod_{n=1}^{\infty} (1 + r^2/|\beta_n|^2),$$

and so it is sufficient to prove 2.12.4 for a function of the form

$$f(z) = \prod_{n=1}^{\infty} (1 + z^2/\alpha_n^2), \qquad \alpha_n > 0,$$

for which $M(r) = f(r)$. Put

$$g(z) = f(z^{1/2}) = \prod_{n=1}^{\infty} (1 + z/\alpha_n^2);$$

then $g(z)$ is at most of order $\frac{1}{2}$, convergence class, and attains its maximum modulus for real positive z. By the second part of 2.11.1, $\int^{\infty} r^{-3/2} \log g(r)\, dr$ converges, that is, $\int^{\infty} r^{-2} \log g(r^2)\, dr = \int^{\infty} r^{-2} \log f(r)\, dr$ converges, the desired conclusion.

From 2.12.4 we have at once the following corollaries.

2.12.9. *Theorem.*[d] *If $f(z)$ and $F(z)$ are even entire functions and $|f(z)| \le F(|z|)$, then $f(z)$ is of genus 0 if $F(z)$ is of genus 0.*

2.12.10. *Theorem.*[d] *If $f(z) = \sum_{n=0}^{\infty} a_{2n} z^{2n}$, $F(z) = \sum_{n=0}^{\infty} b_{2n} z^{2n}$, and $|a_{2n}| \le b_{2n}$, then $f(z)$ is of genus 0 if $F(z)$ is of genus 0.*

We can now show that the behavior described in 2.12.1 cannot occur for even entire functions.

2.12.11. *Theorem.*[e] *If $f_1(z)$ and $f_2(z)$ are even entire functions of genus 0, then $f_1(z) + f_2(z)$ is of genus 0.*

Under the hypothesis of 2.12.11, (2.12.3) holds for f_1 and for f_2, by 2.12.4. Since

$$\log \max_{|z| \le r} |f_1(z) + f_2(z)| \le \log \{M_1(r) + M_2(r)\} \le \log 2 + \log^+ M_1(r) + \log^+ M_2(r),$$

it follows from 2.11.4 that $f_1 + f_2$ is of genus 0.

Let the power series of the even entire function $f(z)$ be

$$f(z) = \sum_{n=0}^{\infty} c_{2n} z^{2n}. \tag{2.12.12}$$

2.12.13. *Theorem.*[f] *The even entire function* (2.12.12) *is of genus* 0 *if and only if*

$$F(z) = \sum_{n=0}^{\infty} |c_{2n}| z^{2n} \tag{2.12.14}$$

is of genus 0.

That $f(z)$ is of genus zero if $F(z)$ is of genus zero follows at once from 2.12.10. The converse, which seems more unexpected, follows from 2.12.10 also. For, let $f(z)$ have the product representation (2.12.8),

$$f(z) = \prod_{n=1}^{\infty} (1 - z^2/\beta_n^2), \qquad \sum 1/|\beta_n| < \infty.$$

The function

$$F^*(z) = \prod_{n=1}^{\infty} (1 + z^2/|\beta_n|^2)$$

is also of genus zero (since $\sum 1/|\beta_n|$ converges), and if the product is multiplied out we obtain a power series for $F^*(z)$, with nonnegative coefficients c_{2n}^*, such that $c_{2n}^* \geq |c_{2n}|$, since the c_{2n} can be obtained by multiplying out the product for $f(z)$. Therefore $F(z)$ is of genus 0 by 2.12.10.

If $f(z) = \sum_{n=0}^{\infty} c_n z^n$ and $f(z)$ is of order less than 1, we have (by (2.2.7)) $|c_n| \leq n^{-n/(\rho+\epsilon)}$ for large n, where $\rho + \epsilon < 1$. Then $|c_n|^{1/n} \leq n^{-1/(\rho+\epsilon)}$ and consequently $\sum |c_n|^{1/n}$ converges. It is not true that this series always converges if $f(z)$ is of order 1 and genus 0, as can be shown by the example quoted in connection with 2.12.1. However, the series does converge if $f(z)$ is an even function of genus zero.

2.12.15.* *Theorem.*[g] *If $f(z)$ is an even entire function of genus* 0 *with the power series* (2.12.12), *then*

$$\sum_{n=0}^{\infty} |c_{2n}|^{1/(2n)} \tag{2.12.16}$$

and even

$$\sum_{n=0}^{\infty} \sup_{k \geq 0} |c_{2n+2k}|^{1/(2n+2k)} \tag{2.12.17}$$

converge.

Notes for Chapter 2

General note: Results not otherwise credited to sources can be found in such books as Titchmarsh [3], Valiron [3], [5], Borel [1].

2.1a. Recent Russian literature uses the term "function of finite degree".

2.1b. The lower order was introduced by J. M. Whittaker [1].

2.1c. For elementary constructions of proximate orders see Valiron [10], p. 25, Shah [18]. Shah [22] has discussed lower proximate orders. Levin [1] has introduced another kind of proximate order: there is always a function $L(r)$ which is positive, continuous, satisfies $k^{-\epsilon} \leq L(kr)/L(r) \leq k^{\epsilon}$ for $r \geq R(\epsilon)$, $k \geq 1$ (R independent of k) and for which $\log M(r) \leq r^{\rho} L(r)$ with equality for a sequence $r_k \to \infty$. This proximate order appears to be worth further study.

2.2a. S. Bernstein ([2], p. 80) originally defined ν of (2.2.9), $\rho = 1$, as the degree of $f(z)$. More recent Russian literature identifies the degree with τ.

2.3a. Shah [19], where further results along this line are to be found.

2.3b. Shah [24].

2.3c. Some of these relationships also hold for $\rho = 0$ or ∞. For proofs and further relationships see Valiron [1], [3], Pólya and Szegö [1], vol. 2, pp. 1–13, Mazurkiewicz [1], Okamura [1], Whittaker [1], Shah [10]–[16], Singh [1]. The notations of these authors are not always consistent either with each other or with this book.

2.4a. For these and other results see Shah [21], [25], Singh [1].

2.5b. For examples see Boas [15], Buck [6].

2.5c. Shah [23], Lakshminarasimhan [1], Singh [2].

2.5d. Examples illustrating the various possibilities are constructed by Shah [4].

2.5e. (2.5.21), with ρ_1 instead of λ_1, was proved by Pólya [1], sharpened by Valiron

[4], and obtained in its present form by Shah [5]. The proof of the text is new. Valiron deduced from Jensen's theorem that lim inf $\{n(r) \log r\}/\{\log M(r) \log\log M(r)\} \leq 1$; this follows from (2.5.21) if $\lambda > 0$. Further results connecting $n(r)$ and $M(r)$ are given by Shah [1], [7], [8], [9]; see also §§2.9, 2.10.

2.6a. For information on infinite products in general see Knopp [1]. The theorems of this section are due to Borel: see for example Valiron [3] or Borel [1].

2.7a. For alternative proofs see Titchmarsh [3], Chandrasekharan [1].

2.7b. Hayman [1]. Hayman also gives lower estimates for $m(r)$.

2.7c. Beurling [1] proves a stronger result.

2.8a. Titchmarsh [3], Borel [1].

2.9a. See Valiron [1], Pólya [1], and especially Valiron [4], where an explicit positive lower bound is obtained for lim sup $n(r)/\log M(r)$. For the proof of the text see Boas [28].

2.9b. Lindelöf [2].

2.10a. 2.10.1 and 2.10.3 are both due to Lindelöf [2]; the first will be used more frequently.

2.10b. Lindelöf [2]. By further developing the idea of this proof, Boas, Buck and Erdős [1] proved that there is a universal constant K, $0 < K \leq (1 + \sigma)^{-1}$, such that the set of points for which $|f(z)| \leq M(r)^{-\sigma}$ has upper planar density at most K; as $\sigma \to \infty$, the lower planar density of the set is $o(1/\sigma)$. In particular, $|f(z)| \leq 1/M(r)$ at most on a set of upper planar density $\frac{1}{2}$. Much stronger results are true for entire functions of small order. Cf. Arima [1].

2.10c. Shah [2], [3], [6]. For a short proof along the lines of that of 2.9.4 see Boas [28].

2.10d. Shah [3].

2.10e. A special case is given by Levin [6] with an unnecessarily complicated proof.

2.11a. Valiron [2].

2.11b. See Valiron [2] for these and further results.

2.11c. Valiron [3].

2.11d. Valiron [2].

2.11e. Chang [1].

2.11f. Valiron [2].

2.12a. Lindelöf [1]. The specific function mentioned is discussed briefly by Valiron ([3], pp. 87–88). Another possibility which it illustrates is that a function of genus 1 can have its derivative of genus 0.

2.12b. Valiron [3], pp. 182 ff.

2.12c. S. Bernstein [16].

2.12d. Videnskii [1].

2.12e. S. Bernstein [14].

2.12f. Videnskii [1].

2.12g. (2.12.16) is due to S. Bernstein [2]. Chang [1] gives two simplified proofs and a generalization to functions of larger order. (2.12.17) and another generalization are given by Mandelbrojt [1], p. 76.

Chapter 3

THE MINIMUM MODULUS

3.1. Functions of order less than ½. The theorems of this chapter give lower bounds for the minimum modulus $m(r)$ of entire functions of (in general) small order. Of course $m(r)$ vanishes whenever $f(z)$ has a zero of modulus r, but we may hope to show that $m(r)$ is sometimes not very small, or even that $m(r)$ is frequently quite large. Such results are important in applications where one has to divide by a given entire function.

We begin by seeing what can be said by rather elementary methods. We have already noticed (2.7.4) that

$$m(r) \neq o(e^{-r^{\rho+\epsilon}})$$

for each positive ϵ; but if $\rho < \frac{1}{2}$ it is easy to get much more.

3.1.1. *Theorem.*[a] *If $f(z)$ is of growth $(\frac{1}{2}, 0)$ and not a constant,*

$$\limsup_{r\to\infty} m(r) = \infty. \tag{3.1.2}$$

Later we shall see that more precise statements can be made both about how large $m(r)$ must be and for how many values of r.

We may assume that $f(0) = 1$ without affecting (3.1.2). Then

$$f(z) = \prod_{n=1}^{\infty} (1 - z/z_n).$$

Let

$$g(z) = \prod_{n=1}^{\infty} (1 + z/r_n). \tag{3.1.3}$$

Then since $|1 - z/z_n| \geq |1 - r/r_n|$,

$$m_f(r) \geq m_g(r) = |g(-r)|, \qquad M_f(r) \leq g(r) = M_g(r). \tag{3.1.4}$$

By theorems 2.5.8, 2.9.1 and 2.9.5, $g(z)$ is also of growth $(\frac{1}{2}, 0)$, and 3.1.2 follows if we prove

3.1.5. *Theorem. An entire function $g(z)$ of growth $(\frac{1}{2}, 0)$ is not bounded on any half-line unless it is a constant.*

To prove this, consider $h(z) = g(z^2)$, which is of growth (1,0). If $g(z)$ is bounded on a half-line, $h(z)$ is bounded on a line, which we may take to be the imaginary axis. Using Theorem 1.4.3, we infer that $h(z)$ is bounded in $x \geq 0$, and in $x \leq 0$, and so everywhere; hence $h(z)$ is a constant.[b] (Cf. § 6.2.)

By using a more precise auxiliary theorem we can prove considerably more than 3.1.1.

3.1.6. *Theorem.*[c] *If $f(z)$ is of growth $(\rho, 0)$, $0 < \rho < \frac{1}{2}$, and $f(z)$ is not a constant, there is a sequence $r_n \to \infty$ such that*

$$\log m(r_n) > \cos \pi\rho \cdot \log M(r_n). \tag{3.1.7}$$

As in proving 3.1.1, it is enough to prove the theorem for $g(z)$ (with real negative zeros). We consider $G(z) = g(z^2)$, for which we have $G(z)$ regular for $x \geq 0$, $|G(iy)| \leq m(y^2)$, and $|G(z)|$ unbounded. Then 1.4.6, with $\phi(y) = m(y^2)$, gives

$$\log M(r_n) < \sec \pi\rho \cdot \log m(r_n)$$

for a sequence $r_n \to \infty$.

For functions of order $\frac{1}{2}$ and positive type we do not even know that $m(r)$ is unbounded, so the following result is of interest.

3.1.8.* *Theorem.*[d] *If $m(r)$ is bounded and*

$$\liminf_{r\to\infty} r^{-1/2} \log M(r) < \infty$$

then

$$\lim_{r\to\infty} r^{-1/2} \log M(r)$$

exists.

Even 3.1.6 does not tell the entire truth about the minimum modulus, since a similar, but not quite so precise, result holds for all entire functions of order less than 1.

3.2. Functions of order less than 1. We prove the following theorem.

3.2.1. *Theorem.*[a] *If $0 < \rho < 1$ and $f(z)$ is of growth $(\rho, 0)$ then*

$$\limsup_{r\to\infty} \frac{\log m(r)}{\log M(r)} \geq \cos \pi\rho. \tag{3.2.2}$$

Because of (3.1.4) we may consider the function $g(z)$ of (3.1.3) instead of $f(z)$. This is obvious when $\rho < \frac{1}{2}$, since replacing $f(z)$ by $g(z)$ makes $m(r)$ smaller and $M(r)$ larger and $\cos \pi\rho$ is positive. When $\rho \geq \frac{1}{2}$, we suppose as before that $f(0) = 1$. If we select a point z' with $|z'| = r$ and $|f(z')| = m(r)$, we have

$$|g(r)g(-r)| \leq |f(z')f(-z')| \leq m(r)M(r),$$

since

$$\left|\left(1 + \frac{r}{r_n}\right)\left(1 - \frac{r}{r_n}\right)\right| = \left|1 - \frac{r^2}{r_n^2}\right| \leq \left|1 - \frac{z'^2}{z_n^2}\right| = \left|\left(1 + \frac{z'}{r_n}\right)\left(1 - \frac{z'}{r_n}\right)\right|.$$

If $|g(-r)| > |g(r)|^{\cos \pi\rho - \epsilon}$ for arbitrarily large values of r, then

$$m(r)M(r) \geq |g(r)g(-r)| > |g(r)|^{1+\cos \pi\rho - \epsilon} \geq M(r)^{1+\cos \pi\rho - \epsilon},$$

and (3.2.2) follows.

We now establish (3.2.2) for $g(z)$. We need the values of two definite integrals; the first transforms by integration by parts into a standard Γ-function formula, and the second may be evaluated by contour integration.

3.2.3. *Lemma. If* $0 < \alpha < 1$,

$$\int_0^\infty x^{-\alpha-1} \log (1 + x)\, dx = \pi\alpha^{-1} \csc \pi\alpha,$$

$$\int_0^\infty x^{-\alpha-1} \log | 1 - x |\, dx = \pi\alpha^{-1} \cot \pi\alpha.$$

Hence we have

$$\int_0^\infty \frac{\log | 1 - t | - \cos \pi\alpha \cdot \log (1 + t)}{t^{1+\alpha}}\, dt = 0. \tag{3.2.4}$$

From this we could deduce that

$$\int_0^\infty \frac{\log | g(-r) | - \cos \pi\alpha \cdot \log g(r)}{r^{1+\alpha}}\, dr = 0, \qquad \alpha > \rho,$$

which implies that the numerator is positive for at least one value of r; however, we need to prove that the numerator is positive for arbitrarily large values of r. We shall therefore prove that

$$\int_x^\infty \frac{\log | g(-r) | - \cos \pi\alpha \cdot \log g(r)}{r^{1+\alpha}}\, dr > 0, \qquad \alpha > \rho, \tag{3.2.5}$$

for all positive x.

To do this, we put $\theta(t)$ equal to the numerator in (3.2.4) and

$$\psi(x) = \int_x^\infty t^{-1-\alpha}\theta(t)\, dt.$$

Then we have $\psi(\infty) = 0$, $\psi(0) = 0$ by (3.2.3), so that

$$\psi(x) = -\int_0^x t^{-1-\alpha}\theta(t)\, dt.$$

A sketch of the graph of $\theta(t)$ shows that $\theta(t)$ is negative near 0 and has just one change of sign in $(0, \infty)$, and so $\psi(x)$ first increases and then decreases; since it vanishes at 0 and ∞ it is strictly positive in $0 < x < \infty$. If we put

$$\lambda(x) = \psi(x)x^\alpha/\log (1 + x),$$

we find that $\lambda(\infty) = \alpha^{-1}(1 - \cos \pi\alpha)$, $\qquad \lambda(0+) = (1 - \alpha)^{-1}(1 + \cos \pi\alpha)$. Since $\lambda(x)$ is continuous, does not vanish for $0 < x < \infty$, and has positive limits at 0 and ∞, it has a positive minimum[b] $k(\alpha)$. Thus

$$\psi(x) \geq k(\alpha)x^{-\alpha} \log (1 + x), \qquad 0 < x < \infty,$$

and so

$$(3.2.6)\qquad \int_{x/r_n}^{\infty} \frac{\log|1-t| - \cos\pi\alpha\cdot\log(1+t)}{t^{1+\alpha}}\,dt \geq k(\alpha)(r_n/x)^{\alpha}\log(1+x/r_n).$$

Now

$$\log|g(-r)| = \sum_{n=1}^{\infty}\log|1-r/r_n|, \qquad \log g(r) = \sum_{n=1}^{\infty}\log(1+r/r_n),$$

and if $\alpha > \rho$ we have $\sum r_n^{-\alpha}$ convergent. We then have

$$\int_x^{\infty} r^{-1-\alpha}\log g(r)\,dr = \sum_{n=1}^{\infty}\int_x^{\infty} r^{-1-\alpha}\log(1+r/r_n)\,dr$$
$$= \sum_{n=1}^{\infty} r_n^{-\alpha}\int_{x/r_n}^{\infty} t^{-1-\alpha}\log(1+t)\,dt,$$

the interchange of limiting operations being justified because everything is positive; and

$$\int_x^{\infty} r^{-1-\alpha}\log|g(-r)|\,dr = \sum_{n=1}^{\infty}\int_x^{\infty} r^{-1-\alpha}\log|1-r/r_n|\,dr$$
$$= \sum_{n=1}^{\infty} r_n^{-\alpha}\int_{x/r_n}^{\infty} t^{-1-\alpha}\log|1-t|\,dt,$$

because the integral is dominated by the convergent integral

$$\int_0^{\infty} r^{-1-\alpha}\,|\log|g(-r)||\,dr = \sum_{n=1}^{\infty}\int_0^{\infty} r^{-1-\alpha}\,|\log|1-r/r_n||\,dr$$
$$= \sum_{n=1}^{\infty} r_n^{-\alpha}\int_0^{\infty} t^{-1-\alpha}\,|\log|1-t||\,dt.$$

Thus

$$\int_x^{\infty}\frac{\log|g(-r)| - \cos\pi\alpha\cdot\log g(r)}{r^{1+\alpha}}\,dr$$
$$(3.2.7)\qquad = \sum_{n=1}^{\infty} r_n^{-\alpha}\int_{x/r_n}^{\infty}\frac{\log|1-t| - \cos\pi\alpha\cdot\log(1+t)}{t^{1+\alpha}}\,dt$$
$$\geq k(\alpha)x^{-\alpha}\sum_{n=1}^{\infty}\log(1+x/r_n) = k(\alpha)x^{-\alpha}\log g(x),$$

by (3.2.6). This is considerably more than (3.2.5). Using only (3.2.5), however, we see that there must be arbitrarily large values of r for which the integrand is positive, i.e. $\log |g(-r)| > \cos \pi\alpha \cdot \log g(r)$, and hence

$$\log m(r) > \cos \pi\alpha \cdot \log M(r). \tag{3.2.8}$$

Since we required only that $\sum r_n^{-\alpha}$ converges, (3.2.8) still holds when $\alpha = \rho$ and $f(z)$ is of order ρ and convergence class, which is more than is asserted in 3.2.1.

Still more than this is true, and we quote some further results.

3.2.9.* *Theorem.*[c] *If* $\epsilon > 0$ *and* $f(z)$ *is of order greater than* $\rho - \epsilon$, *but of growth* $(\rho, 0)$, $0 < \rho < \frac{1}{2}$, *there is a sequence* $r = r_n \to \infty$ *such that*

$$\log m(r) > \cos \pi\rho \cdot \log M(r) > r^{\rho-\epsilon}. \tag{3.2.10}$$

We already know that each of the inequalities in (3.2.10) holds for a sequence of values of r; the point of 3.2.9 is that there are arbitrarily large values of r for which both are satisfied simultaneously, and hence for which $\log m(r)$ has a large lower bound independent of $f(z)$; otherwise it might happen that $\log m(r)$ would exceed $\cos \pi\rho \cdot \log M(r)$ only for values of r at which $\log M(r)$ is itself relatively small.

The fact that $\log m(r) > r^{\rho-\epsilon}$ for a sequence $r_n \to \infty$, which follows from (3.2.10), can be proved by taking $\alpha = \rho$ in (3.2.7), since if we had

$$\log |g(-r)| - \cos \pi\alpha \cdot \log g(r) > r^{\rho-\epsilon}$$

for all $r > x$ we should obtain a contradiction.

Actually, not only is $m(r)$ sometimes large, but it is large for a considerable proportion of the values of r, as the following theorems indicate.

3.2.11.* *Theorem.*[d] *If* $0 < \rho < 1$ *and* $\epsilon > 0$,

$$\log m(r) > (\cos \pi\rho - \epsilon) \log M(r)$$

for $R_n \le r \le R_n + R_n^{1-\rho-\epsilon}$, *where* $R_n \to \infty$.

3.2.12.* *Theorem.*[e] *If* $0 < \rho < \frac{1}{2}$ *and* $\epsilon > 0$, *the upper density of the set of r's for which* $\log m(r) > r^{\rho-\epsilon}$ *is at least* $1 - 2\rho$.

For similar results see also §3.7 and Theorem 4.4.17.

3.3. Functions of order 1. For functions of order 1 there is a theorem which may be regarded as a limiting case of the theorems of §3.2.

3.3.1. *Theorem.*[a] *If* $f(z)$ *is an entire function of order* 1 *then, for each positive* ϵ,

$$\limsup_{r\to\infty} m(r)M(r)^{1+\epsilon} = \infty. \tag{3.3.2}$$

In particular, if $f(z)$ is of exponential type τ,

$$\limsup_{r\to\infty} m(r)e^{r(\tau+\epsilon)} = \infty. \tag{3.3.3}$$

For further results in this direction see §3.7.

Let the zeros of $f(z)$ be z_n, $|z_n| = r_n$; then

$$g(z) \equiv f(z)f(-z) = \prod_{n=1}^{\infty} (1 - z^2/z_n^2),$$

and

$$h(z) = \prod_{n=1}^{\infty} (1 - z^2/r_n^2)$$

is an entire function of the same order as $g(z)$, and of exponential type if $g(z)$ is of exponential type, of zero exponential type if $g(z)$ is of zero exponential type. Indeed, $g(z^{1/2})$ is an entire function of order at most ½; by (2.9.5), whether it is of growth (½, 0) or (½, c), $c > 0$, is determined by its $n(r)$, which is the same as for $h(z^{1/2})$. We shall show that for real x and each positive ϵ.

(3.3.4) $$|h(x)| \neq o\{M_f(x)^{-\epsilon}\}, \qquad |x| \to \infty;$$

from this we have, since $|g(z)| \geq |h(|z|)|$, $|g(z)| \geq M_f(r)^{-\epsilon}$ for a sequence of arbitrarily large values of r, and so

$$|f(z)| = |g(z)/f(-z)| \geq M_f(r)^{-1-\epsilon}$$

for the same values of r; and this is (3.3.2). It remains to prove (3.3.4).

Suppose that (3.3.4) is not true, so that

(3.3.5) $$|h(x)| \leq M_f(x)^{-\epsilon} \qquad -\infty < x < \infty.$$

If $f(z)$ is of zero exponential type, so is $h(z)$, and (3.3.5) is a contradiction even with $\epsilon = 0$ since an entire function of zero exponential type which is bounded on a line is a constant (the same fact was used in proving 3.1.5).

If $f(z)$ is of zero or finite exponential type, we apply Carleman's theorem (1.2.2) to $h(z)$ in the upper half plane, and obtain, using (3.3.5),

$$0 \leq O(1) - \epsilon \int_1^r (x^{-2} - r^{-2}) \log M_f(x)\, dx$$

$$+ (\pi r)^{-1} \int_0^\pi \log |h(re^{i\theta})| \sin\theta\, d\theta$$

$$\leq O(1) - \epsilon \int_1^r x^{-2} \log M_f(x)\, dx + \epsilon r^{-1} \log M(r) + 2(\pi r)^{-1} \log M_h(r).$$

If $f(z)$ is of finite positive type of order 1, the last two terms are bounded while the integral is unbounded, by 2.11.3, and we have a contradiction.

Finally, if $f(z)$ is of order 1 and infinite type, we have

$$\log M_h(r) = \log \prod_{n=1}^{\infty} (1 + r^2/r_n^2) = r^2 \int_0^\infty \frac{n(t)\, dt}{t(t^2 + r^2)}$$

$$\leq r^2\left(\int_0^r + \int_r^\infty\right) \leq N(r) + r^2 \int_r^\infty \frac{dN(t)}{t^2 + r^2}$$

$$= \frac{N(r)}{2} + 2r^2 \int_r^\infty \frac{N(t)t\,dt}{(t^2 + r^2)^2}$$

$$\leq \frac{\log M_f(r)}{2} + 2r^2 \int_r^\infty t^{-3} \log M_f(t)\,dt.$$

Thus we have

$$0 \leq O(1) - \epsilon \int_1^r x^{-2} \log M_f(x)\,dx + (\epsilon + \tfrac{1}{2})r^{-1} \log M_f(r)$$

$$+ (2/\pi)r \int_r^\infty x^{-3} \log M_f(x)\,dx.$$

This is, however, impossible, as we see from the following lemma, in which $\phi(x) = \log M_f(x)$ and δ is arbitrarily small.

3.3.6. *Lemma. If $\phi(x)$ is positive and $\int_1^\infty x^{-2-\delta}\,\phi(x)\,dx$ exists, $\delta > 0$, and a and b are positive constants, then*

$$\limsup_{r\to\infty} \frac{x \int_1^x t^{-2}\phi(t)\,dt}{a\phi(x) + bx^2 \int_x^\infty t^{-3}\phi(t)\,dt} \geq \frac{1 - \delta}{\delta\{(1 - \delta)a + b\}}. \tag{3.3.7}$$

To prove the lemma we suppose that the conclusion is false and obtain a contradiction. Assume then that for $r \geq R$ we have

$$\Phi(x) \equiv \int_1^x t^{-2}\phi(t)\,dt \leq c\left\{ax^{-1}\phi(x) + bx \int_x^\infty t^{-3}\phi(t)\,dt\right\}, \tag{3.3.8}$$

where c is less than the constant on the right of (3.3.7). Then

$$\int_1^\infty x^{-1-\delta}\Phi(x)\,dx = \int_1^\infty t^{-2}\phi(t)\,dt \int_t^\infty x^{-1-\delta}\,dx = \delta^{-1} \int_1^\infty t^{-2-\delta}\phi(t)\,dt,$$

and so the integral on the left converges. From (3.3.8) we then have

$$\int_R^\infty x^{-1-\delta}\Phi(x)\,dx \leq ca \int_R^\infty x^{-2-\delta}\phi(x)\,dx + cb \int_R^\infty x^{-\delta}\,dx \int_x^\infty t^{-3}\phi(t)\,dt$$

$$= ca \int_R^\infty x^{-2-\delta}\phi(x)\,dx + cb \int_R^\infty t^{-3}\phi(t)\,dt \int_R^t x^{-\delta}\,dx$$

$$\leq \left\{ca + \frac{cb}{1 - \delta}\right\} \int_R^\infty x^{-2-\delta}\phi(x)\,dx.$$

Now

$$\int_R^\infty x^{-2-\delta}\phi(x)\,dx = \int_R^\infty x^{-\delta}\,d\Phi(x) = \delta\int_R^\infty x^{-\delta-1}\Phi(x)\,dx,$$

since $\Phi(x) = o(x^\delta)$. Hence

$$\int_R^\infty x^{-1-\delta}\Phi(x)\,dx \le \delta c\left\{a + \frac{b}{1-\delta}\right\}\int_R^\infty x^{-\delta-1}\Phi(x)\,dx,$$

which is a contradiction since the coefficient of the integral on the right is less than 1.

An application of (3.3.3) is

3.3.9.* *Theorem.*[b] *If $f(z)$ is a function of order less than 2, satisfying the hypotheses of Laguerre's theorem* 2.8.1, *then $f'(z)$ is of the same genus* (0 *or* 1) *as $f(z)$.*

3.4. The minimum modulus of a polynomial. We shall need the following result.

3.4.1. *Boutroux-Cartan lemma.*[a] *Let* $P(z) = \prod_{\nu=1}^{n}(z - z_\nu)$; *for any positive H, the inequality*

$$|P(z)| > (H/e)^n \tag{3.4.2}$$

holds outside at most n circles the sum of whose radii is at most $2H$.

We shall prove this on the assumption that all the z_ν are different; the general case follows by continuity.

We group the zeros z_ν into classes as follows.Let λ_1 be the largest integer which is such that there is a circle C_1 of radius $\lambda_1 H/n$ containing exactly λ_1 points z_ν. We must of course show that λ_1 exists. If some circle of radius H contains all the z_ν, then $\lambda_1 = n$. If not, and some circle of radius $H(n-1)/n$ contains $n-1$ of the z_ν, then $\lambda_1 = n - 1$. If not, we continue in this way; in the worst case, we should have no circle of radius H/n containing exactly one z_ν. Suppose that every circle of radius H/n containing one z_ν contains at least two; then the concentric circle of radius $2H/n$ contains the circles of radius H/n about both these and so contains at least four z_ν's; the concentric circle of radius $4H/n$ contains at least eight; and so on; finally the circle of radius $2^p H/n$, where $2^p \le n < 2^{p+1}$, and hence the circle of radius H, contain at least $2^{p+1} > n$ of the z_ν, a contradiction. Hence λ_1 does in fact exist, and we put the λ_1 points z_ν which are in the circle C_1 into the first class and call them of rank λ_1.

Next consider the points (if any) which are not of rank λ_1; there is a largest integer $\lambda_2 \le n - \lambda_1$ which is such that there is a circle C_2 of radius $\lambda_2 H/n$ containing exactly λ_2 of the remaining points z_ν; these form the second class and are called of rank λ_2. We proceed in this way, considering next points which are neither in the first nor the second class, and so on, until we have all the z_ν's grouped into p classes, and $\lambda_1 + \lambda_2 + \cdots + \lambda_p = n$. The sum of the radii of the C_k is thus H.

Now let S be any circle of radius $\lambda H/n$, $\lambda \leq n$ (an integer); if S contains at least λ points z_ν, then S contains a point of rank at least λ. For, if S is of radius $\lambda H/n$, $\lambda > \lambda_2$ (hence $\lambda > n - \lambda_1$), and S contains λ points z_ν, there are λ_1 points of rank λ_1 and hence at least one of them is in S. Moreover, $\lambda_1 > \lambda$ by the definition of λ_1. If $\lambda_3 < \lambda \leq \lambda_2$, then $\lambda > n - \lambda_1 - \lambda_2$, and if S contains λ points, none of rank λ_1, there are $n - \lambda_1$ points remaining, of which λ are in S, and λ_2 points of rank λ_2, so at least one of them is in S. Moreover, $\lambda_2 \geq \lambda$. This process can be continued down to $\lambda = 1$.

Let $\Gamma_1, \Gamma_2, \cdots$ be circles concentric with $C_1, C_2, \cdots$, and of twice the radii, and let z be outside all the Γ_k. Then a circle S of center z and radius $\lambda H/n$ contains at most $\lambda - 1$ points z_ν. For, if z_ν is a point in S of rank λ_j, and a_j is the center of the corresponding C_j,

$$2\lambda_j H/n \leq | z - a_j | < \lambda_j H/n + \lambda H/n,$$

and hence $\lambda_j < \lambda$. Now let the z_ν be arranged in order of increasing distance from z; the first point is at least H/n from z, the second at least $2H/n$, etc.; hence

$$| P(z) | \geq (H/n)^n \, n! \geq (H/e)^n.$$

3.5. Lemmas on functions of small order. The lemmas of this section are of no particular intrinsic interest, but it is convenient to group them together here.

3.5.1. *Lemma. If $f(z)$ is an entire function of genus zero, with $f(0) = 1$, we have*

$$\log M(r) \leq N(r) + Q(r), \tag{3.5.2}$$

where

$$Q(r) = r \int_r^\infty t^{-2} n(t)\, dt. \tag{3.5.3}$$

Since we have $N(r) \leq \log M(r)$ by Jensen's theorem, we thus have

$$N(r) \leq \log M(r) \leq N(r) + Q(r), \tag{3.5.4}$$

so that $Q(r)$ appears as a term measuring the error involved in replacing $\log M(r)$ by $N(r)$.

We have

$$f(z) = \prod_{n=1}^{\infty} (1 - z/z_n),$$

and we increase $| f(z) |$ if we replace $| 1 - z/z_n |$ by $1 + r/r_n$; hence, using the inequality $\log(1 + x) \leq x$ $(x > 0)$ and the fact that $n(x) = o(x)$ (2.5.13, 2.10.3, 2.10.13), we obtain

$$\log M(r) \leq \log \prod_{n=1}^{\infty} (1 + r/r_n) = \int_0^\infty \log(1 + r/t)\, dn(t)$$

$$= r \int_0^\infty \frac{n(t)}{t(t+r)}\, dt = r\left(\int_0^r + \int_r^\infty\right) \frac{n(t)}{t(t+r)}\, dt \leq N(r) + Q(r).$$

3.5.5. *Lemma. If $f(z)$ is of order zero, and not constant,*

$$\limsup_{r\to\infty} N(r)/Q(r) = \infty. \tag{3.5.6}$$

This is, for $\rho = 0$, a strengthening of (2.5.21), which states that

$$\limsup \{\log M(r)\}/n(r) = \infty,$$

since $N(r) \leq \log M(r)$ and

$$Q(r) = r \int_r^\infty t^{-2} n(t)\, dt \geq rn(r) \int_r^\infty t^{-2}\, dt = n(r).$$

Assume that (3.5.6) is not true, so that for some positive α and R,

$$N(r) \leq \alpha Q(r), \qquad r > R. \tag{3.5.7}$$

If $0 < \beta < 1/(1 + \alpha)$, we have $\int^\infty u^{-1-\beta}\, n(u)\, du$ convergent, and so

$$\int_R^\infty t^{-\beta-1} Q(t)\, dt = \int_R^\infty t^{-\beta}\, dt \int_t^\infty u^{-2} n(u)\, du$$

$$= \int_R^\infty u^{-2} n(u)\, du \int_R^u t^{-\beta}\, dt$$

$$\leq (1 - \beta)^{-1} \int_R^\infty u^{-1-\beta} n(u)\, du,$$

so that the left-hand integral is finite; and then, by (3.5.7),

$$\int_R^\infty t^{-\beta-1} Q(t)\, dt \leq (1 - \beta)^{-1} \int_R^\infty t^{-\beta}\, dN(t)$$

$$\leq \beta(1 - \beta)^{-1} \int_R^\infty t^{-\beta-1} N(t)\, dt$$

$$\leq \alpha\beta(1 - \beta)^{-1} \int_R^\infty t^{-\beta-1} Q(t)\, dt.$$

Since $\alpha\beta/(1 - \beta) < 1$ and $Q(t) \not\equiv 0$, this is a contradiction, and so (3.5.7) must fail for arbitrarily large values of R.

3.5.8. *Lemma. If* $\log M(r) = O((\log r)^2)$, $Q(r) = O(\log r)$.

In fact, by (2.5.9), $N(r) = O((\log r)^2)$, and then

$$n(r) \log r = n(r) \int_r^{r^2} t^{-1}\, dt \leq \int_r^{r^2} t^{-1} n(t)\, dt \leq N(r^2) = O((\log r)^2);$$

so $n(r) = O(\log r)$,

$$Q(r) = r \int_r^\infty t^{-2} n(t)\, dt = O\left(r \int_r^\infty t^{-2} \log t\, dt\right)$$

$$= O\left(r^{\frac{1}{2}} \log r \int_r^\infty t^{-3/2}\, dt\right) = O(\log r).$$

It is interesting to note that the converse of 3.5.8 also holds, and that the hypothesis (or conclusion) is equivalent to $n(r) = O(\log r)$. In the first place, $n(r) \leq Q(r)$ and so $n(r) = O(\log r)$ if $Q(r) = O(\log r)$, while we have just seen that $Q(r) = O(\log r)$ if $n(r) = O(\log r)$. Moreover, unless $f(z)$ is a polynomial, $\log M(r)/\log r \to \infty$ and so by (3.5.4),

$$N(r) \sim \log M(r) \text{ if } n(r) = O(\log r) \tag{3.5.9}$$

(since this statement is trivial for polynomials). Finally, if $n(r) = O(\log r)$,

$$N(r) = \int_0^r t^{-1} n(t)\, dt = O\left(\int_1^r t^{-1} \log t\, dt\right) = O((\log r)^2).$$

We next apply the Boutroux-Cartan lemma to obtain a preliminary estimate for the minimum modulus of an entire function of genus zero.

3.5.10. *Lemma. If* $f(z)$ *is of genus* 0, $f(0) = 1$, $\epsilon > 0$, *and* $\sigma > 1$, *then there is a function* $\Delta(r)$, *tending arbitrarily slowly to* ∞, *such that for sufficiently large* R

$$\log |f(z)| \geq \log M(\sigma R) - Q(\sigma R)\Delta(R), \qquad r < R, \tag{3.5.11}$$

outside a set of circles the sum of whose radii is at most ϵR.

Since $\log M(r)$ increases and $Q(r)$ decreases, this implies in particular that

$$\log m(r) \geq \log M(r) - Q(r)\, \Delta(r) \tag{3.5.12}$$

in a set of unit linear density, where $\Delta(r)$ (not the same function as in (3.5.11)) tends to ∞ arbitrarily slowly.

We have, since $g(z)$ is of genus 0,

$$|f(z)| = \prod_{n=1}^{\infty} |1 - z/z_n| \geq \Big| \prod_{r_n \leq \sigma R} (1 - z/z_n)\Big| \prod_{r_n > \sigma R} |1 - z/z_n| = P_1 P_2,$$

where $r_n = |z_n|$, $r = |z| \leq R$, $\sigma > 1$. Now we have

$$|1 - z/z_n| \geq |1 - r/r_n|,$$

so (since $n(r) = o(r)$)

$$\log P_2 \geq \log \prod_{|r_n| \geq \sigma R} (1 - r/r_n) = \int_{\sigma R}^{\infty} \log(1 - r/t)\, dn(t)$$

$$= -n(\sigma R) \log(1 - r/(\sigma R)) - r \int_{\sigma R}^{\infty} \frac{n(t)}{t(t-r)}\, dt$$

$$\geq \frac{rn(\sigma R)}{\sigma R} - \frac{\sigma R}{\sigma - 1} \int_{\sigma R}^{\infty} t^{-2} n(t)\, dt$$

$$= \frac{rn(\sigma R)}{\sigma R} - \frac{r}{R(\sigma - 1)} Q(\sigma R).$$

For P_1, we apply the Boutroux-Cartan lemma (3.4.1). We deduce that, if z is outside circles the sum of whose radii is $2eH$,

$$\log P_1 \geq n(\sigma R) \log H + \log \prod_{r_n \leq \sigma R} r_n^{-1}$$

$$\geq n(\sigma R) \log H - \int_0^{\sigma R} \log t\, dn(t)$$

$$= n(\sigma R) \log (H/(\sigma R)) + N(\sigma R).$$

Combining the results for P_1 and P_2, and using (3.5.2), we have, outside the exceptional circles,

$$\log |f(z)| \geq n(\sigma R) \left(\frac{r}{\sigma R} + \log \frac{H}{\sigma R} \right) + N(\sigma R) - \frac{r}{R(\sigma - 1)} Q(\sigma R)$$

$$\geq n(\sigma R) \left(\frac{r}{\sigma R} + \log \frac{H}{\sigma R} \right) + \log M(\sigma r) - \left(1 + \frac{r}{R(\sigma - 1)} \right) Q(\sigma R).$$

Now we are naturally to suppose that $H(R) = o(R)$ so that there will be something outside the excluded circles; then for large R the coefficient of $n(\sigma R)$ is negative, and since $n(r) \leq Q(r)$ we have

$$\log |f(z)| \geq \log M(\sigma R) - \Delta(r) Q(\sigma R), \qquad r < R,$$

where $\Delta(R) \to \infty$ (arbitrarily slowly), outside a set of circles the sum of whose radii is $o(R)$.

3.6. Functions of order zero. If $f(z)$ is of order zero, and more particularly if it is of such slow growth that $\log M(r) = O((\log r)^2)$, the relation between $m(r)$ and $M(r)$ is especially simple.

3.6.1. *Theorem.*[a] *If* $\log M(r) = O((\log r)^2)$, *or equivalently if* $n(r) = O(\log r)$, *then* $\log m(r) \sim \log M(r)$ *on a set of unit density.*

We have $Q(r) = O(\log r)$ by 3.5.8, and if we suppose that $f(z)$ is not a polynomial (for polynomials, 3.6.1 is trivial), we have $\log M(r)/\log r \to \infty$

and (for a fixed σ) $\Delta(r) \to \infty$ arbitrarily slowly, hence $\Delta(r)Q(\sigma r) = o(\log M(r))$ and so by (3.5.11),

$$\log |f(z)| \geq \log M(\sigma r)\,(1 - o(1)) \geq \log M(r)\,(1 - o(1));$$

the conclusion follows.

For functions of zero order in general we have as a limiting case of the theorems of §3.2,

3.6.2. *Theorem.*[b] *If* $f(z)$ *is of order zero,* $\log m(r) \sim \log M(r)$ *for a sequence* $r = r_n \to \infty$.

By (3.5.5), $Q(t)/N(t) \to 0$ for $t = t_n \to \infty$; since

$$\frac{Q(r)}{N(r)} = \frac{r\int_r^\infty t^{-2}n(t)\,dt}{N(r)} \geq \frac{r\int_{2r}^\infty t^{-2}n(t)\,dt}{N(r)} = \frac{Q(2r)}{2N(r)},$$

$Q(t)/N(t) \to 0$ in the intervals $(t_n, 2t_n)$, which (for large t_n) contain values of r belonging to the set of unit density for which (3.5.12) holds. By making $\Delta(r) \to \infty$ slowly enough, we can make $\Delta(r)Q(r)/N(r) \to 0$ for these values of r, and since $N(r) \leq \log M(r)$, $\Delta(r)Q(r) = o(\log M(r))$ for these values of r, and the theorem follows from (3.5.12).

3.7. Functions of larger order. If $0 < \rho < 1$, or even if $f(z)$ is of order 1, convergence class, we can deduce immediately from (3.5.12) that $\log m(r) > -o(r^\rho)$ on a set of unit density; this is of course much weaker than the theorems of §3.2 as far as concerns the order of $\log m(r)$, but much stronger as far as concerns the density. The following theorems show that a result of the same kind is true for unrestricted values of ρ.

3.7.1. *Theorem.*[a] *If* $f(z)$ *is an entire function of growth* $(\rho, 0)$, *then for every positive* ϵ *and* η *and for sufficiently large* R *we have*

$$\log |f(z)| > -\epsilon R^\rho, \qquad |z| < R, \tag{3.7.2}$$

except in a set of circles the sum of whose radii is at most ηR.

We may express the content of 3.7.1 more picturesquely when $\rho = 1$ by saying that if $f(z)$ is of zero exponential type there are annuli of arbitrarily large width on which $\log\{1/|f(z)|\} = o(r^\epsilon)$; and, for every ray $\arg z = \theta_0$, the same inequality holds if the ray is modified by indentations whose total length out to $|z| = r$ is $o(r)$. Similarly, if $f(z)$ is of (some) exponential type, there are annuli of arbitrarily large width on which $1/|f(z)|$ is at most exponentially large; and for any ray $\arg z = \theta_0$ there is a path, whose length out to $|z| = r$ is $O(r)$ and which stays in a specified angle

$$|\theta - \theta_0| < \epsilon,$$

on which $1/|f(z)|$ is at most exponentially large.

3.7.3. *Corollary. If $f(z)$ is of growth* (1, 0), log $m(r) > -o(r)$ *on a set of unit density.*

3.7.1 is a special case of

3.7.4. *Theorem. If $f(z)$ is of growth (ρ, σ), then for every positive ϵ and η, and $k > 1$, for sufficiently large R we have*

$$\log |f(z)| > -H(\sigma + \epsilon)(kR)^\rho, \qquad |z| < R, \tag{3.7.5}$$

except in a set of circles the sum of whose radii is at most $2\eta kR$; H depends only on η and k.

We may suppose as usual that $f(0) = 1$, since we are interested only in large R. We shall show that there is an H such that

$$\log |f(z)| > -H \log M(R) \tag{3.7.6}$$

for $|z| \leq R/k$ except at most in circles the sum of whose radii is $2\eta R$. This implies the theorem, since it means in the first place (replace R by kR) that

$$\log |f(z)| > -H \log M(kR)$$

for $|z| \leq R$ except at most in circles the sum of whose radii is $2\eta kR$. Now

$$\log M(kR) \leq (\sigma + \epsilon)(kR)^\rho$$

for sufficiently large R, and (3.7.5) follows.

The inequality (3.7.6) results from combining the Boutroux-Cartan lemma with Carathéodory's inequality (1.3.1). Take k_1 so that $1 < k_1 < k$ and write $f(z) = f_1(z)f_2(z)$, where

$$f_2(z) = \prod_{|z_n| \leq R/k_1} (1 - z/z_n),$$

z_n being the zeros of $f(z)$. If m is the number of factors in $f_2(z)$,

$$\prod |1 - z/z_n| = |z_1 \cdots z_m|^{-1} |\prod (z - z_n)| > |z_m|^{-m}(\eta R/e)^m$$

outside circles the sum of whose radii is $2\eta R$. Now

$$n(R/k_1) \log k_1 = n(R/k_1) \int_{R/k_1}^{R} t^{-1}\, dt \leq \int_{R/k_1}^{R} n(t)t^{-1}\, dt$$
$$\leq N(R) \leq \log M(R);$$

so since $m = n(R/k_1)$ and $|z_m| \leq R/k_1$, outside the excluded circles we have

$$\log |f_2(z)| > m \log \frac{\eta R}{e|z_m|} > \frac{\log M(R)}{\log k_1} \log \frac{\eta k_1}{e}. \tag{3.7.7}$$

If η is small (as we may assume without loss of generality), we can find

k_2 between k_1 and k so that the circle $|z| = R/k_2$ avoids all the excluded circles and thus so that (3.7.7) holds for $|z| = R/k_2$. Choose θ so that $|f_1(Re^{i\theta}/k_2)| = M_1(R/k_2)$, where $M_1(r)$ is the maximum modulus of $f_1(z)$. Then since $f_1 = f/f_2$ and $M(R/k_2) \leq M(R)$, we have

$$\begin{aligned}\log M_1(R/k_2) &\leq \log M(R/k_2) - \log |f_2(Re^{i\theta}/k_2)| \\ &\leq \log M(R) \left\{1 - \frac{\log(\eta k_1/e)}{\log k_1}\right\}.\end{aligned} \tag{3.7.8}$$

Now $f_1(z)$ has no zeros inside $|z| \leq R/k_2$, and $f_1(0) = 1$, so we have by (1.3.2) (a consequence of Carathéodory's inequality)

$$\log |f_1(z)| \geq -\frac{2r}{(R/k_2) - r} \log M_1(R/k_2), \qquad |z| = r < R/k_2. \tag{3.7.9}$$

We now write, for $|z| = r \leq R/k$, and z outside the excluded circles,

$$\log |f(z)| = \log |f_1(z)| + \log |f_2(z)|,$$

and use (3.7.9), then (3.7.8) to estimate $\log M_1(R/k_2)$, and (3.7.7) to estimate $\log |f_2(z)|$. We obtain

$$\log |f(z)| \geq -\log M(R) \left\{\left[1 - \frac{\log(\eta k_1/e)}{\log k_1}\right] \frac{2r}{R/k_2 - r} - \frac{\log(\eta k_1/e)}{\log k_1}\right\}.$$

The coefficient of $-\log M(R)$ does not exceed

$$\frac{2k_2}{k - k_2} + \frac{k_1 + k}{k - k_1} \frac{\log(e/(\eta k_1))}{\log k_1},$$

and this is an H of the required kind.

Notes for Chapter 3

3.1a. Wiman [1], Titchmarsh [3], p. 274. See also Arima [1].

3.1b. The explicit statement that an entire function of zero exponential type is a constant if it is bounded on a line seems to be due to S. Bernstein [1], [2]. His proof is given in §11.2.

3.1c. For this proof see Kjellberg [1].

3.1d. Heins [2].

3.2a. The theorem was conjectured by Littlewood, who proved a weaker result, and proved by Valiron and by Wiman. The proof given here is that of Pólya, as simplified by Denjoy. [1]. See also Valiron [9] for further developments. Many proofs have been given; Kjellberg [1] gives a survey. See also Huber [1].

3.2b. Denjoy states that $k(\alpha)$ is the smaller of $\lambda(\infty)$, $\lambda(0+)$, but he gives no proof and I have been unable to verify this. However, all that is required is that $k(\alpha) > 0$.

3.2c. Valiron [7], Kjellberg [1]; actually 3.2.10 still holds for order ρ, minimum type.

3.2d. Amirà [1].

3.2e. Besicovitch [1], Pennycuick [1]. For further results in this direction see these papers and Cartwright [5], Shah [20], Inoue [1], [2], [3], Tsuji [1]. These theorems are closely connected with the theory of "flat regions" of entire functions, for which see Whittaker [2], Pennycuick [2], A. J. Macintyre [2], Maitland [1], Mandelbrojt and Ulrich [1], Noble [4], and references given by these authors.

3.3a. Cartwright [5] proves this and gives some results on the density of the set where $m(r) > M(r)^{-1-\epsilon}$. Titchmarsh ([3], p. 276) gives another proof of (3.3.1) for functions of exponential type. The proof of the text is still different.

3.3b. Borel [1], p. 32; Valiron [3], p. 90.

3.4a. A less precise result is due to Boutroux, and the form given here to H. Cartan; for the proof see Valiron [10].

3.6a. Valiron [3].

3.6b. Littlewood; cf. Valiron [3].

3.7a. For this and the next theorem see Chebotarëv and Meiman [1], pp. 73–75. Theorems of this kind are due, in various forms, to Valiron [7], V. Bernstein, Cartwright, and Pfluger; see Pfluger [3] for one version and references. See also Boas [20] and Gelfond [4]. The constants involved in the conclusion of 3.7.4 can be estimated more precisely: see especially Gelfond [4].

CHAPTER 4

FUNCTIONS WITH REAL NEGATIVE ZEROS

4.1. Direct theorems. If all the zeros of an entire function are on, or close to, a given half-line (which we may take to be the negative real axis) there are especially simple connections between the distribution of the zeros and the rate of growth of the function. We shall consider only functions of order less than 1, although there are similar theorems for any nonintegral order. When the order is an integer the situation is more complicated, as usual; some results for functions of order 1, belonging to the same circle of ideas, but with the zeros distributed along a whole line, will be discussed in Chapter 8; some, but not all, of these can be deduced from properties of functions of order ½. We begin with a theorem in which the rate of growth is inferred from the distribution of the zeros.

4.1.1. *Theorem.*[a] *If $f(z)$ is an entire function of order ρ, $0 < \rho < 1$, with real negative zeros, $f(0) = 1$, and $n(t) \sim \lambda t^\rho$ ($\lambda > 0$) as $t \to \infty$, then*

$$\log f(x) \sim \pi\lambda(\csc \pi\rho)x^\rho, \qquad x \to \infty; \tag{4.1.2}$$

and more generally

$$\log f(re^{i\theta}) \sim e^{i\rho\theta}\pi\lambda(\csc \pi\rho)r^\rho, \qquad r \to \infty, \tag{4.1.3}$$

for each fixed θ in $-\pi < \theta < \pi$; the branch of the logarithm is that which is real on the positive real axis.

We have

$$f(z) = \prod_{n=1}^{\infty} (1 + z/r_n), \qquad r_n > 0,$$

$$\begin{aligned} \log f(z) &= \sum_{n=1}^{\infty} \log (1 + z/r_n) = \int_0^\infty \log (1 + z/t)\, dn(t) \\ &= n(t) \log (1 + z/t) \Big|_0^\infty + z \int_0^\infty \frac{n(t)\, dt}{t(t+z)} \\ &= z \int_0^\infty \frac{n(t)\, dt}{t(t+z)}. \end{aligned} \tag{4.1.4}$$

We may carry out the calculation first for real positive z, when there is no problem about which determination of $\log(1 + z/r_n)$ to take; since $n(t) \equiv 0$ in a neighborhood of 0, the integrated term vanishes; and since the final integral and $\log f(z)$ are both regular except on the negative real axis, and coincide on the positive real axis, they coincide in the plane cut along the negative real axis.

If $\epsilon > 0$, we have $(\lambda - \epsilon)\, t^\rho < n(t) < (\lambda + \epsilon)t^\rho$ for $t > T$. If we replace $n(t)$ by λt^ρ, we commit an error of at most

$$\left| z \int_0^\infty \frac{n(t) - \lambda t^\rho}{t(t+z)}\, dt \right| \le |z| \int_0^T \frac{n(t) + (\lambda + \epsilon)t^\rho}{t\,|t+z|}\, dt + |z|\,\epsilon \int_0^\infty \frac{t^\rho\, dt}{t\,|t+z|}. \tag{4.1.5}$$

Now

$$z \int_0^\infty \frac{t^\rho\, dt}{t(t+z)} = \pi z^\rho \csc \pi\rho; \tag{4.1.6}$$

for, using a well-known definite integral we have

$$x \int_0^\infty \frac{t^\rho\, dt}{t(t+x)} = \pi x^\rho \csc \pi\rho,$$

and if $z = re^{i\theta}$, $-\pi < \theta < \pi$, we can turn the line of integration in (4.1.6) to the ray $\arg z = \theta$ with the same result. Also $|t + z| \ge t + x$ if $x > 0$, and

$$|t + z| \ge (r + t)\sin(\pi - |\theta|)/2 \tag{4.1.7}$$

if $x \le 0$. So with a fixed θ the first term on the right of (4.1.5) is $O(1)$ and the second is $\epsilon O(r^\rho)$, and (4.1.3) follows.

There is a similar result even for points on the negative real axis; here, of course, we can expect to determine the asymptotic behavior of $f(z)$ only outside neighborhoods of the zeros.

4.1.8.* *Theorem.*[b] *Under the hypotheses of 4.1.1, if $\epsilon > 0$ we have*

$$\log|f(-x)| < (\pi\lambda \cot \pi\rho + \epsilon)x^\rho, \qquad x > x_0(\epsilon), \tag{4.1.9}$$

and if $\eta > 0$ we also have

$$\log|f(-x)| > (\pi\lambda \cot \pi\rho - \epsilon)x^\rho, \tag{4.1.10}$$

for $0 < x < X$ except in a set of measure ηX, provided that X is sufficiently large.

Thus, in particular,

$$\log|f(-x)| \sim \pi\lambda(\cot \pi\rho)x^\rho \text{ in a set of density 1.} \tag{4.1.11}$$

Another sense in which the asymptotic formula in (4.1.11) is true "on the average" is given in 4.4.14.

4.2. Converse theorems. The converse of 4.1.1 is also true, but is rather more difficult to prove.

4.2.1. *Theorem.*[a] *If $f(z)$ is of order less than 1, if all its zeros are real and* *negative, if* $f(0) = 1$ *and if*

$$\log f(r) \sim \pi \csc \pi\rho \cdot r^{\rho}, \qquad r \to \infty, \tag{4.2.2}$$

then

$$n(r) \sim r^{\rho}, \qquad r \to \infty. \tag{4.2.3}$$

Our first step[b] is to show that

$$\log f(z) \sim \pi z^{\rho} \csc \pi\rho \tag{4.2.4}$$

uniformly in any angle $|\arg z| \leq \pi - \delta < \pi$.

In the first place, (4.2.2) implies that there is a constant A such that

$$\log f(r) \leq Ar^{\rho}, \qquad r > r_0. \tag{4.2.5}$$

Setting $\phi(z) = z^{-\rho} \log f(z)$, which is regular in $|\arg z| < \pi$ (with $z^{-\rho}$ positive for real positive z), we shall show that

$$\phi(z) \text{ is bounded in } |\arg z| \leq \pi - \delta/2. \tag{4.2.6}$$

As soon as we have (4.2.6), (4.2.4) follows from Montel's theorem (1.4.8) applied to $\phi(z)$. Now by (4.1.4) and (4.1.7),

$$|\log f(z)| \leq r \int_0^{\infty} \frac{n(t)\, dt}{t\,|t + z|} \leq r \csc(\delta/4) \cdot \int_0^{\infty} \frac{n(t)\, dt}{t(t + r)}$$

$$= \csc(\delta/4) \cdot \log f(r) = O(r^{\rho})$$

by (4.2.5), and (4.2.6) follows. Hence we have (4.2.4).

The most straightforward way of proving (4.2.3) is now to start from Jensen's theorem, which gives us

$$N(r) = \int_0^r t^{-1} n(t)\, dt = (2\pi)^{-1} \int_{-\pi}^{\pi} \log |f(re^{i\theta})|\, d\theta. \tag{4.2.7}$$

We have, by (4.2.4),

$$\log f(z) \sim \pi r^{\rho} \csc \pi\rho\, e^{i\rho\theta}, \qquad |\arg z| \leq \pi - \delta < \pi,$$

and if we can take the limit under the integral sign in (4.2.7) we shall have

$$r^{-\rho} N(r) \to 1/\rho, \tag{4.2.8}$$

from which $r^{-\rho}\, n(r) \to 1$ will follow by an elementary Tauberian argument. Now

$$r^{-\rho} \log |f(re^{i\theta})| \to \pi \csc \pi\rho \cos \rho\theta,$$

uniformly outside every neighborhood of $-\pi$, and is bounded above for $|\theta| \leq \pi$; if we could show that it is bounded below we could take the limit under the integral sign[c] in (4.2.7) by "bounded convergence." Of course this expression is not bounded below for r's in the neighborhood of the moduli of the zeros, but we can find a sufficiently large set of r's on which it is bounded below and take the limit over this set. To do this, we note that, by 3.7.4, for any positive η we have $\log |f(z)| \geq -H\eta r^\rho$ with a finite H (depending on η), except in circles the sum of whose diameters out to $|z| = R$ is at most ηR (we have changed the notation of 3.7.4 somewhat). Hence when $|z| = r$ does not intersect the excluded circles, $r^{-\rho} \log |f(re^{i\theta})|$ is bounded, uniformly for $|\theta| \leq \pi$. Therefore we can take the limit under the integral sign in (4.2.7) when $r \to \infty$ through any sequence of values avoiding a collection of intervals whose total length for $r \leq R$ is at most ηR, with η arbitrarily small.

Accordingly we have

$$\lim_{r\to\infty} r^{-\rho} N(r) = 1/\rho, \tag{4.2.9}$$

where r is to avoid the excluded intervals.

It is now easy to show that the excluded intervals can be included in (4.2.9). In fact, if (r_1, r_2) is an excluded interval and $r_1 < r < r_2$, we have

$$r^{-\rho}N(r) \leq r^{-\rho}N(r_2) = (r_2/r)^\rho r_2^{-\rho}N(r_2) \leq (1 - \eta)^{-\rho} r_2^{-\rho}N(r_2),$$

and since $n(t) = O(t^\rho)$,

$$\begin{aligned} r^{-\rho}N(r) &= r^{-\rho}N(r_2) - r^{-\rho} \int_r^{r_2} t^{-1} n(t)\, dt \\ &\geq r_2^{-\rho}N(r_2) - Ar^{-\rho} \int_r^{r_2} t^{\rho-1}\, dt \\ &\geq r_2^{-\rho}N(r_2) - A\eta/(1 - \eta), \end{aligned}$$

so that

$$\rho^{-1} - A\eta/(1 - \eta) \leq \liminf_{r\to\infty} r^{-\rho}N(r) \leq \limsup_{r\to\infty} r^{-\rho}N(r) \leq (1 - \eta)^{-\rho}/\rho;$$

since η is arbitrary, (4.2.9) holds as $r \to \infty$ without restriction.

Finally, we show that (4.2.9) implies (4.2.3).[d] If $\lambda > 1$, we have, writing (4.2.9) for r and for λr and subtracting,

$$r^{-\rho} \int_r^{\lambda r} t^{-1} n(t)\, dt \to \rho^{-1}(\lambda^\rho - 1),$$

and since $n(t)$ increases, the left-hand side is at least $r^{-\rho} n(r) \log \lambda$, so that

$$\limsup_{r\to\infty} r^{-\rho} n(r) \leq \rho^{-1}(\lambda^\rho - 1)/\log \lambda,$$

and letting $\lambda \to 1$ we have

$$\limsup_{r\to\infty} r^{-\rho}n(r) \leq 1.$$

Similarly we find $\liminf r^{-\rho}n(r) \geq 1$, and the proof is complete.

4.3. Generalizations. In the theorems of §4.1 and §4.2 the assumption that the zeros are actually on a half line is stronger than is necessary; it is sufficient[a] to have $\arg z_n \to \pi$ as $n \to \infty$. This is easy for 4.1.1, and 4.2.1 with the general hypothesis can be reduced to the theorem with real zeros by a straightforward argument. It is less simple to show[b] that the hypothesis (4.2.2) in 4.2.1 can be replaced by $\log f(w_n) \sim \pi(\csc \pi\rho)w_n^\rho$ if $|w_n| \uparrow \infty$, $|\arg w_n| \leq \pi - \delta < \pi$, and $w_{n+1}/w_n \to 1$.

There is also a converse of 4.1.8.

4.3.1.* *Theorem.*[c] *Let $f(z)$ be an entire function of genus 0 with real negative zeros, let, for each $\epsilon > 0$,*

$$\log |f(-x)| < \pi x^\rho(\cot \pi\rho + \epsilon)$$

for all sufficiently large positive x, and

$$\log |f(-x_n)| > \pi x_n^\rho(\cot \pi\rho - \epsilon)$$

for sufficiently large x_n of a sequence for which $x_n \to \infty$, $x_{n+1}/x_n \to 1$. Then

$$\log f(x) \sim \pi x^\rho \csc \pi\rho, \qquad x \to +\infty,$$

if either $0 < \rho < \frac{1}{2}$ and $f(z)$ is of growth $(\frac{1}{2}, 0)$, or $\frac{1}{2} < \rho < 1$.

A very general theorem along similar lines is due to Pfluger. To state it, we first require some terminology.

4.3.2. *Definition. The entire function $f(z)$ of (arbitrary finite) order ρ is of regular asymptotic behavior if, for each θ, $\lim_{r\to\infty} r^{-\rho} \log |f(re^{i\theta})|$ exists as $r \to \infty$ through a certain set of density* 1.

4.3.3. *Definition. The set of zeros of $f(z)$ is said to be measurable with respect to r^ρ if there is a nondecreasing $N(\theta)$ such that for every two points θ', θ'' of continuity of $N(\theta)$, the number of zeros with $|z| \leq r$, $\theta' \leq \theta < \theta''$, is equal to* $\{N(\theta'') - N(\theta')\}r^\rho + o(r^\rho)$.

Then Pfluger's theorem is as follows.[d]

4.3.4.* *Theorem. The entire function $f(z)$ of nonintegral order ρ is of regular asymptotic behavior if and only if its zeros are measurable with respect to r^ρ, and its indicator function is given by*

$$h(\theta) = \pi \csc \pi\rho \int_0^{2\pi} \cos(\rho t - \rho\pi)\, dN(\theta + t);$$

moreover, in any zero-free angle, $\lim_{r\to\infty} r^{-\rho} \log |f(re^{i\theta})|$ exists.

Though much more general in some ways than the theorems of §4.2, this theorem does not imply them, because before we can use it to draw conclusions about the zeros we must know the behavior of the function not only on one ray but on all rays, including those containing zeros.

4.4. Another kind of theorem. We now prove another result connecting the behavior of $f(z)$ along the negative real axis with the distribution of the zeros. Here instead of working with anything as precise as (4.1.9), (4.1.10), we use an integral which effectively smooths out the irregularities resulting from the presence of the zeros. The situation is especially simple in the case $\rho = \frac{1}{2}$ on which 4.3.1 gives no information, and this case is particularly interesting because it is equivalent to an important theorem about functions of order 1 (§8.2).

4.4.1. *Theorem.*[a] *Let $f(z)$ be entire, of order less than* 1, *with negative real zeros, and let $f(0) = 1$. Then*

$$\int_0^r x^{-3/2} \log |f(-x)| \, dx \text{ is bounded} \tag{4.4.2}$$

if and only if

$$x^{-1/2} \log |f(x)| \text{ is bounded for } 0 \leq x < \infty; \tag{4.4.3}$$

the existence of

$$\int_0^{\to\infty} x^{-3/2} \log |f(-x)| \, dx = \alpha \tag{4.4.4}$$

and of

$$\lim_{x\to\infty} x^{-1/2} \log f(x) = \beta \tag{4.4.5}$$

are equivalent, and $\alpha = -\pi\beta$.

We start by proving that (4.4.2) and (4.4.4) imply (4.4.3) and (4.4.5), respectively. Consider the integral

$$I = \int \frac{r}{r - z} \frac{\log f(z) \, dz}{z^{3/2}}$$

over the contour consisting of the circle $|z| = R > r$ with a cut from $-R$ to 0 and back again (initially the cut has indentations to avoid the zeros of $f(z)$ and the origin, but the contributions of the indentations tend to 0 with their diameters). The branches of the multiple-valued functions are fixed by being taken positive for large positive values of z. The pole at $z = r$ makes

$$I = -2\pi i r^{-1/2} \log f(r). \tag{4.4.6}$$

We next show that the contribution to I from the integral around $|z| =$

R tends to 0 as $R \to \infty$, at least if R is restricted to a suitable sequence of values. We have $\log^+ |f(z)| = O(r^\lambda)$, $\lambda < 1$; hence $n(t) = O(t^\lambda)$ (2.5.13), and so

$$\text{(4.4.7)} \qquad \arg f(z) = \Im \log f(z) = y \int_0^\infty \frac{n(t)\,dt}{(t+x)^2 + y^2} = O(r^\lambda);$$

we may verify (4.4.7) as follows (A denotes various constants):

$$\int_0^\infty \frac{n(t)\,dt}{(t+x)^2+y^2} \le \int_0^{2|x|} \frac{n(t)\,dt}{(t+x)^2+y^2} + A \int_{2|x|}^\infty \frac{t^\lambda\,dt}{(t+x)^2+y^2}$$
$$\le n(2|x|) \int_0^{2|x|} \frac{dt}{(t+x)^2+y^2} + A \int_{2|x|}^\infty \frac{t^\lambda\,dt}{(t/2)^2+y^2}$$
$$\le (A|x|^\lambda + A) \int_0^\infty \frac{du}{u^2+y^2} = (A|x|^\lambda + A)/|y|.$$

In addition, by (2.6.18) we have $\log |f(z)| > -R^{\lambda+\epsilon}$ for $0 < \epsilon < 1 - \lambda$ and a sequence of values of R tending to ∞. Hence

$$|\log f(z)| \le |\log |f(z)|| + |\arg f(z)| = O(R_n^{\lambda+\epsilon}), \quad \lambda + \epsilon < 1,$$

for $|z| = R_n \to \infty$. We therefore have, combining the integrals along the two sides of the cut and taking imaginary parts in (4.4.6),[b]

$$\text{(4.4.8)} \qquad -\pi r^{-1/2} \log f(r) = \lim_{n\to\infty} \int_0^{R_n} \frac{r}{r+x} x^{-3/2} \log |f(-x)|\,dx.$$

Suppose first that (4.4.2) is true, and let

$$J(r) = \int_0^r x^{-3/2} \log |f(-x)|\,dx;$$

then

$$-\pi r^{-1/2} \log f(r) = \lim_{n\to\infty} \int_0^{R_n} r(r+x)^{-1}\,dJ(x)$$
$$= \lim_{n\to\infty} \int_0^{R_n} J(x) r(r+x)^{-2}\,dx,$$

and this is bounded if $J(x)$ is bounded, i.e., (4.4.3) is true.

Next suppose that (4.4.4) is true. Then

$$\lim_{n\to\infty} \int_0^{R_n} r(r+x)^{-1} x^{-3/2} \log |f(-x)|\,dx$$
$$= \alpha - \lim_{n\to\infty} \int_0^{R_n} x(r+x)^{-1} x^{-3/2} \log |f(-x)|\,dx,$$

and it remains only to show that the second term on the right approaches zero as $r \to \infty$. Now we have, if $n > m$,

$$\int_{R_m}^{R_n} \frac{x}{r+x} x^{-3/2} \log | f(-x) | \, dx$$

$$\frac{R_n}{r+R_n} \int_{R}^{R_n} x^{-3/2} \log | f(-x) | \, dx, \qquad R_m < R < R_n,$$

since $x/(r+x)$ increases; and

$$\int_0^{R_n} \frac{x}{r+x} x^{-3/2} \log | f(-x) | \, dx = \int_0^{R_m} \frac{x}{r+x} x^{-3/2} \log | f(-x) | \, dx$$

$$+ \frac{R_n}{r+R_n} \int_R^{R_n} x^{-3/2} \log | f(-x) | \, dx.$$

If we first take m so large that the second integral on the right is small (by the convergence of (4.4.4)), and then with a fixed m take r large, we can make the first integral on the right small also (since $x^{-3/2} \log | f(-x) |$ is absolutely integrable over a finite range); this gives us the desired conclusion.

We turn now to the proof that (4.4.3) and (4.4.5) imply (4.4.2) and (4.4.4), respectively.

We showed in the proof of 4.2.1 that (4.4.3) implies that

$$| z |^{-1/2} | \log f(z) |$$

is bounded in $| \arg z | \le \pi - \delta < \pi$, and that (4.4.5) implies that

$$z^{-1/2} \log f(z) \to \beta$$

uniformly in $| \arg z | \le \pi - \delta < \pi$. In addition, in either case $f(z)$ is at most of finite type of order $\frac{1}{2}$, and so $n(r) = O(r^{1/2})$. Now consider the integral

$$\int z^{-3/2} \log f(z) \, dz$$

taken around the contour consisting of the circle $| z | = r$ with a cut from $-r$ to 0 and back again (again, initially with indentations at the zeros of $f(z)$ and at 0, which contribute nothing in the limit). Since the integrand is regular inside the contour, the integral is zero, so that we have

$$\int_{|z|=r} z^{-3/2} \log f(z) \, dz = -2i \int_0^r x^{-3/2} \log f(-x) \, dx,$$

and, taking imaginary parts,

$$\Im \int_{|z|=r} z^{-3/2} \log f(z) \, dz = -2 \int_0^r x^{-3/2} \log | f(-x) | \, dx.$$

The left-hand side is

$$\Im\left\{i\int_{-\pi}^{\pi} z^{-1/2}\log f(z)\,d\theta\right\}, \tag{4.4.9}$$

and to complete the proof we have to show that (4.4.9) is bounded or approaches a limit as $|z| \to \infty$ if $x^{-1/2}\log f(x)$ is bounded or approaches a limit as $x \to \infty$.

The case when $x^{-1/2}\log f(x)$ is bounded is quite simple, since $|f(z)|$ is largest for real positive z, so that $f(z)$ is at most of order $\frac{1}{2}$, mean type, $N(r) = O(r^{1/2})$, and by Jensen's theorem

$$r^{-1/2}\int_{-\pi}^{\pi}\log|f(re^{i\theta})|\,d\theta = 2\pi r^{-1/2}N(r) = O(1). \tag{4.4.10}$$

Indeed, (4.4.9) is

$$r^{-1/2}\int_{-\pi}^{\pi}\{\log|f(re^{i\theta})|\cos\theta/2 + \arg f(re^{i\theta})\sin\theta/2\}\,d\theta; \tag{4.4.11}$$

as we saw in the first part of the proof (4.4.7), $\arg f(re^{i\theta}) = O(r^{1/2})$, and we are left with

$$r^{-1/2}\int_{-\pi}^{\pi}\log|f(re^{i\theta})|\cos\theta/2\,d. \tag{4.4.12}$$

Since $\log|f(z)| \leq O(r^{1/2})$, which means that

$$r^{-1/2}\int_{-\pi}^{\pi}\log^{+}|f(re^{i\theta})|\,d\theta = O(1),$$

we also have, by (4.4.10),

$$r^{-1/2}\int_{-\pi}^{\pi}\log^{-}|f(re^{i\theta})|\,d\theta = O(1),$$

and inserting the factor $\cos\theta/2$ into each of the last two integrals does not affect their boundedness (since the integrands have a fixed sign); the boundedness of (4.4.12) follows, and the proof that (4.4.3) implies (4.4.2) is complete.

To show that (4.4.5) implies (4.4.4), we have to appeal to 4.2.1. We know that $z^{-1/2}\log f(z)$ approaches a limit uniformly in $|\arg z| \leq \pi - \delta$, so that in view of (4.4.11) it is necessary only to show that

$$r^{-1/2}\int_{\pi-\delta}^{\pi}\{\log|f(re^{i\theta})|\cos\theta/2 + \arg f(re^{i\theta})\sin\theta/2\}\,d\theta$$

is small with δ (and the same for the integral over $(-\pi, -\pi+\delta)$). The second part is $O(\delta)$ as $\delta \to 0$ because $\arg f(re^{i\theta}) = O(r^{1/2})$, and we are left with

$$r^{-1/2} \int_{\pi-\delta}^{\pi} \log | f(re^{i\theta}) | \cos \theta/2 \, d\theta. \tag{4.4.13}$$

But we also have, again because $r^{-1/2} \log f(z) \to 0$ uniformly in $| \arg z | \leq \pi - \delta$, and because $r^{-1/2} N(r) \to \beta/\pi$, that

$$r^{-1/2} \int_{\pi-\delta}^{\pi} \log | f(re^{i\theta}) | \, d\theta .= O(\delta),$$

and by the same argument as before, the same is true of (4.4.13).

The generalization of 4.4.1 to orders other than ½ is not entirely obvious, but may be found by attempting to generalize the proof.

4.4.14.* Theorem.[c] *If $f(z)$ is an entire function of order less than 1, with real negative zeros, and $f(0) = 1$, the conditions*

$$\log f(r) \sim \lambda\pi \csc \pi\rho \, r^{\rho}, \qquad r \to \infty, \lambda > 0, \tag{4.4.15}$$

and (for any σ, $0 < \sigma < 1$)

$$\int_0^r x^{-1-\sigma} \{\log | f(-x) | - \pi (\cot \pi\sigma) \, n(x)\} \, dx \sim \pi\lambda \frac{\cot \pi\rho - \cot \pi\sigma}{\rho - \sigma} r^{\rho-\sigma} \tag{4.4.16}$$

are equivalent; the right-hand side is to be interpreted as $-\pi^2\lambda \csc^2 \pi\rho$ when $\sigma = \rho$.

In particular, (4.4.15), or its equivalent $n(r) \sim \lambda r^{\rho}$, implies

$$\int_0^r x^{-1-\rho} \log | f(-x) | \, dx \sim \pi\lambda \, (\cot \pi\rho) \log r, \qquad \rho \neq \tfrac{1}{2},$$

$$\int_0^r x^{-1-\sigma} \log | f(-x) | \, dx \sim \frac{\pi\lambda r^{\rho-\sigma} \cot \pi\rho}{\rho - \sigma}, \qquad \rho > \sigma,$$

$$\int_r^{\infty} x^{-1-\sigma} \log | f(-x) | \, dx \sim \frac{\pi\lambda r^{\rho-\sigma} \cot \pi\rho}{\sigma - \rho}, \qquad \sigma > \rho;$$

these are just the relations which would follow from 4.1.8 if (4.1.11) were true everywhere.

There is a less precise version of 4.4.14, corresponding to the first part of 4.4.1, which asserts that $f(z)$ (with real negative zeros) is of finite type of order ρ if and only if the left-hand side of (4.4.16) is $O(r^{\rho-\sigma})$. From this (and 2.11.1) one can deduce the following result.

4.4.17.* Theorem. *If $f(z)$ is an entire function of order ρ, $0 < \rho < \frac{1}{2}$, and $m(r)$ is its minimum modulus, then $\int_1^R r^{-\rho-1} \log m(r) dr$ is bounded $(1 < R < \infty)$ if and only if $f(z)$ is of convergence class.*

One half of this result also follows from the convergence of the integral (3.2.5) with $\alpha = \rho$.

Notes for Chapter 4

4.1a. Valiron [1], Titchmarsh [1].

4.1b. Titchmarsh [1], Ikehara [1].

4.2a. 4.2.1 has a long history. The experienced reader will recognize it as a Tauberian theorem in which the Tauberian condition is that $n(t)$ increases. It was proved by Valiron [1] by a rather long Tauberian argument; rediscovered by Titchmarsh [1] and proved by reduction to an equivalent Tauberian theorem of Hardy and Littlewood; and deduced by Paley and Wiener [1] from Wiener's general Tauberian theorem. Function-theoretical proofs were given by Pfluger [3], Delange [1], [2], Heins [2] and Bowen [1]; the last two proofs are relatively short. Several of these authors give generalizations to functions of arbitrary nonintegral order and generalizations in other directions. See also Bowen and Macintyre [1]. Hypothesis (4.2.2) can be replaced by (4.1.3) with a single θ if $|\theta| < \pi/2$; or, if we assume to begin with that $f(z)$ is of order at most ρ, it is enough to have $|\theta| < \pi/(2\rho)$ if $\rho > \frac{1}{2}$, $|\theta| < \pi$ if $\rho \leq \frac{1}{2}$; for, in all these cases (4.1.3) implies (4.2.5) by a Phragmén-Lindelöf argument using the angle $|\arg z| \leq \theta$, since $f(z)$ satisfies (4.1.3) for $-\theta$ if for θ. Cf. Delange [2], where there is an example showing that $|\theta| < \pi/(2\rho)$ cannot be replaced by $|\theta| < \pi$.

4.2b. This is the essential step in the proofs of Bowen and Heins.

4.2c. This is done directly by Heins, but it is shorter for us to appeal to 3.7.4; the proof obtained in this way would be longer if written out to include 3.7.4 as part of the proof of 4.2.1. Bowen writes (4.2.4) in the form

$$\int_0^\infty t^{-1}(t + z)^{-1}n(t)\, dt \sim \pi z^\rho \csc \pi\rho ,$$

takes the imaginary part of both sides, and so reduces the theorem to a Tauberian theorem of an elementary kind.

4.2d. This well-known argument seems to have originated with Landau.

4.3a. Valiron [1], Bowen [1]. For an analogous theorem with the zeros clustering around the (whole) imaginary axis, see Noble [1].

4.3b. Bowen [1].

4.3c. Titchmarsh [1], Bowen and Macintyre [2]. Titchmarsh proved the first part and the theorem was completed by Bowen and Macintyre, who give further generalizations. See also Ikehara [1].

4.3d. Pfluger [3]. Even more generally, regular asymptotic behavior implies measurability of zeros even for functions of integral order and even in an angle; and the theorem can be further generalized in terms of proximate orders. See §8.1 for further discussion of the case $\rho = 1$.

4.4a. The equivalent theorem for entire functions of order 1 is given by Paley and Wiener [1]. For the proof of the text see Boas [27].

4.4b. By using the more precise theorem 3.7.4 instead of 2.6.18 we could, as in the proof of 4.2.1, replace R_n by a variable R tending continuously to ∞; but it is unnecessary to do so.

4.4c. Boas [27].

CHAPTER 5

GENERAL PROPERTIES OF FUNCTIONS OF EXPONENTIAL TYPE

5.1. Properties of the indicator. In the rest of this book we shall be concerned almost exclusively with the functions of exponential type, and usually with entire functions of exponential type. To begin with we collect some material which will be needed frequently. We shall constantly have to refer to properties of the Phragmén-Lindelöf indicator function, and it is useful to have as many as possible of its properties stated in such a form that they can be used for functions which are not necessarily entire.

In this section we suppose that $f(z)$ is regular and of exponential type in a closed angle, which we take for convenience to be $|\theta| \leq \alpha$. The behavior of $f(z)$ near the vertex is unimportant for our purposes, although it is sometimes necessary to apply the theorems when (for example) $f(z)$ is regular in the open angle and continuous in the closed angle. Such generalizations can be dealt with by limiting processes and are left to the reader. The indicator function of $f(z)$ is defined as

$$h(\theta) = \limsup_{r \to \infty} r^{-1} \log |f(re^{i\theta})|, \qquad |\theta| \leq \alpha, \tag{5.1.1}$$

and since we require that $f(z)$ is of exponential type, $h(\theta)$ is finite or $-\infty$. We write $h_f(\theta)$ when it is necessary to call attention to the particular function $f(z)$ which is under consideration.

There is a corresponding definition for functions of growth (ρ, τ), $\rho \neq 1$, and a more general one in which the growth of $f(z)$ is compared with a proximate order; but we shall not need these, and we prove the necessary properties for the simplest case only. For *entire* functions of exponential type it is possible to give proofs which, once we have established some theorems which will be needed in any case, are shorter (cf. §5.4).

5.1.2. *Theorem.*[a] *If* $|\theta_1| \leq \alpha$, $|\theta_2| \leq \alpha$, $0 < \theta_2 - \theta_1 < \pi$, $h(\theta_1) \leq h_1$, $h(\theta_2) \leq h_2$, *and*

$$H(\theta) = \frac{h_1 \sin(\theta_2 - \theta) + h_2 \sin(\theta - \theta_1)}{\sin(\theta_2 - \theta_1)}$$

is the (unique) sinusoid (i.e., function of the form $a \cos \theta + b \sin \theta$*) which takes the values* h_1, h_2 *at* θ_1, θ_2, *respectively, then*

$$h(\theta) \leq H(\theta), \qquad \theta_1 \leq \theta \leq \theta_2 .$$

Let $\delta > 0$ and let $H_\delta(\theta) = a_\delta \cos\theta + b_\delta \sin\theta$ be the sinusoid which takes the values $h_1 + \delta$, $h_2 + \delta$ at θ_1, θ_2, respectively. Let

(5.1.3) $$F(z) = f(z) \exp \{-(a_\delta - ib_\delta)z\};$$

then

$$| F(z) | = | f(z) | \exp \{-H_\delta(\theta)r\},$$

and so $F(z)$ is bounded on the rays $\arg z = \theta_1, \theta_2$. Since $0 < \theta_2 - \theta_1 < \pi$, this implies by (1.4.2) that $F(z)$ is bounded in $\theta_1 \leq \theta \leq \theta_2$, and so by (5.1.3), $f(z) = O\{\exp(H_\delta(\theta)r)\}$, uniformly in this angle. Hence $h(\theta) \leq H_\delta(\theta)$ and since $H_\delta(\theta) \to H(\theta)$ as $\delta \to 0$, 5.1.2 follows.

With slight modifications the same proof shows that if $h(\theta_1)$ or $h(\theta_2)$ is $-\infty$ then $h(\theta) = -\infty$ for $\theta_1 < \theta < \theta_2$. Hence $h(\theta) = -\infty$ for some θ only if $h(\theta) = -\infty$ for all θ in the angle under consideration, with the possible exception of $\theta = \pm\alpha$.

5.1.4. *Theorem. Unless* $h(\theta) \equiv -\infty$, $h(\theta)$ *is continuous in* $-\alpha < \theta < \alpha$.

For later use we remark that the proof will use only the property established in 5.1.2.

To prove 5.1.4 we begin by proving

5.1.5. *Lemma. If* $h(\theta_1)$ *and* $h(\theta_2)$ *are finite,* $-\alpha < \theta_1 < \theta_2 < \theta_3 < \alpha$, $\theta_3 - \theta_1 < \pi$, *and* $H(\theta)$ *is a sinusoid such that* $h(\theta_1) \leq H(\theta_1)$ *and* $h(\theta_2) \geq H(\theta_2)$, *then* $h(\theta_3) \geq H(\theta_3)$.

First, $h(\theta_3) > -\infty$ by the remark made just before 5.1.4. If there is a positive δ such that $h(\theta_3) < H(\theta_3) - \delta$, put

$$H_\delta(\theta) = H(\theta) - \delta \sin(\theta - \theta_1) \csc(\theta_3 - \theta_1).$$

Then

$$H_\delta(\theta_1) = H(\theta_1), \qquad H_\delta(\theta_2) < H(\theta_2), \qquad H_\delta(\theta_3) = H(\theta_3) - \delta,$$

and so

$$h(\theta_1) \leq H_\delta(\theta_1), \qquad h(\theta_3) \leq H_\delta(\theta_3),$$

whence by 5.1.2,

$$h(\theta_2) \leq H_\delta(\theta_2) < H(\theta_2),$$

a contradiction.

To prove 5.1.4, let $-\alpha < \theta_1 < \theta_2 < \theta_3 < \alpha$ and let $H_{1,2}(\theta)$ be the sinusoid which takes the values $h(\theta_1)$, $h(\theta_2)$ at θ_1, θ_2, respectively; and similarly for $H_{2,3}(\theta)$. Then for $\theta_1 \leq \theta \leq \theta_2$, $h(\theta) \leq H_{1,2}(\theta)$ by 5.1.2, and $H_{2,3}(\theta) \leq h(\theta)$ by 5.1.5. Similarly $H_{1,2}(\theta) \leq h(\theta) \leq H_{2,3}(\theta)$ for $\theta_2 \leq \theta \leq \theta_3$. Hence

(5.1.6) $$\frac{H_{1,2}(\theta) - H_{1,2}(\theta_2)}{\theta - \theta_2} \leq \frac{h(\theta) - h(\theta_2)}{\theta - \theta_2} \leq \frac{H_{2,3}(\theta) - H_{2,3}(\theta_2)}{\theta - \theta_2},$$

and as $\theta \to \theta_2$ the right-hand and left-hand sides approach limits. Hence

$h(\theta)$ is continuous at θ_2, and even has finite right-hand and left-hand derivatives.

If θ_2 is one of the endpoints, say α, take $\theta_3 = \alpha$ and $\theta_1 < \theta < \theta_3$; then we still have $h(\theta) \leq H_{1,3}(\theta)$, $\theta_1 \leq \theta \leq \theta_3$, so

$$\limsup_{\theta \to \alpha-} h(\theta) \leq H_{1,3}(\theta_3) = h(\alpha);$$

similarly

$$\limsup_{\theta \to -\alpha+} h(\theta) \leq h(-\alpha).$$

5.1.7. *Theorem.* If $-\alpha \leq \theta_1 < \theta_2 < \theta_3 \leq \alpha$, $\theta_2 - \theta_1 < \pi$, $\theta_3 - \theta_2 < \pi$, *then* $h(\theta)$ *satisfies*

$$h(\theta_1) \sin(\theta_3 - \theta_2) + h(\theta_2) \sin(\theta_1 - \theta_3) + h(\theta_3) \sin(\theta_2 - \theta_1) \geq 0. \tag{5.1.8}$$

When $\theta_3 - \theta_1 < \pi$, this is just another way of writing the conclusion of 5.1.2. To prove (5.1.8) in general, note first that the left-hand side is identically zero for a sinusoid $H(\theta)$. Choose θ_4 so that $\theta_2 < \theta_4 < \theta_1 + \pi$; let $H(\theta)$ be the sinusoid coinciding with $h(\theta)$ at θ_1 and θ_2; then by 5.1.5, $h(\theta_4) \geq H(\theta_4)$. Next apply 5.1.5 to θ_2, θ_4, θ_3; it follows that $h(\theta_3) \geq H(\theta_3)$. Since

$$h(\theta_1) \sin(\theta_3 - \theta_2) + h(\theta_2) \sin(\theta_1 - \theta_3) + H(\theta_3) \sin(\theta_2 - \theta_1) = 0$$

and $\sin(\theta_2 - \theta_1) > 0$, replacing $H(\theta_3)$ by $h(\theta_3)$ (which is not smaller) makes the left-hand side larger (if it changes it), and (5.1.8) follows.

5.1.9. *Theorem. If* ϵ *is a given positive number and* $h(\theta) \not\equiv -\infty$, *then for* $r > r(\epsilon)$ *we have*

$$\log |f(re^{i\theta})| \leq r\{h(\theta) + \epsilon\}, \tag{5.1.10}$$

uniformly in any closed subinterval of $(-\alpha, \alpha)$, *uniformly in* $-\pi \leq \theta \leq \pi$ *if* $f(z)$ *is an entire function, and uniformly in* $(-\alpha, \alpha)$ *if* $h(\theta)$ *is continuous at* $\theta = \pm\alpha$.

The definition of $h(\theta)$ shows only that (5.1.10) holds for each θ.

Since $h(\theta)$ is continuous, it is uniformly continuous and hence varies by less than $\epsilon/3$ in each of a finite number of intervals $\theta_1 \leq \theta \leq \theta_2$ covering the closed interval in question. If these intervals are small enough, and $H_{1,2}(\theta)$ is the sinusoid coinciding with $h(\theta) + \epsilon/3$ at θ_1, θ_2, we shall have $H_{1,2}(\theta) \leq h(\theta) + 2\epsilon/3$ in (θ_1, θ_2). Then by the proof of 5.1.2,

$$\log |f(re^{i\theta})| \leq r\{H_{1,2}(\theta) + \epsilon/3\} \leq r\{h(\theta) + \epsilon\}, \quad r > r(\epsilon),$$

uniformly in each interval and hence uniformly in their union.

The following variant of 5.1.9 is sometimes useful.

5.1.11. *Theorem. If* ϵ *is a given positive number and* $h(\theta) \leq H(\theta)$, $-\alpha \leq \theta$

$\le \alpha$, *where* $H(\theta)$ *is a sinusoid, then for* $r > r(\epsilon)$ *we have*

$$\log |f(re^{i\theta})| \le r\{H(\theta) + \epsilon\},$$

uniformly in $-\alpha \le \theta \le \alpha$.

5.1.12. *Theorem. Unless* $f(z) \equiv 0$, *we cannot have either* (a) $h(\theta) \le 0$ *for* $\theta = \theta_1, \theta_2, \theta_3$ *with* $h(\theta) < 0$ *at one of these points if* $\theta_3 - \theta_1 > \pi$ *and* $\theta_1 < \theta_2 < \theta_3$; *or* (b) $h(\theta) \le 0$ *for* $\theta = \theta_1, \theta_2$ *with* $h(\theta) < 0$ *at one of these points and* $\theta_2 - \theta_1 = \pi$; *or* (c) $h(\theta) \le 0$ *for* $\theta = \theta_1, \theta_3$, $h(\theta) \ge 0$ *for* $\theta = \theta_2$, *with strict inequality at one point, if* $\theta_3 - \theta_1 < \pi$ *and* $\theta_1 < \theta_2 < \theta_3$.

If there are values of θ satisfying (a), (b), or (c), none of the terms on the left of (5.1.8) is positive, at least one is negative, and we have a contradiction. The next three statements are corollaries of (a), (b), (c), respectively.

5.1.13. *Corollary. Unless* $f(z) \equiv 0$, $h(\theta) < 0$ *at most in an interval of length* π.

5.1.14. *Corollary.*[b] *A function of exponential type in a half-plane cannot approach zero exponentially along the boundary unless it vanishes identically.*

5.1.15. *Corollary.*[c] *A function of exponential type cannot be of zero type on the boundary of an angle of opening less than* π *and of positive type on some ray in the interior.*

5.1.16. *Theorem.*[d] *If* $\alpha > \pi/2$ *and* $h(\theta) \le a$ *for* $|\theta| \le \alpha$, *then* $h(\theta) \ge -a|\sec\alpha|$ *for* $|\theta| \le \pi - \alpha$.

5.1.17. *Corollary. If* $f(z)$ *is an entire function of exponential type,*

$$\min h(\theta) \ge -\max h(\theta).$$

We give another proof of 5.1.17 in §5.4. To prove 5.1.16, write (5.1.8) in the form

$$2h(\theta_2) \sin \frac{\theta_3 - \theta_1}{2} \cos \frac{\theta_1 - \theta_3}{2} \le h(\theta_1) \sin(\theta_3 - \theta_2) + h(\theta_3) \sin(\theta_2 - \theta_1)$$

$$\le 2\, a \sin \frac{\theta_3 - \theta_1}{2} \cos \frac{\theta_3 - 2\theta_2 + \theta_1}{2},$$

whence $h(\theta_2) \cos(\theta_1 - \theta_3)/2 \le a$, provided that $0 < \theta_2 - \theta_1 < \pi$ and $0 < \theta_3 - \theta_2 < \pi$. If $|\theta_2| < \pi - \alpha$ we can take $\theta_1 = -\alpha$, $\theta_3 = \alpha$, and the conclusion of 5.1.16 follows.

If we take $\theta_3 = \theta_1 + \pi$ in (5.1.8) we see that if $-\alpha \le \theta < \theta + \pi \le \alpha$ then $h(\theta) + h(\theta + \pi) \ge 0$ (cf. 5.4.4). A little more than this is obtainable if $\alpha = \pi/2$ by letting $\theta_1 \to -\pi/2$, $\theta_2 \to 0$, $\theta_3 \to \pi/2$.

5.1.18. *Theorem. If* $\alpha = \pi/2$,

$$\limsup_{\theta \to -\pi/2} h(\theta) + \limsup_{\theta \to \pi/2} h(\theta) \ge 0.$$

In view of the remark made at the end of the proof of 5.1.5, this implies, for example, that

$$\limsup_{\theta \to -\pi/2} h(\theta) + h(\pi/2) \geq 0.$$

5.2. Convex sets. We shall see that the growth properties of an entire function of exponential type are conveniently described by the geometrical properties of a convex set associated with the function. This section contains the necessary preliminary material about convex sets. We shall not go into any more detail than is necessary for the applications.[a]

Let K be a nonempty bounded closed set of points $z = x + iy$. We call K convex if, whenever K contains z_1 and z_2, it also contains the line segment joining z_1 and z_2, that is, all points $tz_1 + (1 - t)z_2$, $0 \leq t \leq 1$. If $0 \leq \phi \leq 2\pi$, we project K on the ray $\arg z = \phi$ and denote by $k(\phi)$ the distance from the origin to the most remote point of this projection in the positive direction ($k(\phi)$ is negative if the projection of K lies completely on $\arg z = \phi + \pi$). Analytically, then, $k(\phi)$ is the maximum of $\Re(ze^{-i\phi}) = x \cos \phi + y \sin \phi$ as z ranges over K.

Since K is closed, the line $x \cos \phi + y \sin \theta - k(\phi) = 0$ contains at least one point of K, and K lies entirely on the side of this line where $x \cos \phi + y \sin \phi - k(\phi) \leq 0$ (which we may also write as

$$r \cos (\theta - \phi) \leq k(\phi),$$

if $z = re^{i\theta}$); this line is called a supporting line of K, and $k(\phi)$ is called the supporting function of K. Points z not in K are characterized by the fact that $x \cos \phi + y \sin \phi - k(\phi) > 0$ for some ϕ.

It is not always quite easy to calculate the supporting function for a convex set given, for example, by the equation of its boundary. The calculation is facilitated by the fact that some simple operations on convex sets correspond to simple operations on their supporting functions. By the sum of two convex sets K_1 and K_2 we shall mean the set of all points $z_1 + z_2$, where $z_1 \epsilon K_1$, $z_2 \epsilon K_2$; we denote it by $K_1 + K_2$. (For the set-theoretical sum of K_1 and K_2 we use the term union and the notation $K_1 \cup K_2$.) The set $K_1 + K_2$ is convex and its supporting function is $k_1(\phi) + k_2(\phi)$. If K_2 is a point z_0, $K_1 + K_2$ is the set K_1 translated by the vector $(0, z_0)$. If K_2 is the circle $|z| \leq \epsilon$, $K_1 + K_2$ has supporting function $k_1(\phi) + \epsilon$, and all the points of K_1 are interior points of $K_1 + K_2$. By way of illustration, and for reference, we tabulate the supporting functions for some special sets. (Here $a, b > 0$.)

Set K	Supporting function $k(\phi)$
point $x_0 + iy_0$	$x_0 \cos \phi + y_0 \sin \phi$
segment $(-a, a)$	$a\lvert \cos \phi \rvert$

segment $(-ia, ia)$	$a\|\sin\phi\|$
circle $\|z\| \leq R$	R
K translated so that the point originally at 0 is at $x_0 + iy_0$	$k(\phi) + x_0 \cos\phi + y_0 \sin\phi$
K rotated about 0 through angle γ	$k(\phi - \gamma)$
$K_1 + K_2$	$k_1(\phi) + k_2(\phi)$
smallest convex set containing $K_1 \cup K_2$	$\max\{k_1(\phi), k_2(\phi)\}$
rectangle, vertices at $\pm a \pm ib$	$a\|\cos\phi\| + b\|\sin\phi\|$
semicircle $r \leq 1$, $\|\theta\| \leq \pi/2$	1, $\|\phi\| \leq \pi/2$; $\|\sin\phi\|$, $\pi/2 \leq \phi \leq 3\pi/2$

The definition of a supporting function can be extended to unbounded convex sets; in that case $k(\phi)$ has to be allowed to take infinite values.

A supporting function has period 2π, but not all functions of period 2π are supporting functions. Since it is important to be able to recognize a supporting function when we see one, we establish the following characterization.

5.2.1. *Theorem. The function $k(\phi)$ is the supporting function of some nonempty bounded closed convex set if and only if it is of period 2π and, for $\phi_1 < \phi_2 < \phi_3$, $\phi_2 - \phi_1 < \pi$, $\phi_3 - \phi_2 < \pi$, we have*

$$k(\phi_1)\sin(\phi_3 - \phi_2) + k(\phi_2)\sin(\phi_1 - \phi_3) + k(\phi_3)\sin(\phi_2 - \phi_1) \geq 0. \tag{5.2.2}$$

The inequality (5.2.2) implies (as we shall show) that $k(\phi)$ is continuous and we could if we preferred add continuity as a requirement on $k(\phi)$ and prove directly (it is very easy to do so) that a supporting function is necessarily continuous.

We prove first that a supporting function satisfies (5.2.2). For every z in K, and arbitrary ϕ_1, ϕ_3,

$$x\cos\phi_1 + y\sin\phi_1 - k(\phi_1) \leq 0,$$

$$x\cos\phi_3 + y\sin\phi_3 - k(\phi_3) \leq 0.$$

If $0 < \phi_3 - \phi_2 < \pi$ and $0 < \phi_2 - \phi_1 < \pi$, we can multiply the first inequality by $\sin(\phi_3 - \phi_2)$ and the second by $\sin(\phi_2 - \phi_1)$, and add. There results, on combining terms,

$$-\{x\cos\phi_2 + y\sin\phi_2\}\sin(\phi_1 - \phi_3) - k(\phi_1)\sin(\phi_3 - \phi_2) - k(\phi_3)\sin(\phi_2 - \phi_1) \leq 0,$$

and since $x\cos\phi_2 + y\sin\phi_2 = k(\phi_2)$ for some z in K, (5.2.2) follows.

Next we prove that a (finite) solution of (5.2.2) is a supporting function.[b] First, we have to prove that it is continuous. This, in fact, we have already done, since (5.2.2) is, when $\phi_3 - \phi_1 < \pi$, merely another way of writing (with $k(\phi)$ instead of $h(\theta)$) the conclusion of 5.1.2, and so the continuity of $k(\phi)$ is established by 5.1.4.

Now let K be the set of points z for which

$$x \cos \phi + y \sin \phi - k(\phi) \leq 0, \qquad 0 \leq \phi \leq 2\pi \tag{5.2.3}$$

Then K is bounded: for, choosing $\phi = 0, \phi = \pi$, we have $-k(\pi) \leq x \leq k(0)$ if $x + iy$ is in K; choosing $\phi = \pi/2, 3\pi/2$, we have $-k(3\pi/2) \leq y \leq k(\pi/2)$. Also K is closed: (5.2.3) for a sequence $\{z_n\}$ implies (5.2.3) for $\lim z_n$, since $k(\phi)$ is continuous. K is convex, since the two inequalities (5.2.3) for z_1 and z_2 can be multiplied by t and $1 - t$, $0 < t < 1$, and added to give (5.2.3) for $tz_1 + (1 - t)z_2$. The crux of the proof consists in showing that K is not empty.

Let $k(\phi)$ assume its maximum at ϕ_0. We show that it is impossible to have $k(\phi_1) < k(\phi_0) \cos(\phi_1 - \phi_0)$ for any ϕ_1, $\phi_0 - \pi \leq \phi_1 < \phi_0 + \pi$. Consider the case $\phi_1 < \phi_0$; the other is similar. Let $H(\phi)$ be the sinusoid which coincides with $k(\phi)$ at ϕ_0 and ϕ_1. Then $H(\phi) - k(\phi_0) \cos(\phi - \phi_0)$ is another sinusoid which vanishes at ϕ_0, and so is of the form $C \sin(\phi_0 - \phi)$; and $C < 0$ since the sinusoid is negative at ϕ_1. Hence $H'(\phi_0) = -C > 0$, so that $H(\phi)$ is increasing at ϕ_0. That is, for small positive δ,

$$H(\phi_0) < H(\phi_0 + \delta).$$

Now $H(\phi_0) = k(\phi_0)$ and, by 5.1.5, whose proof uses only the property (5.2.2) (in the form of the conclusion of 5.1.2), $k(\phi_0 + \delta) \geq H(\phi_0 + \delta)$, so $k(\phi_0) < k(\phi_0 + \delta)$, contradicting the assumption that $k(\phi)$ attains its maximum at ϕ_0. Hence we have $k(\phi) \geq k(\phi_0) \cos(\phi - \phi_0)$ for all ϕ. That is, the point $z = k(\phi_0)e^{i\phi_0}$ belongs to K.

Since we showed in 5.1.7 that $h(\theta)$ for an entire function of exponential type satisfies (5.2.2), we have as an immediate application of 5.2.1:

5.2.4. *Theorem. If $f(z)$ is an entire function of exponential type, its indicator $h(\theta)$ is a supporting function.*

The following theorem states a fact which we shall need.

5.2.5.* *Theorem. The boundary of a nonempty closed bounded convex set is a point, a line segment, or a rectifiable Jordan curve.*

The proof is left to the reader, who may either proceed directly or appeal to properties of convex functions.

By an extreme point of a convex set K we mean a boundary point which is not an interior point of a line segment belonging to the boundary of K. Extreme points are characterized by the following property.

5.2.6. *Theorem. If p is an extreme point of K, then after an arbitrarily small neighborhood of p is removed from K the remainder of K is contained in a proper convex subset of K.*

For orientation, compare the effects of removing from a convex polygon (a) a neighborhood of a vertex, (b) a neighborhood of an interior point of one of its edges.

The property is obvious when K is a point or a line segment. Otherwise there is a supporting line through p (as follows from the definition), and p is either the only point of K on the supporting line, or an endpoint of the line segment which the supporting line has in common with K. In either case, a small circle drawn with center at p meets the boundary of K in two points which are not the ends of a diameter, and the intersection of K and a half plane determined by the line through these points has the required property.

5.3. The indicator diagram.[a] We are now ready to associate a convex set with each entire function of exponential type.

5.3.1. *Theorem. The function*

$$f(z) = \sum_{n=0}^{\infty} a_n z^n \tag{5.3.2}$$

is an entire function of exponential type if and only if

$$F(z) = \sum_{n=0}^{\infty} n! a_n / z^{n+1} \tag{5.3.3}$$

is convergent for some (finite) z. If the radius of convergence of the series in (5.3.3) is σ, $f(z)$ is of order 1 and type σ if $\sigma > 0$, of exponential type 0 if $\sigma = 0$.

If $f(z)$ is of order 1 and type τ, $\limsup_{n\to\infty} |f^{(n)}(0)|^{1/n} = \tau$ by (2.2.12) and so the radius of convergence of $\sum n!\, a_n/w^{n+1}$ is τ. On the other hand, if the radius of convergence of the last series is τ, the converse follows by (2.2.12) again.

The function $F(z)$ is often called the Borel transform of $f(z)$. It is the Laplace transform of $f(z)$ for z of sufficiently large positive real part, and the analytic continuation of this for the other values of z for which it is regular. Indeed, if $z = x + iy$ and $x > \tau$,

$$\int_0^\infty f(t) e^{-zt}\, dt = \sum_{n=0}^{\infty} a_n \int_0^\infty t^n e^{-zt}\, dt = \sum_{n=0}^{\infty} n! a_n z^{-n-1} = F(z), \tag{5.3.4}$$

the integration term by term being justified by Fubini's theorem because

$$\int_0^\infty \sum_{n=0}^{\infty} |a_n| t^n e^{-xt}\, dt = \sum_{n=0}^{\infty} n!\, |a_n|\, x^{-n-1};$$

the right-hand series converges for $x > \tau$ because $\limsup (n!\, |a_n|)^{1/n} = \tau$.

Let $f(z)$ be again an entire function of exponential type defined by (5.3.2), and let D denote the interesection of all closed bounded convex sets outside which $F(z)$ is regular; D is then the smallest closed convex set outside which $F(z)$ is regular; it is called the conjugate indicator diagram of $f(z)$, for reasons which will appear shortly. One useful property of D can be established immediately.

5.3.5. *Theorem. If $f(z)$ is an entire function of exponential type, D is the conjugate indicator diagram and C is a contour containing D in its interior, then*

$$(5.3.6) \qquad f(z) = (2\pi i)^{-1} \int_C F(w) e^{zw} \, dw.$$

The representation (5.3.6), which we shall refer to as the Pólya representation, is one of the most useful tools for investigating entire functions of exponential type. The contour C can always be taken to be a circle $|w| = \tau + \epsilon$, $\epsilon > 0$, where τ is the type of $f(z)$. Later (§§5.5, 6.7) we shall consider some cases where C can be shrunk to the boundary of D.

Let Γ be a circle $|w| = \tau + \epsilon$, $\epsilon > 0$, such that C is inside Γ. Then

$$\int_C F(w)e^{zw}\, dw = \int_\Gamma F(w)e^{zw}\, dw = \int_\Gamma \left\{ \sum_{n=0}^{\infty} n! a_n w^{-n-1} \sum_{k=0}^{\infty} z^k w^k / k! \right\} dw$$

$$= 2\pi i \sum_{n=0}^{\infty} a_n z^n = 2\pi i f(z),$$

the integration term by term now being justified by the uniform convergence of the series on Γ. This establishes (5.3.6).

We can now characterize D more directly in terms of $f(z)$.

5.3.7. *Theorem. The supporting function of D is $h(-\theta)$, where $h(\theta)$ is the indicator function* (5.1.1) *of $f(z)$.*

In other words, $h(\theta)$ is the supporting function of a convex set, the indicator diagram of $f(z)$, and D is the reflection of the indicator diagram with respect to the real axis.

To prove 5.3.7, we begin by supposing that z is real and positive in (5.3.4). We may then put $zt = u$ in the integral, and obtain

$$(5.3.8) \qquad F(z) = z^{-1} \int_0^{\infty} f(u/z) e^{-u}\, du.$$

The integral in (5.3.8) is now a regular function of z for $|z|$ sufficiently large, and so is in fact equal to $F(z)$ whenever it is regular. Now let $z = re^{i\phi}$ in the region of convergence of the integral, and put $u = rt$. Then

$$F(re^{i\phi}) = e^{-i\phi} \int_0^{\infty} f(te^{-i\phi}) e^{-rt}\, dt,$$

where the integral now converges if $r > h(-\phi)$. Finally, replace r by a complex variable $w = \rho e^{i\theta}$; the function $F(we^{i\phi})$ is regular for large $|w|$ and coincides with the integral

$$(5.3.9) \qquad e^{-i\phi} \int_0^\infty f(te^{-i\phi})e^{-wt}\,dt$$

for large positive values of w. But the integral converges for $\Re(w) = \rho \cos \theta > h(-\theta)$, and so $F(we^{i\phi})$ is regular in this half plane, or $F(\rho e^{i\psi})$ is regular for $\rho \cos (\psi - \phi) > h(-\phi)$. Since $h(-\phi)$ is a supporting function, we see that $F(z)$ is regular outside the closed convex set of which $h(-\phi)$ is the supporting function, and since D is the interesection of all closed convex sets outside which $F(z)$ is regular, we have $h(-\phi) \geq k(\phi)$, where $k(\phi)$ is the supporting function of D.

On the other hand, from the Pólya representation (5.3.6) we have

$$(5.3.10) \qquad |f(re^{i\theta})| \leq (2\pi)^{-1}L \max_C |F(w)| \max_C \exp \{r\Re(we^{i\theta})\},$$

where L is the length of C, and so

$$(5.3.11) \qquad h(\theta) \leq \max_C \Re(we^{i\theta}).$$

Let C be the boundary of the sum of D and the circle $|z| \leq \epsilon$; the supporting function of this set is $k(\phi) + \epsilon$, and so by (5.3.11) and the definition of a supporting function, $h(\theta) \leq k(-\theta) + \epsilon$. Since we already have $h(\theta) \geq k(-\theta)$, the theorem follows on letting $\epsilon \to 0$.

5.3.12. *Theorem. The extreme points of the conjugate indicator diagram of $f(z)$ are singular points of the Borel transform $F(z)$ of $f(z)$.*

If the extreme point p were not a singular point of $F(z)$, we could remove a neighborhood of p from D and obtain a smaller convex set outside which $F(z)$ is regular (by 5.2.6), thus contradicting the definition of D as the smallest such convex set.

5.4. Properties of the indicator diagram. The interpretation of $h(-\theta)$ as the supporting function of the convex set D, and the representation (5.3.6), lead to simple proofs of several properties of $h(\theta)$, some of which have already been proved in other ways.

In the first place, we have (using only the definition of $h(\theta)$)

5.4.1. *Theorem. If $f(z)$ is an entire function of order 1, its type is the maximum of $h(\theta)$.*

Next, $h(\theta)$, being a supporting function, can never be $-\infty$ (unless $f(z) \equiv 0$). Stated without reference to $h(\theta)$, this means

5.4.2. *Theorem. If $f(z)$ is an entire function of exponential type and, for some θ, $|f(re^{i\theta})| \leq e^{-r\omega(r)}$, where $\omega(r) \to \infty$, then $f(z) \equiv 0$.*

This also follows from 3.7.4. Compare 5.4.5 and 5.4.7.

5.4.3. *Theorem. Unless $f(z) \equiv 0$, $h(\theta) < 0$ at most in an interval of length π.*

This is 5.1.13. To prove it by our present methods, we note that if the origin is inside the indicator diagram, $h(\theta) \geq 0$ for all θ. If the origin is outside the diagram, the diagram is completely on one side of some line through the origin, so $h(\theta)$ will be negative only for rays in the half plane determined by this line and not containing the diagram.

5.4.4. *Theorem. For all* θ, $h(\theta - \pi/2) + h(\theta + \pi/2) \geq 0$.

In fact, the sum is the distance between the two supporting lines which have the direction θ.

5.4.5. *Theorem. For all* θ, $h(\theta) \geq -\max h(\theta)$.

This is 5.1.17; it is now geometrically evident. A still more precise result is

5.4.6. *Theorem. If* $h(\theta) \leq \tau$ *and* $h(0) = -\tau$ *then* $f(z) = g(z)e^{-\tau z}$ *with* $g(z)$ *an entire function of exponential type zero.*

Of course $h(0) = -\tau$ could be replaced by $h(\theta_0) = -\tau$ with a corresponding change in the form of $f(z)$.

If we put $g(z) = f(z)e^{\tau z}$, we have $h_g(\theta) = h_f(\theta) + \tau \cos\theta \leq \tau(1 + \cos\theta)$. Thus we have $h_g(0) = 0$ and $h_g(\pi) \leq 0$, so that the indicator diagram of $g(z)$ is a segment of the imaginary axis (or a point), and hence $h_g(\theta)$ is of the form $a\,|\sin\theta| + b\sin\theta$, $a \geq 0$. Thus we have

$$a\,|\sin\theta| + b\sin\theta \leq \tau(1 + \cos\theta),$$

and letting $\theta \to \pi$ we infer that $a = b = 0$. Hence the indicator diagram of $g(z)$ is the point 0, which means that $g(z)$ is of zero exponential type.

5.4.7. *Corollary.*[a] *If* $f(z)$ *is an entire function satisfying* $|f(z)| \leq e^{\tau r}$ *and* $f(re^{i\theta}) = O(e^{-\tau r})$ *for some one* θ *then* $f(z) \equiv Ae^{\lambda z}$, $|\lambda| = \tau$.

Compare 5.4.5, 3.7.4 and 2.7.6. The next two theorems give some further corollaries of the properties of $h(\theta)$.

5.4.8.* *Theorem.*[b] *If* $f(z)$ *is an entire function of exponential type* τ, $0 < \alpha < \pi/2$, *and* $h(\pm\alpha) < -\tau\cos\alpha$, *then* $f(z) \equiv 0$.

5.4.9.* *Theorem.*[c] *If* $f(z)$ *is an entire function of exponential type* τ, $0 < \beta < \alpha < \pi/2$, $h(\pm\alpha) \geq -\gamma\tau\cos\alpha$ *and* $h(\pm\beta) < -\gamma\tau\cos\beta$, $0 < \gamma \geq 1$, *then* $f(z) \equiv 0$.

5.4.10. *Theorem. The indicator diagram of the sum of two functions is contained in the smallest convex set containing the union of the indicator diagrams of the summands.*

5.4.11. *Theorem. The indicator diagram of the product of two functions is contained in the sum (in the sense of §5.2) of the indicator diagrams of the factors.*

The proofs are left to the reader.

In certain cases the indicator diagram of a product is equal to the sum of the diagrams of the factors, but the proofs are more difficult. In §10.4 we shall prove

5.4.12. *Theorem. If the indicator diagram of $f_1(z)$ is a point, the indicator diagram of $f_1(z)f_2(z)$ is the sum of the indicator diagrams of the factors.*

The difficulty which keeps us from proving this now is as follows. It is enough, by 5.4.11, to show that (with self-explanatory notation) $h_{12}(\theta) \geq h_1(\theta) + h_2(\theta)$. To do this it is necessary to be able to find, for each θ, arbitrarily large values of r for which not only do we have

$$\log |f_1(re^{i\theta})| \geq r\{h_1(\theta) - \epsilon\}, \qquad \log |f_2(re^{i\theta})| \geq r\{h_2(\theta) - \epsilon\},$$

but both inequalities hold simultaneously.

Another case in which the indicator diagram of a product is the sum of the indicator diagrams of the factors is given in §7.6.

On the other hand, the indicator diagram of a product can be a point even when the indicator diagram of neither factor is a point.[d]

5.4.13. *Theorem. The indicator diagram of $f'(z)$ is a subset of that of $f(z)$; it is the same as that of $f(z)$ except when $w = 0$ is an extreme point of the diagram and is a simple pole of the Borel transform of $f(z)$.*

In fact, the conjugate indicator diagram of $f(z)$ is the smallest convex set outside which $F(w)$ is regular, and that of $f'(z)$ is the smallest convex set outside which $wF(w)$ is regular.

5.4.14. *Theorem. If $\epsilon > 0$ then for $r > r(\epsilon)$ we have*

$$\log |f(re^{i\theta})| \leq r\{h(\theta) + \epsilon\},$$

uniformly in θ.

This is 5.1.9. For entire functions the proof is immediate because of the estimate (5.3.10).

5.5. The Borel transform on the boundary of the conjugate indicator diagram. Under some conditions the contour C in the Pólya representation (5.3.6) of an entire function of exponential type can be replaced by the boundary B of the conjugate indicator diagram. A general condition of this kind will be given here, and some conditions involving more special restrictions on the function and on D will be given in the next chapter.

5.5.1. *Theorem.*[a] *If $\psi(r)$ is a positive function such that*

$$|f(z)| \leq \psi(r)e^{rh(\theta)} \tag{5.5.2}$$

for all θ, and

$$\int_1^\infty \psi(r)\, dr < \infty \tag{5.5.3}$$

then

$$f(z) = (2\pi i)^{-1} \int_B F(w)e^{zw}\, dw. \tag{5.5.4}$$

We shall show that (5.5.2) and (5.5.3) imply that $F(z)$ is bounded outside D; we can then let C in (5.3.6) shrink to B and (5.5.4) follows by "bounded convergence." (The details of this last process are left to the reader.) To show that $F(z)$ is bounded, we use the representation (5.3.9),

$$F(we^{i\phi}) = e^{i\phi} \int_0^\infty f(te^{i\phi})e^{-wt}\,dt, \qquad \Re(w) > h(-\phi).$$

By (5.5.2),

$$|\,F(we^{i\phi})\,| \leq \int_0^\infty \psi(t) \exp\,[\{h(-\phi) - \Re(w)\}t]\,dt,$$

so that if $\Re(w) > h(-\,\phi)$ the right-hand side is bounded. That is, $|\,F(\rho e^{i\lambda})\,|$ is uniformly bounded for $\rho \cos{(\lambda - \phi)} > h(-\,\phi)$, in other words on the positive side of the supporting line of D which is perpendicular to the direction ϕ. Since this is true for all λ, $F(z)$ is bounded outside D.

5.6. Functions of exponential type in an angle.[a] The Pólya representation (5.3.6) of an entire function of exponential type by means of a contour integral can be generalized to apply to a function which is merely of exponential type in an angle. A consequence of this generalization is a lemma which enables many results to be generalized from entire functions of exponential type (for which proofs are often easier) to functions which are only of exponential type in an angle.

5.6.1. *Theorem. If $f(z)$ is regular and of exponential type c in the angle $|\arg z| \leq \alpha \leq \pi/2$, then its Laplace transform*

$$F(z) = \int_0^\infty e^{-zt}f(t)\,dt \tag{5.6.2}$$

defines a function which is regular outside the unbounded convex set whose supporting function is $h(-\,\phi)$ for $|\,\phi\,| \leq \alpha$. If $0 < \beta < \alpha$, B is a sufficiently large positive number, and C is the contour composed of two lines meeting at B and making angles $\pm(\beta + \pi/2)$ with the real axis, then

$$f(z) = (2\pi i)^{-1} \int_C e^{zw}F(w)\,dw, \qquad |\arg z| < \beta. \tag{5.6.3}$$

The region of regularity of $F(z)$ includes at least the region to the right of the curve composed of the arc $|\arg z| \leq \alpha$ of the circle $|z| = c$ and the left-hand halves of the tangents to this arc at its ends.

If $F(z)$ is defined by (5.6.2) we initially have $F(z)$ regular for $x > h(0)$. Proceeding as in §5.3 we see that $F(z)$ can be continued into any half plane $\Re(ze^{-i\phi}) > h(-\,\phi)$ provided that $|\,\phi\,| < \alpha$, with the representation

(5.6.4) $$F(z) = \int_0^{\infty e^{-i\phi}} f(w)e^{-zw}\, dw,$$

where the integral is taken along the ray $\arg w = -\phi$. Thus $F(z)$ is regular in the region U outside the intersection of all the half planes

$$\Re(ze^{-i\phi}) > h(-\phi), \qquad |\phi| < \alpha,$$

an unbounded convex set. From the representation (5.6.4) we have the estimate

(5.6.5) $$|F(\rho e^{i\psi})| \le \int_0^{\infty} |f(te^{-i\phi})|\, e^{-t\rho\cos(\psi-\phi)}\, dt.$$

If $|\phi| \le \beta < \alpha$ we have $|f(te^{-i\phi})| \le e^{t\{h(-\phi)+\epsilon\}}$, $r > r(\epsilon)$, uniformly in ϕ (for a fixed β), and so

$$|F(\rho e^{i\psi})| \le \int_0^{r(\epsilon)} |f(te^{-i\phi})|\, e^{-t\rho\cos(\psi-\phi)}\, dt$$
$$+ \int_{r(\epsilon)}^{\infty} e^{-t\{\rho\cos(\psi-\phi)-h(-\phi)-\epsilon\}}\, dt, \qquad |\psi| < \alpha + \pi/2.$$

If $\Re(ze^{-i\phi}) > \max(h(-\phi), 0)$, the exponent is negative in the first integral, which therefore is $O(\rho^{-1})$, uniformly in $|\psi| \le \beta + \pi/2 < \alpha + \pi/2$. The same estimate applies to the second integral if $\rho\cos(\psi - \phi) - h(-\phi) > 2\epsilon$ (say), that is, if z is at least a fixed positive distance inside the region U in which $F(z)$ is regular. Hence $F(z) = O(1/|z|)$ at least in the part of U defined by $\Re(ze^{-i\phi}) > \max(h(-\phi), 0)$. Call this region V.

Now let z satisfy $|\arg z| < \alpha$ and let $0 < \beta < \gamma < \alpha$. Then let B be a real number greater than $c \sec \gamma$, and let C_1 be a curve starting at $w = B$, lying in the region V, and asymptotic to the ray $\arg w = \beta + \pi/2$; let C_2 be the reflection of C_1 in the real axis. Consider the function

$$f^*(z) = (2\pi i)^{-1} \int_{C_1} e^{zw} F(w)\, dw - (2\pi i)^{-1} \int_{C_2} e^{zw} F(w)\, dw,$$

where the line of integration starts at $w = B$ in each integral. On C_1 and C_2 we have the representation (5.6.4) with $\phi = -\gamma$ and γ, respectively. We shall show that the iterated integral

(5.6.6) $$\int_{C_1} e^{zw}\, dw \int_0^{\infty e^{-i\gamma}} e^{-wt} f(t)\, dt$$

converges absolutely and uniformly for $|\arg z| \le \beta$; this (together with the corresponding fact for the other integral) implies both that $f^*(z)$ is regular in $|\arg z| < \beta$ and that the order of integration can be reversed;

since the inner integral can then be evaluated, we obtain

$$(5.6.7)\quad f^*(z) = (2\pi i)^{-1}\int_0^{\infty e^{-i\gamma}} \frac{e^{B(z-t)}}{t-z} f(t)\,dt - (2\pi i)^{-1}\int_0^{\infty e^{i\gamma}} \frac{e^{B(z-t)}}{t-z} f(t)\,dt.$$

On the other hand, we have for real positive x,

$$f(x)e^{-Bx} = (2\pi i)^{-1}\int_\Gamma \frac{e^{-Bt}f(t)}{t-x}\,dt,$$

where Γ is the sector bounded by the lines $\arg t = \pm\gamma$ and the arc of $|z| = R > x$ where $|\arg t| \le \gamma$.

Since $B > c \sec \gamma$, and $|f(t)| \le e^{c|t|}$ for large $|t|$, the integral over the arc tends to 0 as $R \to \infty$, and so

$$f(x) = (2\pi i)^{-1}\int_0^{\infty e^{-i\gamma}} \frac{e^{B(x-t)}}{t-x} f(t)\,dt - (2\pi i)^{-1}\int_0^{\infty e^{i\gamma}} \frac{e^{B(x-t)}}{t-x} f(t)\,dt = f^*(x).$$

In other words, $f^*(x) = f(x)$ for real positive x, and so throughout the angle $|\arg z| < \beta$.

We now verify the absolute convergence of (5.6.6). In the inner integral, $|f(t)| < e^{|t|h(-\gamma)}$ for large $|t|$, while $|e^{-wt}| = \exp\{-|t|\,\Re\,(we^{-i\gamma})\}$; since $\Re(we^{-i\gamma}) > h(-\gamma) + \epsilon$, $\epsilon > 0$, for large $|w|$ on C_1, the inner integral converges absolutely and the integral of the absolute value has an upper bound independent of w. Along C_1,

$$|e^{zw}| = \exp\{-|z|\,|w| \sin(\arg z + \beta)\},$$

and so the outer integral converges absolutely in any angle contained in $|\arg z| < \beta$.

5.6.8. *Macintyre's lemma. Let $f(z)$ be regular and of exponential type in $|\arg z| \le \alpha \le \pi/2$, $h(\pm\alpha) < \pi \sin \alpha$, $h(0) \le 0$. Then $f(z) = f_1(z) + f_2(z)$, where $f_2(z)$ is an entire function of exponential type less than π and*

$$f_1(z) = O(r^{-1}e^{\pi r|\sin\theta|}), \qquad z = re^{i\theta},$$

uniformly for $|\phi| \le \beta < \alpha$.

We represent $f(z)$ by (5.6.3). The type c of $f(z)$ is less than $\pi \sin \alpha$; then if we choose $\beta < \gamma < \alpha$ and $B = c \sec \gamma$, and β and γ are close enough to α, the contour C of 5.6.1 will cross the imaginary axis at $\pm ib$, $b < \pi$. Since $F(w)$ will be regular in the triangle between the imaginary axis and the right-hand part of C, we can deform this part of C into a segment of the vertical line $x = \epsilon > 0$ together with short pieces of C to connect it with $\pm ib$. Let $f_2(z)$ be the part of (5.6.3) with integration over this deformed path; $f_2(z)$ will be an entire function of exponential type $b < \pi$, if ϵ is small enough. Let $f_1(z)$ be the rest of (5.6.3). Taking the integral over the upper half of the path as representative, we can estimate this part of $f_1(z)$ by using the fact that $F(w)$ is bounded; and we have, on the path, $w = B + \rho e^{i(\beta+\pi/2)}$, $z = re^{i\theta}$,

$$\begin{aligned}\int | e^{zw} | \, | dw | &= \int_{B\csc\beta}^{\infty} e^{rB\cos\theta - r\rho\sin(\beta+\theta)} \, d\rho \\ &= \frac{e^{rB\cos\theta}}{r \sin(\beta+\theta)} e^{-rB\csc\beta\sin(\beta+\theta)} \\ &= r^{-1} \csc(\beta+\theta) e^{-rB\cot\beta\sin\theta} = O(r^{-1} e^{\pi r |\sin\theta|})\end{aligned}$$

if $\theta > -\beta$ and β and γ are close enough to α.

Notes for Chapter 5

Most of the results of Chapter 5 (except as noted below) are to be found in Pólya [2], together with remarks on their history. For the properties of $h(\theta)$ see also Phragmén and Lindelöf's original paper [1].

5.1a. Theorem 5.1.2 shows that $h(\theta)$ is "sub-sinusoid;" that is, it has the same property with respect to sinusoids that convex functions have with respect to linear functions. The remaining theorems of this section can be obtained as consequences of this remark and the general theory of convex functions.

5.1b. Carlson; see Titchmarsh [3], p. 185; cf. Riesz [1], Hoheisel [1]. Case (b) of 5.1.12 shows more generally that a function of exponential type in a half-plane cannot approach 0 exponentially along one half of the boundary and be of less than exponential increase on the other half, unless it vanishes identically. A still more precise result will be given in 6.3.6.

5.1c. Dinghas [1].

5.1d. Srivastava [3].

5.2a. For more details see Pólya [2], and for convex sets in general see Bonnesen and Fenchel [1].

5.2b. For this proof see Chebotarëv and Meiman [1], p. 115.

5.3a. For methods of proof cf. A. J. Macintyre [1], where a corresponding discussion is carried out for functions of any finite order, and references are given to earlier and less successful attempts in the same direction.

5.4a. Srivastava [1], generalizing a theorem of Cramér; Titchmarsh [2], p. 175.

5.4b. Ahiezer [1]; cf. Pólya [2], p. 596.

5.4c. Ahiezer [1], where the statement of the theorem contains an evident misprint.

5.4d. Pólya [2], p. 596. Examples are readily constructed using the results of Chapter 8. See Wilson [1] for further results.

5.5a. Schmidli [1]. The proof of the text is different.

5.6a. A. J. Macintyre [1], where a more general result is given.

CHAPTER 6

FUNCTIONS OF EXPONENTIAL TYPE, RESTRICTED ON A LINE. I. THEOREMS IN THE LARGE

6.1. Introduction. We are now ready to consider the properties of a function of exponential type which follow from the imposition of additional restrictions on the function. In this chapter we shall study the effect on the general behavior of the function of imposing a condition like boundedness along a line. Later chapters will deal on the one hand with conditions imposed on the function only along a sequence of points, and on the other hand with deeper results concerning the influence which restrictions on the function have on the distribution of its zeros.

6.2. Functions bounded on a line. We shall begin by assuming a condition of the form

$$|f(x)| \leq \omega(x), \qquad -\infty < x < \infty, \tag{6.2.1}$$

where $\omega(x) = O(e^{\epsilon|x|})$, $|x| \to \infty$, for every positive ϵ. This implies that $h(0) \leq 0$, $h(\pi) \leq 0$; and so, according to (5.1.2), if $h(\pm\pi/2) \leq c$ we have $|h(\theta)| \leq c|\sin\theta|$, whence, for large r,

$$|f(re^{i\theta})| \leq e^{r(c|\sin\theta|+\epsilon)}. \tag{6.2.2}$$

We can, however, say much more than this. For example, an entire function of exponential type which is bounded on the real axis is bounded on every parallel to the real axis (6.2.4), and a quite precise evaluation of the bound can be given; similarly, a function belonging to L^p on the real axis belongs to L^p on each parallel to the real axis (6.7.1). Results of this kind are, of course, not true for entire functions in general.

We begin with the case where $\omega(x)$ is a constant. There are several variants, according to how much we assume, and we state several related theorems before proving any of them.

6.2.3. *Theorem. If $f(z)$ is regular and of exponential type in the first quadrant, $|f(iy)| \leq Ae^{cy}$ $(0 \leq y < \infty)$, and $|f(x)| \leq M$ $(0 \leq x < \infty)$, then*

$$|f(x + iy)| \leq \max(A, M)e^{c|y|}, \qquad 0 \leq x < \infty, 0 < y < \infty$$

6.2.4. *Theorem.*[a] *If $f(z)$ is regular and of exponential type in the upper half plane, $h(\pi/2) \leq c$, and $|f(x)| \leq M$, $-\infty < x < \infty$, then*

$$|f(x + iy)| \leq Me^{cy}, \qquad -\infty < x < \infty,\ 0 \leq y < \infty \tag{6.2.5}$$

Notice the difference between the hypotheses of 6.2.3 and 6.2.4: in 6.2.4 we assume only that $|f(iy)| \leq A(\epsilon)e^{(c+\epsilon)y}$ for every positive ϵ.

Theorem 6.2.4 is a best possible result, in the sense that neither M nor c can be replaced by a smaller constant in (6.2.5), as, for example, $f(z) = \sin z$ shows. Even assuming that $f(z)$ is an entire function leads to no improvement. Nevertheless, if $f(z)$ is entire and is also real on the real axis, there is a sharper result.

6.2.6.* *Theorem.*[b] *If $f(z)$ is an entire function of exponential type, real for real z, $h(\pm\pi/2) \leq c$, and $|f(x)| \leq M$, $-\infty < x < \infty$, then*

$$(6.2.7) \qquad |f(x + iy)| \leq M \cosh cy,$$

and there is equality at some nonreal point only if $f(z) \equiv M \cos (cz + b)$, where b is real.

6.2.8. *Theorem. If $f(z)$ is regular and of exponential type c in the first quadrant, and $f(x) \to 0$ as $x \to \infty$, then $f(x + iy) \to 0$ as $x \to \infty$ for each fixed positive y (and hence, by Montel's theorem (1.4.9), uniformly in any bounded y-range). Moreover, for $0 \leq \theta < \pi/2$, $f(re^{i\theta}) = o(e^{cr\sin\theta})$.*

The first part of 6.2.8 is a special case of

6.2.9. *Theorem. If $f(z)$ is regular and of exponential type in the first quadrant, $h(\pi/2) \leq c$, and $\limsup_{x\to\infty} |f(x)| = K$ then*

$$\limsup_{x\to\infty} |f(x + iy)| \leq Ke^{c|y|}, \qquad 0 < y < \infty.$$

Another natural generalization of (6.2.8) would be that if $|f(x)|$ approaches a limit as $x \to \infty$ then $|f(x + iy)|$ approaches a limit for each y; but it is not known whether or not this is true.

6.2.10.* *Theorem.*[c] *If $f(z)$ is regular and of exponential type c in the right-hand half plane and*

$$(6.2.11) \qquad M[f(x)] \equiv \limsup_{T\to\infty} T^{-1} \int_0^T |f(t)|\, dt = K < \infty,$$

then

$$(6.2.12) \qquad M[f(x + iy)] \leq Ke^{c|y|}, \qquad -\infty < y < \infty.$$

The finiteness of the mean value (6.2.11) is a generalization of boundedness, and a significant generalization because (6.2.11) can hold for an entire function of exponential type which is unbounded on the positive real axis.[d] For the corresponding generalization for other mean values see 6.7.24.

6.2.13. *Theorem.*[e] *If $f(z)$ is an entire function of exponential type,*

$$h(\pm\pi/2) = 0$$

and $f(x) = O(|x|^p)$, $x \to \pm\infty$, then $f(z)$ is a polynomial of degree not exceeding p.

This is a corollary of 6.2.4: applying 6.2.4 to $g(z) = z^{-k-1}\{f(z) - P_k(z)\}$, where $P_k(z)$ is the polynomial consisting of the terms in the Maclaurin series of $f(z)$ through z^k, $k = [p]$, we have $g(z)$ bounded in the entire plane and so a constant; the constant must be zero since $g(z) \to 0$ on the real axis.

A less immediate corollary (not an immediate corollary of 6.2.10) is

6.2.14.* *Theorem. If $f(z)$ is an entire function of exponential type zero and $M[f(x)]$ is finite then $f(z)$ is a constant.*

We now turn to the proofs of these theorems.

In proving 6.2.3 we may suppose that $M = 1$. Consider the function $g(z) = f(z)e^{icz}$ in the first quadrant. We have $|g(z)|$ bounded (by 1) on the positive real axis, and $g(z)$ bounded on the positive imaginary axis; hence by the Phragmén-Lindelöf theorem (1.4.2), $g(z)$ is bounded in the first quadrant. If $|g(z)| \le N$, $|f(z)| \le Ne^{cy}$.

Theorem 6.2.4 is a special case of 1.4.3 (with a different half plane).

To prove 6.2.8 and 6.2.9, we note first that since $f(x)$ is bounded,

$$|f(z)| \le A(\epsilon)e^{(c+\epsilon)y}$$

for every positive ϵ, by 6.2.3. We now modify the line of reasoning used in 1.4.4 by considering the auxiliary function

$$g(z) = \frac{z}{z+\lambda} f(z+x_0)e^{i(c+\epsilon)z},$$

where ϵ (small) and λ (large) are positive numbers and

$$|f(x)| < K + \delta, \qquad \delta > 0$$

for $x > x_0$, so that $|g(x)| \le K + \delta$, $0 \le x < \infty$. Then

$$|f(z+x_0)e^{i(c+\epsilon)z}| \to 0$$

for $z = iy$, $y \to \infty$; hence by choosing λ large enough we can make

$$|g(iy)| < K + \delta, \qquad y \ge 0.$$

Therefore $|g(z)| \le K + \delta$ throughout the first quadrant and so

$$|f(z+x_0)| \le |(z+\lambda)/z|\, e^{(c+\epsilon)y}(K+\delta), \qquad x \ge 0,\ y \ge 0.$$

Since the left-hand side is independent of ϵ we have

$$(6.2.15) \qquad |f(z+x_0)| \le |(z+\lambda)/z|\, e^{cy}(K+\delta), \qquad x \ge 0,\ y \ge 0.$$

Finally, if y is fixed and $x \to \infty$, $(z+\lambda)/z \to 1$ and so

$$|f(z+x_0)| \le e^{cy}(1+\delta)\,(K+\delta)$$

for sufficiently large x; since δ can be arbitrarily small this proves 6.2.9 and so (with $K = 0$) the first part of 6.2.8. For the second part of 6.2.8, we return to (6.2.15), take $K = 0$, and let $z = re^{i\theta} - x_0$, $0 < \theta < \pi/2$, and $r \to \infty$. Then $e^{-cr\sin\theta}|f(re^{i\theta})| \le \delta(1+\delta)$ if r is large enough, and so $e^{-cr\sin\theta}|f(re^{i\theta})| \to 0$.

6.3. Relations among integrals, mean values, and distribution of zeros. We collect here some theorems connecting integrals along the real axis, mean values on semicircles, and the distribution of the zeros of a function which is regular and of exponential type in the upper half plane. These are all deductions from Carleman's formula, and quite simple except for some elementary Tauberian arguments. Some of them will be needed as lemmas in this chapter, while some will not be wanted until Chapter 7; since they all come from a common source it is convenient to discuss them together even though they are not obviously connected with the main theme of this chapter.

Throughout this section $f(z)$ will be regular in $y > 0$, continuous in $y \geq 0$, with zeros having no finite limit point, and will satisfy the condition, less restrictive than that of being of exponential type,

(6.3.1) $$\alpha = \liminf_{r\to\infty} r^{-1} \log M(r) < \infty.$$

We set

(6.3.2) $$J(R) = R^{-1} \int_0^\pi \log | f(Re^{i\theta}) | \sin \theta \, d\theta,$$

(6.3.3) $$I(R) = \int_1^R \left(\frac{1}{x^2} - \frac{1}{R^2}\right) \log | f(x)f(-x) | \, dx,$$

(6.3.4) $$\Sigma(R) = \sum_{r_k \leq R} \left(\frac{1}{r_k} - \frac{r_k}{R^2}\right) \sin \theta_k ,$$

(6.3.5)
$$I^*(R) = \int_1^R x^{-2} \log | f(x)f(-x) | \, dx,$$
$$I^*(\infty) = \int_1^{\to\infty} x^{-2} \log | f(x)f(-x) | \, dx,$$

where $r_k \exp(i\theta_k)$ are the zeros of $f(z)$ in $y > 0$. Our hypothesis (6.3.1) thus implies that $J(R)$ is bounded above for a sequence $R_n \to \infty$.

6.3.6. *Theorem.*[a] *If $f(z)$ satisfies* (6.3.1) *and*

(6.3.7) $$\int_{-\infty}^{\infty} \frac{\log^+ | f(x) |}{1 + x^2} dx < \infty$$

then

(6.3.8) $$\int_{-\infty}^{\infty} \frac{| \log | f(x) || \, dx}{1 + x^2} < \infty.$$

This includes 5.1.14 as a special case.

6.3.9. *Theorem. If $f(z)$ satisfies* (6.3.1) *and if $I^*(R)$ is bounded above, then $\Sigma(R)$ approaches a finite limit and the series $\sum r_k^{-1} \sin \theta_k$ converges.*

6.3.10. *Theorem. If $f(z)$ satisfies* (6.3.1) *and $I^*(\infty)$ exists then $J(R)$ approaches a finite limit as $R \to \infty$.*

6.3.11. *Theorem. If $f(z)$ is of exponential type and $I^*(R)$ is bounded above, then $I^*(R)$ and $J(R)$ are bounded.*

6.3.12. *Theorem. If $J(R)$ is bounded and $\sum r_k^{-1} \sin \theta_k$ converges, $I^*(R)$ is bounded.*

6.3.13. *Theorem. If $f(z)$ is of exponential type and $\lim_{R\to\infty} J(R)$ exists then $I^*(R)$ approaches a limit if it is bounded.*

For the purposes of this chapter we need only 6.3.6, and a weaker form of 6.3.9 and 6.3.10 (which would be no easier to prove than the more general results).

Note that by (2.10.22) $J(R)$ is always bounded for an entire function of exponential type. If we use this fact and compare 6.3.9, 6.3.11, and 6.3.12, we have the following result.

6.3.14. *Theorem. If $f(z)$ is an entire function of exponential type,*

$$\sum r_k^{-1} \mid \sin \theta_k \mid$$

(summed over all the zeros of $f(z)$) converges if and only if $I^*(R)$ is bounded (or bounded above).

We may write Carleman's theorem in the form

$$\Sigma(R) = \pi^{-1} J(R) + (2\pi)^{-1} I(R) + A, \tag{6.3.15}$$

where A is a constant and Σ, J, I are defined in (6.3.2)—(6.3.4).

To prove 6.3.6, note that $\Sigma(R) \geq 0$ and so by (6.3.1) and (6.3.15) we have for a sequence $R_n \to \infty$

$$-\int_1^{R_n/2} (x^{-2} - R_n^{-2}) \log^- \mid f(x)f(-x) \mid dx$$

$$\leq -\int_1^{R_n} (x^{-2} - R_n^{-2}) \log^- \mid f(x)f(-x) \mid dx \leq O(1).$$

In the left-hand integral, $-1/R_n^2 \geq -1/(2x)^2$, so that

$$\frac{3}{4}\int_1^{R_n/2} x^{-2}\{ - \log^- \mid f(x)f(-x) \mid\} \, dx \leq O(1).$$

The integrand is not negative, so

$$\int_1^R x^{-2}\{-\log^- \mid f(x)f(-x) \mid\} \, dx = O(1). \tag{6.3.16}$$

By (6.3.6),

$$\int_1^R x^{-2} \log^+ | f(x)f(-x) | \, dx \le \int_1^R x^{-2} \log^+ | f(x) | \, dx$$

$$+ \int_1^R x^{-2} \log^+ | f(-x) | \, dx + O(1) = O(1);$$

combining this with (6.3.16), we have

$$\int_1^R x^{-2} | \log | f(x)f(-x) || \, dx = O(1).$$

which is equivalent to (6.3.8).

Now suppose that the hypothesis of 6.3.9 is satisfied. By hypothesis, $J(R_n)$ is bounded above for a sequence $R_n \to \infty$. By the second mean value theorem,

$$I(R) = \int_1^R (1 - x^2/R^2)x^{-2} \log | f(x)f(-x) | \, dx$$

$$= (1 - R^{-2}) \int_1^{R'} x^{-2} \log | f(x)f(-x) | \, dx, \qquad 1 < R' < R,$$

so $I(R_n)$ is bounded above. Therefore $\Sigma\,(R_n)$ is bounded above, from (6.3.15). But $\Sigma(R)$ is a nondecreasing function of R, since when R increases we add nonnegative terms and subtract less than before from what we had. Hence $\Sigma(R)$ approaches a finite limit. Since

$$\sum_{r_k \le R} \left(\frac{1}{r_k} - \frac{r_k}{R^2}\right) \sin \theta_k \ge \sum_{r_k \le R/2} \left(\frac{1}{r_k} - \frac{r_k}{R^2}\right) \sin \theta_k \ge \frac{3}{4} \sum_{r_k \le R/2} r_k^{-1} \sin \theta_k ,$$

this means that $\sum r_k^{-1} \sin \theta_k$ converges. This proves 6.3.9.

The proof of 6.3.10 is similar. By the proof of 6.3.9, $\Sigma(R)$ approaches a finite limit. It follows that $J(R)$ approaches a finite limit provided that we show that the convergence of $I^*(R)$ implies the convergence of $I(R)$. To do this, put

$$\psi(t) = \int_t^\infty x^{-2} \log | f(x)f(-x) | \, dx;$$

then

$$-R^{-2} \int_1^R \log | f(x)f(-x) | \, dx = R^{-2} \int_1^R x^2 \, d\psi(x)$$

$$= \psi(R) - R^{-2}\psi(1) - 2R^{-2} \int_1^R x\psi(x) \, dx.$$

Since $\psi(x) = o(1)$ as $x \to \infty$, we thus have $I(R) = I^*(R) + o(1)$ if $I^*(\infty)$ exists, and this proves 6.3.10.

For 6.3.11, we have $J(R)$ bounded above because $f(z)$ is of exponential type. As in the proof of 6.3.9, $\Sigma(R)$ approaches a limit and $I(R)$ is bounded above. Then (6.3.15) shows that $I(R)$ is bounded below, and therefore that $J(R)$ is bounded. To complete the proof we have to show that $I^*(R)$ is bounded if $I(R)$ is bounded (and $f(z)$ is of exponential type, so that

$$\log | f(x)f(-x) | \leq O(x)).$$

This follows from the first case of the following lemma.

6.3.17. *Lemma.*[b] *If $\omega(x)$ has an absolutely continuous first derivative in $(1, \infty)$, then, as $x \to \infty$, $\omega(x) = O(x)$ and $\omega''(x) \leq O(1/x)$ imply*

$$\omega'(x) = O(1);$$

$\omega(x) \sim Ax$ and $\omega''(x) \leq O(1/x)$ imply that $\omega'(x) \to A$.

Let $0 \leq \delta(x) \leq x$; then by Taylor's theorem with remainder,

$$\omega(x \pm \delta(x)) = \omega(x) \pm \delta(x)\omega'(x) + \delta(x)^2\omega''(x \pm \theta\delta(x)), \quad 0 \leq \theta \leq 1,$$

or

$$\mp\omega'(x) = \frac{\omega(x) - \omega(x \pm \delta(x))}{\delta(x)} + \delta(x)\omega''(x \pm \theta\delta(x)).$$

For the first part of the lemma, take $\omega(x) = x$; then $\mp\omega'(x) \leq O(1)$, for both choices of the sign, as required. For the second part, we may consider $\omega(x) - Ax$ instead of $\omega(x)$ and thus reduce the lemma to the case when $A = 0$, i.e. $\omega(x) = o(x)$. Take $0 \leq \delta(x) \leq x/2$. Then if

$$\sigma(x) = \max_{x/2 \leq t \leq 3x/2} | \omega(t) |, \qquad \tau(x) = \max_{x/2 \leq t \leq 3x/2} \omega''(t) > 0,$$

we have

$$\mp\omega'(x) \leq 2\sigma(x)/\delta(x) + \delta(x)\tau(x),$$

and by choosing $\delta(x) = \{\sigma(x)/\tau(x)\}^{1/2}$ we have

$$\mp\omega'(x) \leq 3\{\sigma(x)\tau(x)\}^{1/2} = o(1).$$

If $\tau(x) \leq 0$, we have only to take $\delta(x) = x/2$.

We now prove 6.3.12. If $\sum r_k^{-1} \sin\theta_k$ converges, so does $\Sigma(R)$, since $\Sigma(R) \leq \sum_{r_k \leq R} r_k^{-1} \sin\theta_k$. Then if $J(R)$ is bounded, so is $I(R)$, by (6.3.15). We showed in the proof of 6.3.11 that $I^*(R)$ is bounded if $I(R)$ is bounded and $f(z)$ is of exponential type, so 6.3.12 is proved.

For 6.3.13, if $J(R)$ approaches a limit it is bounded; then if $I^*(R)$ is bounded, $\Sigma(R)$ approaches a limit as in the proof of 6.3.9. Hence $I(R)$ approaches a limit and it then follows from 6.3.17 that $I^*(R)$ approaches a limit.

6.4. Generalizations of boundedness. If we assume only that $f(z)$ is regular and of exponential type in the upper half plane and

(6.4.1) $$|f(x)| \le \omega(x), \qquad -\infty < x < \infty,$$

with a specified $\omega(x)$, there is no very obvious procedure for estimating the size of $|f(z)|$ unless $\omega(x)$ is the boundary function of a function $\omega(z)$ whose reciprocal is of exponential type in the upper half plane, or at least is dominated by such a function. For example, if $|f(x)| \le |x|^\alpha$, $\alpha > 0$, we can consider $f(z)(z+i)^{-\alpha}$ (with a fixed branch of $(z+i)^{-\alpha}$ if α is not an integer) and infer for instance that

$$|f(x+iy)| \le (|x|+y)^\alpha e^{cy}, \qquad y > 0.$$

It is not always easy to decide whether a given $\omega(x)$ is dominated by the reciprocal of the boundary function of a function of exponential type. One case in which an answer can be given is that in which $\omega(z)$ is an entire function of exponential type with no zeros for $y \ge 0$. In this case a necessary and sufficient condition for $1/\omega(z)$ to be of exponential type in the upper half plane is that $-\log|\omega(x)| \le O(|x|)$ as $x \to \pm\infty$. This is a special case of the following result, which will be needed in Chapter 7.

6.4.2. *Theorem.*[a] *Let $f(z)$ and $g(z)$ be entire functions of exponential type, let $f(z)$ have no zeros for $y > 0$ and let $|g(x)| \le |f(x)|\phi(x)$, $-\infty < x < \infty$. Then $\psi(z) = g(z)/f(z)$ is of exponential type if and only if* $\log\phi(x) \le O(|x|)$ *as $x \to \pm\infty$.*

The necessity of the condition is immediate. To prove the sufficiency we note first that by (2.6.18), $\log|\psi(z)| \le |z|^{1+\epsilon}$ on arbitrarily large semicircles, while $\log|\psi(x)| \le O(|x|)$ for real x. We shall show that

(6.4.3) $$\limsup_{y\to\infty} y^{-1}\log|1/f(iy)| < \infty;$$

this implies that $h_\psi(\pi/2) < \infty$. By the Phragmén-Lindelöf theorem (1.4.2), since $\psi(z)$ is at most of order 1 in the first and second quadrants, and of finite exponential type on their boundaries, $\psi(z)$ is of exponential type in the upper half plane.

To prove (6.4.3), let the Hadamard factorization of $f(z)$ be

$$f(z) = Az^m e^{cz} \prod_{n=1}^{\infty} (1 - z/z_n)\exp(z/z_n).$$

Since the sums $\sum_{|z_n|\le r} 1/z_n$ are bounded, by Lindelöf's theorem (2.10.1), and the z_n all have negative imaginary parts, $\sum \Im(1/z_n)$ converges and we can write

$$(6.4.4) \qquad f(z) = Az^m e^{kz} \prod_{n=1}^{\infty} (1 - z/z_n) \exp \{z\Re(1/z_n)\},$$

with $k = c + \sum_{n=1}^{\infty} i\Im(1/z_n)$. Since $\Im(1/z_n) \geq 0$, we have $|1 - iy/z_n| \geq 1$ for $y > 0$. The last exponential factor in (6.4.4) has modulus 1 when $z = iy$. Hence

$$\log |f(iy)| \geq \log |A| + m \log y - y\,\Im(k),$$

and (6.4.3) follows.

In connection with the problem raised at the beginning of this section, and in the light of 6.4.2, it is natural to ask what entire functions of exponential type can be dominated on the real axis by entire functions of exponential type which have no zeros in the upper half plane. The following theorem gives a complete answer.

6.4.5. *Theorem.*[b] *An entire function $f(z)$ of exponential type satisfies $|f(x)| \leq |\omega(x)|$, $-\infty < x < \infty$, for some entire function $\omega(z)$ of exponential type, having no zeros in the upper half plane, if and only if*

$$(6.4.6) \qquad \sum |\Im(1/z_n)| = \sum r_n^{-1} |\sin \theta_n|$$

converges, where $z_n = r_n \exp(i\theta_n)$ are the zeros of $f(z)$. If this condition is satisfied, $\omega(z)$ can be taken so that $|f(x)| = |\omega(x)|$.

See also §7.8. The possibility of replacing an entire function $f(z)$ by $\omega(z)$ which has the same modulus for real z and has a zero-free half-plane is useful in a number of problems.

Suppose first that there is a function $\omega(z)$ with the specified properties; we shall prove that (6.4.6) holds. If w_n are the zeros of $\omega(z)$, then as in the proof of 6.4.2, $\sum \Im(1/w_n)$ converges. By 6.3.14, the integral $I^*(R)$ in (6.3.5) is bounded; and since the integral is the same for $\omega(z)$ and $f(z)$, the series (6.4.6) converges, again by (6.3.14).

Suppose on the other hand that (6.4.6) converges. Let w_n be the zeros of $f(z)$ in the (open) upper half plane, and ζ_n the rest of the zeros together with $\overline{w}_n$. Let

$$(6.4.7) \qquad \phi(z) = \prod_{n=1}^{\infty} (1 - z/\overline{w}_n)/(1 - z/w_n);$$

the product converges because $\sum \Im(1/w_n)$ converges, and this is implied by the convergence of (6.4.6). In fact, we have the following lemma (in which the w_n are not necessarily confined to a half plane).

6.4.8. *Lemma. If $\sum |\Im(1/w_n)|$ converges, the product (6.4.7) converges uniformly in any bounded set which is at a positive distance from all the w_n and $\overline{w}_n$.*

Let $w_n = r_n \exp(i\theta_n)$. If n is so large that $|z/w_n| < \frac{1}{2}$, we have

$$\log \frac{1 - z/w_n}{1 - z/\bar{w}_n} = 2i \sum_{k=1}^{\infty} k^{-1} z^k \Im(1/w_n^k) = -2i \sum_{k=1}^{\infty} k^{-1} z^k \mid w_n \mid^{-k} \sin k\theta_n ,$$

$$\left| \log \frac{1 - z/w_n}{1 - z/\bar{w}_n} \right| \leq 2 \sum_{k=1}^{\infty} k^{-1} \frac{\mid z \mid \mid \sin \theta_n \mid}{\mid w_n \mid} \left| \frac{z}{w_n} \right|^{k-1} \left| \frac{\sin k\theta_n}{\sin \theta_n} \right|$$

$$\leq 4 \mid z \mid \mid \Im(1/w_n) \mid,$$

and the lemma follows.

Let $\omega(z) = \phi(z)f(z)$; then $\omega(z)$ is entire and has no zeros in the upper half-plane. Let the Hadamard factorization of $f(z)$ be

$$f(z) = Az^m e^{bz} \prod_{n=1}^{\infty} (1 - z/z_n) \exp (z/z_n),$$

and write $\phi(z)$ in the form

$$\phi(z) = \prod_{n=1}^{\infty} \frac{(1 - z/\bar{w}_n)e^{z/\bar{w}_n}}{(1 - z/w_n)e^{z/w_n}} \prod_{n=1}^{\infty} \exp \{z(w_n^{-1} - \bar{w}_n^{-1})\},$$

which we can do because $\sum \Im(1/w_n)$ converges and so the last product converges. We then have

$$\omega(z) = Az^m e^{cz} \prod_{n=1}^{\infty} (1 - z/\zeta_n) \exp (z/\zeta_n)$$

with some c, so that $\omega(z)$ is of order at most 1 (2.6.5). Again appealing to the fact that the sums $\sum_{|z_n|<r} 1/z_n$ are bounded (Lindelöf's theorem), we have the sums $\sum_{|\zeta_n|<r} 1/\zeta_n$ bounded, since $\sum \Im(1/w_n)$ converges; therefore $\omega(z)$ is of exponential type (Lindelöf's theorem again). Since $| \phi(x) | = 1$ for real x, $| \omega(x) | = | f(x) |$ for real x.

When $f(z)$ and $\omega(z)$ are related as in 6.4.5, we can estimate $f(x + iy)$, $y > 0$, by applying 6.2.4 to $f(z)/\omega(z)$. When $\omega(x)$ is not of this form, it is possible in some cases to obtain an inequality by applying the Poisson integral for semicircles.[c]

6.5. Representation of functions in a half plane. In this section we establish an integral representation for functions which are regular and of exponential type in the upper half plane and satisfy a condition which is less restrictive than, but of the same general nature as, the condition of being bounded on the real axis. We shall then use the representation to deduce a number of results about these functions, which we may conveniently think of as being "almost bounded" on the real axis.

Three conditions will prove to be particularly useful. These are

(6.5.1) $$\int_{-\infty}^{\infty} \frac{\log^+ | f(x) | \, dx}{1 + x^2} < \infty,$$

$$(6.5.2)\qquad \int_1^{\to\infty} x^{-2}\log|f(x)|\,dx \quad\text{and}\quad \int_1^{\to\infty} x^{-2}\log|f(-x)|\,dx \text{ exist,}$$

$$(6.5.3)\qquad \int_1^{\to\infty} x^{-2}\log|f(x)f(-x)|\,dx \text{ exists} \quad\text{and}\quad \limsup_{x\to\infty} x^{-1}\log|f(x)f(-x)| \le 0.$$

Of these, only (6.5.1) is strictly a generalization of boundedness, since (6.5.2) and (6.5.3) impose a restriction on the negative values of $\log|f(x)|$. However, when $f(z)$ is of exponential type in the upper half plane, (6.5.1) implies (6.5.2), by 6.3.6; when $f(z)$ is an entire function of exponential type, (6.5.2) implies (6.5.3), as we shall show later (6.6.5). On the other hand, (6.5.2) does not imply (6.5.1), since (6.5.2) allows large positive and small negative values of $\log|f(x)|$ to cancel each other; and (6.5.3) is still less restrictive since it allows large positive values of $\log|f(x)|$ to be cancelled by small negative values of $\log|f(-x)|$. Connections among (6.5.1), (6.5.2), (6.5.3), related conditions, and the distribution of the zeros of $f(z)$ will be discussed in Chapters 7 and 8; here we shall be concerned only with their effect on the growth of the function. The essential fact for the purposes of this chaper is that either (6.5.1) or (6.5.2) implies an upper bound for $\log|f(z)|$ throughout the upper half plane. We shall obtain this from the following representation theorem.

6.5.4. *Theorem.*[a] *If $f(z)$ is regular in the upper half plane, continuous in the closed half plane, its zeros in the upper half plane have no finite limit point, and it satisfies*

$$(6.5.5)\qquad \alpha = \liminf_{r\to\infty} r^{-1}\log M(r) < \infty,$$

and (6.5.1) or (6.5.2), then

$$(6.5.6)\qquad \log|f(z)| = \log\left|\prod_{n=1}^{\infty}\frac{1-z/z_n}{1-z/\bar{z}_n}\right| + y\pi^{-1}\int_{-\infty}^{\infty}\frac{\log|f(t)|}{(t-x)^2+y^2}\,dt + cy,$$

where z_n are the zeros of $f(z)$ in the upper half plane and

$$(6.5.7)\qquad c = \lim_{r\to\infty}(2/\pi)r^{-1}\int_0^{\pi}\log|f(re^{i\theta})|\sin\theta\,d\theta.$$

We have, in particular,

$$(6.5.8)\qquad c \le 4\alpha/\pi$$

and[b]

(6.5.9) $$\log|f(z)| \le y\pi^{-1}\int_{-\infty}^{\infty}\frac{\log|f(t)|}{(t-x)^2+y^2}\,dt + cy.$$

The existence of c in (6.5.7) follows from 6.3.10; by 6.3.6, (6.5.1) implies (6.5.2). In (6.5.6), the integral is to be interpreted as $\int_0^{\to\infty} + \int_{\to-\infty}^{0}$. The formula is the analogue for a half plane of the Poisson-Jensen formula for a circle, but it is not obtainable trivially from it by conformal mapping because of the complications arising from the unboundedness of the region; in other words, the functions which most naturally present themselves for study in a half plane do not arise by conformal mapping from the functions usually studied in a circle.

It would be considerably simpler to prove just the inequality (6.5.9), and also to use (6.5.1) instead of (6.5.2); these simpler cases suffice in many applications. Our starting point is the Poisson formula for a semicircle (1.2.3):

(6.5.10) $$\begin{aligned}\log|f(re^{i\theta})| &= \sum_{|z_n|<R}\log\left|\frac{z-z_n}{z-\bar{z}_n}\right| + \frac{y}{\pi}\int_{-R}^{R}P_1(R,z,t)\log|f(t)|\,dt\\ &+ \frac{2Ry}{\pi}\int_0^{\pi}P_2(R,z,\phi)\log|f(Re^{i\phi})|\,d\phi = S + I_1 + I_2 + I_3,\end{aligned}$$

where

$$P_1 = \frac{1}{t^2-2tx+r^2} - \frac{R^2}{R^4-2tR^2x+r^2t^2},$$

$$P_2 = \frac{(R^2-r^2)\sin\phi}{|R^2e^{2i\phi}-2Rxe^{i\phi}+r^2|^2},$$

and I_1 and I_2 involve respectively the first and second parts of P_1. Formula (6.5.6) is what we get by the formal process of letting $R\to\infty$ in (6.5.10), and we must now justify this formal process. This requires the following steps. (i) S approaches a limit. (ii) I_1 approaches a limit (i.e., the integral in (6.5.6) converges) and $I_2\to 0$. (iii)

$$2Ry\pi^{-1}\int_0^{\pi}\{P_2-\sin\phi\}\log|f(Re^{i\phi})|\,d\phi\to 0.$$

Since the limit denoted by c in (6.5.7) exists, this shows that I_3 approaches the same limit.

(i) We know from 6.4.9 that $\sum|\Im(1/z_n)|$ converges. If z is in a bounded set at a positive distance from all z_n and $\bar{z}_n$, and n is so large that $|z/z_n|<$

½, by 6.4.8 the product

$$\prod_{n=1}^{\infty} \frac{1 - z/z_n}{1 - z/\bar{z}_n}, \tag{6.5.11}$$

converges and S approaches a limit.

(ii) We assume (6.5.2), which is implied by (6.5.1) (it is here that the proof is easier if we work with (6.5.1)). Write

$$I_1 = \frac{y}{\pi} \int_{-R}^{R} \frac{\log |f(t)|}{t^2} \frac{t^2}{t^2 - 2tx + r^2}\, dt, \tag{6.5.12}$$

and consider the limits of $\int_0^R$ and $\int_{-R}^0$ separately; the first is typical and we give the details for this one only. For $x < 0$, $t^2/(t^2 - 2tx + r^2)$ increases, so by an application of the second mean-value theorem,

$$\int_R^S \frac{\log |f(t)|}{t^2} \frac{t^2}{t^2 - 2tx + r^2}\, dt = \frac{S^2}{S^2 - 2Sx + r^2} \int_{R'}^{S} t^{-2} \log |f(t)|\, dt, \tag{6.5.13}$$

and so the convergence of (6.5.2) implies the convergence of the integral in (6.5.6) for $x < 0$. For $x > 0$, $t^2/(t^2 - 2tx + r^2)$ decreases if $t > r^2/x$, and if we take R and S greater than this value the left-hand side of (6.5.13) is equal to

$$\frac{R^2}{R^2 - 2Rr + r^2} \int_R^{S'} t^{-2} \log |f(t)|\, dt,$$

which approaches zero as $R, S \to \infty$, and so the integral in (6.5.6) converges for $x > 0$ also. This disposes of I_1.

For I_2, write

$$\int_{-R}^{R} \frac{R^2 \log |f(t)|}{R^4 - 2tR^2x + r^2t^2}\, dt = \int_{-R}^{-M} + \int_{-M}^{M} + \int_{M}^{R} = J_1 + J_2 + J_3 .$$

We proceed to show that if M is large and $R > M$ then J_1 and J_3 are small, independently of R. Since $J_2 \to 0$ for a fixed M, this shows that $I_2 \to 0$. Consider J_3 (J_1 can be handled in the same way), writing

$$J_3 = R^2 \int_M^R \frac{\log |f(t)|}{t^2} \frac{t^2}{R^4 - 2tR^2x + r^2t^2}\, dt.$$

We may suppose that $R > x$; then since $R > t$, $t^2/(R^4 - 2tR^2x + r^2t^2)$ is positive and increasing, so that by the second mean-value theorem

$$J_3 = \frac{R^4}{R^4 - 2R^3x - r^2R^2} \int_M^R t^{-2} \log |f(t)|\, dt, \qquad M < M' < R.$$

The integral can be made arbitrarily small, independently of R, by taking M large, and (with a fixed z) its coefficient is bounded.

(iii) We now know that S, I_1 and I_2 approach limits as $R \to \infty$, and therefore I_3 approaches a limit, which we must show has the value (6.5.7). We may write Carleman's theorem in the form

$$\Sigma(R) = (2\pi)^{-1}I(R) + \frac{1}{\pi R}\int_0^\pi \log^+ |f(Re^{i\theta})| \sin\theta\, d\theta$$
$$+ \frac{1}{\pi R}\int_0^\pi \log^- |f(Re^{i\theta})| \sin\theta\, d\theta + A$$

(notation as in (6.3.3), (6.3.4)); under our hypotheses, for a sequence $R_n \to \infty$, the second term on the right is bounded, $\Sigma(R)$ is bounded (6.3.9), and $I(R)$ is bounded; hence the third term on the right is bounded. Then if

$$D = I_3 - \frac{2y}{\pi R}\int_0^\pi \log |f(Re^{i\phi})| \sin\phi\, d\phi,$$

we have

$$|D| \le \frac{2y}{\pi}\int_0^\pi \left| \frac{R(R^2 - r^2)}{|R^2e^{2i\phi} - 2Rxe^{i\phi} + r^2|^2} - \frac{1}{R}\right|$$
$$\times \{\log^+ |f(Re^{i\phi})| - \log^- |f(Re^{i\phi})|\} \sin\phi\, d\phi.$$

If $R > 4r$ we have

$$|R^2e^{2i\phi} - 2Rxe^{i\phi} + r^2|^2 = |(Re^{i\phi} - r)^2 + 2rR(1 - \sin\theta)e^{i\phi}|^2$$
$$\ge |(R - r)^2 - 2rR|^2 \ge R^2(R - 4r)^2;$$

hence

$$|D| \le 2y\pi^{-1}O(R^{-2})\int_0^\pi \{\log^+ |f(Re^{i\phi})| - \log^- |f(Re^{i\phi})|\} \sin\phi\, d\phi,$$

and so $D \to 0$ as $R \to \infty$ through the sequence R_n. This completes step (iii) of the proof.

6.6. Consequences of the representation theorem. As we stated at the beginning of §6.5, if $f(z)$ is an entire function of exponential type the condition that

(6.6.1) $$\int_1^{\to\infty} x^{-2} \log |f(x)|\, dx \quad \text{and} \quad \int_1^{\to\infty} x^{-2} \log |f(-x)|\, dx \text{ exist}$$

implies that $h(0) \leq 0$ and $h(\pi) \leq 0$. This appears to be difficult to prove directly but can be deduced from Theorem 6.5.4. We require the following lemma.

6.6.2. *Lemma.*[a] *If* $f(z)$ *satisfies* (6.6.1) *and*

$$v(z) = y \int_{-\infty}^{\infty} \frac{\log |f(t)| \, dt}{(t - x)^2 + y^2} \tag{6.6.3}$$

then

$$\lim_{r \to \infty} r^{-1} |v(z)| = 0, \quad \text{uniformly in } \delta < \theta < \pi - \delta,\ \delta > 0. \tag{6.6.4}$$

That the integral (6.6.3) converges was shown in the proof of 6.5.4. Since $\int_0^\infty$ and $\int_{-\infty}^0$ converge separately, it is enough to consider one of them, say the first. Write it as

$$I = y \int_0^\infty \frac{\log |f(t)|}{t^2} \frac{t^2}{(t - x)^2 + y^2} \, dt,$$

and note again that $t^2/\{(t - x)^2 + y^2\}$ increases if $x < 0$ or if $x > 0$ and $t < r^2/x$, and decreases if $x > 0$ and $t > r^2/x$. If $x < 0$,

$$I = y \int_0^M + y \int_M^\infty = y \int_0^M \frac{\log |f(t)|}{(t - x)^2 + y^2} \, dt$$
$$+ y \int_{M'}^\infty t^{-2} \log |f(t)| \, dt, \qquad M < M' < \infty,$$

by the second mean-value theorem. Then I/r can be made uniformly small by first taking M large, since $\int^\infty t^{-2} \log |f(t)| \, dt$ converges, and then taking r large. If $x > 0$,

$$I = y \int_0^M + y \int_M^{r^2/x} + y \int_{r^2/x}^\infty = y \int_0^M \frac{\log |f(t)|}{(t - x)^2 + y^2} \, dt$$
$$+ \frac{r^2}{y} \int_M^{M'} t^{-2} \log |f(t)| \, dt$$
$$+ \frac{r^2}{y} \int_{r^2/x}^R t^{-2} \log |f(t)| \, dt, \qquad M' > M,\ R > r^2/x.$$

If $\delta < \theta < \pi - \delta$, $r^2 - x^2 = r^2 \sin^2 \theta \geq r^2 \sin^2 \delta$, and so by taking M large and then r large with a fixed M, we can again make I/r uniformly small.

We can now prove the following theorem, which establishes the statement made at the beginning of this section.

6.6.5. *Theorem.*[b] *If* $f(z)$ *is an entire function of exponential type* c *and satisfies* (6.6.1) *then* $h(0) \leq 0$.

This also shows that $h(\pi) \leq 0$ (consider $f(-z)$), so that the indicator diagram of $f(z)$ is a vertical line segment (or a point).

By 6.5.4,

$$\log | f(z) | \leq \frac{y}{\pi} \int_{-\infty}^{\infty} \frac{\log | f(t) |}{(t - x)^2 + y^2} dt + cy,$$

and by 6.6.2, if $z = re^{i\delta}$, $\delta > 0$,

$$h(\delta) = \limsup_{r\to\infty} r^{-1} \log | f(re^{i\delta}) | \leq c \sin \delta.$$

Applying the same reasoning to $f(z)$ in the lower half plane we have

$$h(-\delta) \leq c \sin \delta.$$

Hence $h(0) \leq c \sin \delta$ and since δ can be arbitrarily small, $h(0) \leq 0$.

As another application of 6.5.4 we prove the following sharpened form of the Phragmén-Lindelöf principle, which we shall later sharpen still further.

6.6.6. *Theorem.*[c] *If* $f(z)$ *is regular in the upper half plane and continuous in the closed half plane, if* $| f(z) |$ *is bounded on the real axis, and if*

$$\liminf_{r\to\infty} r^{-1} \log M(r) = \alpha < \infty \tag{6.6.7}$$

then

$$\beta = \limsup_{r\to\infty} r^{-1} \log M(r) \leq 4\alpha/\pi. \tag{6.6.8}$$

By 6.5.4 we have

$$\log | f(z) | \leq \frac{y}{\pi} \int_{-\infty}^{\infty} \frac{\log | f(t) |}{(t - x)^2 + y^2} dt + 4\alpha y/\pi,$$

and if $\log | f(t) | \leq M$, we infer that

$$\log M(r) \leq M + 4\alpha r/\pi;$$

(6.6.8) follows.

Actually 6.6.6 does not tell the whole truth, which is that in fact $\beta = \alpha$ and that $\alpha < 0$ can occur only when $f(z) \equiv 0$. This will be proved, as a special case of a still more general result, in §7.4.

Still another application is the following extension of 6.2.13.

6.6.9.* *Theorem. If* $f(z)$ *is an entire function of zero exponential type, and* $\log | f(x) | \leq O(| x |^{\rho})$, $\rho < 1$, *then* $f(z)$ *is of order not exceeding* ρ.

6.7. Integrability on a line. The condition that $f(x)$ is bounded on

$(-\infty, \infty)$ may be thought of as the limiting case $p = \infty$ of the condition that

$$\left\{\int_{-\infty}^{\infty} | f(x) |^p \, dx\right\}^{1/p} < \infty,$$

that is, that $f(z)$ belongs to L^p on the real axis. For entire functions in general, the property of belonging to L^p on one line does not carry with it the same property for any other line or for any other value of p, but for functions of exponential type it does. We have, in fact, the following result.

6.7.1. *Theorem.*[a] *If $f(z)$ is an entire function of exponential type τ, and if for some positive number p*

$$\int_{-\infty}^{\infty} | f(x) |^p \, dx < \infty, \tag{6.7.2}$$

then

$$\int_{-\infty}^{\infty} | f(x + iy) |^p \, dx \le e^{p\tau|y|} \int_{-\infty}^{\infty} | f(x) |^p \, dx; \tag{6.7.3}$$

moreover, $f(x) \to 0$ as $|x| \to \infty$ and consequently (6.7.2) holds for every larger value of p.

It is no harder to prove a more general result, of which 6.7.1 is the special case $\phi(t) = t^p$.

6.7.4. *Theorem.*[b] *If $f(z)$ is an entire function of exponential type τ, $\phi(t)$ is a nonnegative nondecreasing convex function of* log t, *and*

$$\int_{-\infty}^{\infty} \phi(| f(x) |) \, dx < \infty \tag{6.7.5}$$

then

$$\int_{-\infty}^{\infty} \phi\{e^{-\tau|y|} | f(x + iy) |\} \, dx \le \int_{-\infty}^{\infty} \phi\{ | f(x) | \} \, dx; \tag{6.7.6}$$

moreover, $\phi\{ |f(x)| \}$ is bounded, and if $\phi(t) > 0$ for $t > 0$, $f(x) \to 0$ as $|x| \to \infty$.

A still more general result is

6.7.7.* *Theorem.*[b] *If $f(z)$ is regular and of exponential type τ for $y \ge 0$, then under the same hypotheses on $\phi(t)$ as in 6.7.4, (6.7.5) implies (6.7.6); and more generally*

$$\int_a^b \phi\{e^{-\tau y} | f(x + iy + s) |\} \, ds \le \sup_{-\infty < x < \infty} \int_a^b \phi\{ | f(x + s) | \} \, ds.$$

It is sometimes useful to have a weaker form of 6.7.1 which holds under a weaker hypothesis.

6.7.8.* *Theorem.*[b] *If $f(z)$ is regular and of exponential type for $x \geq 0$ and $\int_0^\infty |f(x)|^p dx < \infty$ for some positive p, then $\int_0^\infty |f(x+iy)|^p dx < \infty$ for every y, and $f(x) \to 0$ as $x \to \infty$.*

A corollary of 6.7.1 is a generalization of part of 6.2.13, to which 6.7.1 reduces it.

6.7.9. *Theorem.*[c] *If $f(z)$ is an entire function of zero exponential type and $\int_{-\infty}^\infty |f(x)|^p dx < \infty$ for some positive p then $f(z) \equiv 0$.*

In the same way, the following result is a corollary of 6.7.8.

6.7.10.* *Theorem.*[d] *If $f(z)$ is an entire function of exponential type and $\int_0^\infty |f(re^{i\theta})|^p dr < \infty$ for three values of θ, no two of which differ by as much as π, then $f(z) \equiv 0$.*

The condition of belonging to L^p is actually rather artificial in 6.7.9 and 6.7.10; a more general result is given in §10.6.

The following related result will be of interest in connection with §7.2 (cf. note 7.2c).

6.7.11.* *Theorem. If $f(z)$ is regular and of exponential type for $y \geq 0$, the condition*

$$\int_{-\infty}^\infty \frac{\log^+ |f(x)|}{1+x^2} dx < \infty$$

implies that

$$\int_{-\infty}^\infty \frac{\log^+ |f(x+iy)|}{x^2+(1+y)^2} dx$$

is bounded, $0 < y < \infty$.

To prove 6.7.4, we need the following lemma, which is a first step in showing that $\phi(|f(x)|)$ is bounded.

6.7.12. *Lemma. If $\phi(t)$ is a nonnegative nondecreasing convex function of $\log t$ and $\int_0^\infty \phi\{|f(x)|\} dx$ exists then*

$$\int_0^\infty \frac{\log^+ |f(x)|}{1+x^2} dx \tag{6.7.13}$$

converges.

If $f^*(x) = f(x)$ when $|f(x)| \geq 1$, and $f^*(x) = 0$ otherwise, the hypothesis implies that $\int_0^\infty \phi\{|f^*(x)|\} dx$ exists; we may write this in the form

$$\int_0^\infty \psi\{\log |f^*(x)|\} dx = \int_0^\infty \psi\{\log^+ |f(x)|\} dx < \infty,$$

where $\psi(u)$ is a nonnegative nondecreasing convex function of u. Then by Jensen's inequality,

$$\psi\left\{\frac{2}{\pi}\int_0^\infty \frac{\log^+ |f(x)|\,dx}{1+x^2}\right\} \le \frac{2}{\pi}\int_0^\infty \frac{\psi\{\log^+ |f(x)|\}\,dx}{1+x^2}$$

$$\le \frac{2}{\pi}\int_0^\infty \psi\{\log^+ |f(x)|\}\,dx < \infty,$$

and since $\psi(t) \uparrow \infty$ as $t \uparrow \infty$ we have (6.7.13).

We now have $f(z)$ entire, of exponential type, and satisfying (6.5.1); by 6.6.5, $h(0) \le 0$, $h(\pi) \le 0$, and so by 5.1.9 (or 5.4.14),

$$\log |f(re^{i\theta})| \le (c|\sin\theta| + \epsilon)r, \qquad \epsilon > 0,$$

for sufficiently large r. This shows that

$$\limsup_{r\to\infty} r^{-1}\int_0^\pi \log |f(re^{i\theta})| \sin\theta\, d\theta \le \tau\pi/2,$$

and hence that c in (6.5.7) does not exceed τ.

An alternative method for reaching the same conclusion would be to notice that the existence of $\int_{-\infty}^\infty \phi\{|f(x)|\}\,dx$ implies the convergence of $\sum_{n=-\infty}^\infty \int_n^{n+1} \phi\{|f(x)|\}\,dx$ and so the boundedness of $f(\lambda_n)$ for some sequence $\{\lambda_n\}$ with $n < \lambda_n < n+1$; by appealing to 10.5.1 we can again conclude that $h(0) \le 0$, $h(\pi) \le 0$.

We now write (6.5.9) in the form

$$\log\{|f(x+iy)|e^{-\tau|y|}\} \le \pi^{-1}\int_{-\infty}^\infty \log|f(t)| \frac{y}{(x-t)^2+y^2}\,dt.$$

Then if $\psi(\log t) = \phi(t)$, with $\psi(u)$ convex, by Jensen's inequality we have

$$\phi\{|f(x+iy)|e^{-\tau|y|}\} \le \pi^{-1}\int_{-\infty}^\infty \phi\{|f(t)|\}\frac{y}{(x-t)^2+y^2}\,dt$$

and

$$\int_{-\infty}^\infty \phi\{|f(x+iy)|e^{-\tau|y|}\}\,dx$$

$$\le \pi^{-1}\int_{-\infty}^\infty \phi\{|f(t)|\}\,dt \int_{-\infty}^\infty \frac{y}{(x-t)^2+y^2}\,dx = \int_{-\infty}^\infty \phi\{|f(t)|\}\,dt.$$

This establishes (6.7.6). To complete the proof of 6.7.4 we can either appeal to 10.6.1, where a much more general result is established, or argue directly as follows.[e] Since ϕ increases, it follows from (6.7.6) that

(6.7.14) $$\int_{-\infty}^{\infty} \phi\{\,|\,f(x+iy)\,|\,e^{-\tau\delta}\}\,dx \leq \int_{-\infty}^{\infty} \phi\{\,|\,f(t)\,|\,\}\,dt, \qquad |\,y\,| \leq \delta.$$

Since $\phi(t)$ is a convex function of $\log t$, $\phi\{e^{-\tau\delta}\,|\,f(z)\,|\,\}$ is a subharmonic function; the property of subharmonic functions which is relevant here is that the value of a subharmonic function at a point does not exceed its mean value over the area of a circle with center at the point, and so in particular

$$\phi\{e^{-\tau\delta}\,|\,f(x)\,|\,\} \leq \frac{1}{\pi\delta^2} \iint_{s^2+y^2\leq\delta^2} \phi\{e^{-\tau\delta}\,|\,f(x+iy+s)\,|\}\,ds\,dy$$

$$\leq \frac{1}{\pi\delta^2} \int_{-\delta}^{\delta}\int_{-\delta}^{\delta} \phi\{e^{-\tau\delta}\,|\,f(x+iy+s)\,|\,\}\,ds\,dy$$

$$= \frac{1}{\pi\delta^2} \int_{-\delta}^{\delta}\left\{\int_{x-\delta}^{x+\delta} \phi\{e^{-\tau\delta}\,|\,f(s+iy)\,|\,\}\,ds\right\}dy,$$

and this is bounded by (6.7.14). Since $\phi(u) \to \infty$ as $u \to \infty$, it then follows that $|\,f(x)\,|$ is bounded. Furthermore, if $\{x_n\}$ is any sequence such that $x_n \uparrow \infty$ and $x_{n+1} - x_n > 2\delta$, we have

$$\sum_{n=1}^{\infty} \phi\{e^{-\delta\tau}\,|\,f(x_n)\,|\,\} \leq \frac{1}{\pi\delta^2}\int_{-\delta}^{\delta}\left\{\sum_{n=1}^{\infty}\int_{x_n-\delta}^{x_n+\delta} \phi\{e^{-\tau\delta}\,|\,f(s+iy)\,|\,\}\,ds\right\}dy$$

$$\leq \frac{1}{\pi\delta^2}\int_{-\delta}^{\delta}\left\{\int_{-\infty}^{\infty} \phi\{e^{-\tau\delta}\,|\,f(s+iy)\,|\,\}\,ds\right\}dy,$$

so that the series on the left converges, its terms approach zero, and consequently $f(x_n) \to 0$ if $\phi(t) > 0$ for $t > 0$. This implies that $f(x) \to 0$ as $x \to \infty$, and similarly as $x \to -\infty$.

The last part of the proof has incidentally established a result of independent interest.

6.7.15. *Theorem.*[f] *If $f(z)$ is an entire function of exponential type τ and $\int_{-\infty}^{\infty} |\,f(x)\,|^p\,dx < \infty$, $p > 0$, then for any real increasing sequence $\{\lambda_n\}$ such that $\lambda_{n+1} - \lambda_n \geq 2\delta$,*

(6.7.16) $$\sum_{n=-\infty}^{\infty} |\,f(\lambda_n)\,|^p \leq 2\pi^{-1}\delta^{-1}e^{p\tau\delta}\int_{-\infty}^{\infty} |\,f(x)\,|^p\,dx.$$

The chief interest in (6.7.16) is not in the precise value for the coefficient on the right (which is not the best possible) but in the fact that it depends only on p, δ and τ. A converse result, in which the convergence of

$$\int |\,f(x)\,|^p\,dx$$

is inferred from the convergence of $\sum |f(\lambda_n)|^p$, with a suitable $\{\lambda_n\}$, will be given in Chapter 10 (10.6.4, 10.6.8).

The following sharper form of the boundedness of $f(x)$ in 6.7.1 is also true.

6.7.17.* *Theorem.*[g] *If $f(z)$ is an entire function of exponential type τ, if $p \geq 1$ and if $\int_{-\infty}^{\infty} |f(x)|^p dx = M^p < \infty$ then*

$$|f(x+iy)|^p \leq A_p M^p y^{-1} \sinh p\tau y,$$

with $A_p < 1/\pi$ for $p > 1$ and $A_2 = 1/(2\pi)$.

An improvement in another direction is given by

6.7.18.* *Theorem.*[h] *If $p \geq 1$ and $f(z)$ is an entire function of exponential type τ such that $\int_{-\infty}^{\infty} |f|^p dx = M^p < \infty$ then for $q > p$*

$$\left\{ \int_{-\infty}^{\infty} |f(x)|^q \, dx \right\}^{1/q} \leq e^{1/e} \tau^{1/p - 1/q} M \tag{6.7.19}$$

and

$$\left\{ \int_{-\infty}^{\infty} |f(x+iy)|^q \, dx \right\}^{1/q} \leq e^{\tau y} |\pi y|^{1/q - 1/p} M. \tag{6.7.20}$$

The limiting case $q = \infty$ of (6.7.20) is less precise than 6.7.17. The constants in (6.7.19) and (6.7.20) can be improved, but the power of τ in (6.7.19) is asymptotically correct (for large τ).

It is also possible to deduce from 6.5.4 inequalities for integrals of $|f(z)|$ along rays instead of along parallels to the real axis. More precise results can be obtained by other methods, and we quote one of them.

6.7.21.* *Theorem.*[i] *If $f(z)$ is regular and of exponential type in the upper half plane and belongs to $L^p(0, \infty)$ and to $L^q(-\infty, 0)$, where $p \geq 1$, $q \geq 1$, then for $0 < \theta < \pi$ we have*

$$\left\{ \int_0^{\infty} |f(re^{i\theta})|^s e^{-\tau r s \sin\theta} \, dr \right\}^{1/s} \tag{6.7.22}$$

$$\leq \left\{ \int_0^{\infty} |f(x)|^p \, dx \right\}^{\lambda/p} \left\{ \int_0^{\infty} |f(-x)|^q \, dx \right\}^{\lambda'/q},$$

where

$$s = \frac{pq}{\lambda q + \lambda' p}, \qquad \lambda = \frac{\pi - \theta}{\pi}, \qquad \lambda' = \theta/\pi.$$

In particular, if $p = q$,

$$(6.7.23)\quad \int_0 \mid f(re^{i\theta}) \mid^p e^{-\tau r p \sin\theta}\, dr$$

$$\leq \left\{ \int_0^{\infty} \mid f(x) \mid^p dx \right\}^{(\pi-\theta)/\pi} \left\{ \int_0^{\infty} \mid f(-x) \mid^p dx \right\}^{\theta/\pi}$$

(*generalization of* 6.2.4).

There are similar results for mean values, for example

6.7.24.* *Theorem.*[j] *If* $f(z)$ *is an entire function of exponential type* τ *and* $p > 0$ *then*

$$\limsup_{T\to\infty} \frac{1}{2T} \int_{-T}^{T} \mid f(x+iy) \mid^p dx \leq e^{p\tau |y|} \limsup_{T\to\infty} \frac{1}{2T} \int_{-T}^{T} \mid f(x) \mid^p dx.$$

6.8. Representations for functions which are bounded on a line. Entire functions of exponential type which satisfy conditions of the kind we have been discussing can be studied from another point of view. Their indicator diagrams collapse into line segments on the imaginary axis, and if we can shrink the contour in the Pólya representation until it coincides with the conjugate indicator diagram we shall have a representation for the function as an integral over a line segment, and in fact (after a change of variable) as a finite Fourier transform. In practice such representations can be more conveniently established by a different approach. They provide an easy way of establishing some of the theorems of §6.7 if they are established independently of that section.

The most elegant and generally useful representation deals with a function belonging to L^2 on the real axis.

6.8.1. *Theorem of Paley and Wiener.*[a] *The entire function* $f(z)$ *is of exponential type* τ *and belongs to* L^2 *on the real axis if and only if*

$$(6.8.2)\qquad f(z) = \int_{-\tau}^{\tau} e^{izt}\phi(t)\, dt,$$

where

$$(6.8.3)\qquad \phi(t) \in L^2(-\tau, \tau).$$

That (6.8.2) belongs to L^2 on the real axis under (6.8.3) is immediate from the theory of Fourier transforms; it can also be established without appealing to this theory by using the theory of Fourier series and 10.6.6. That (6.8.2) represents an entire function of exponential type is left for the reader to verify.

We can state 6.8.1 in a number of alternative equivalent forms, which appear to be weaker or stronger, depending on the point of view. The hypothesis that $f(z)$ is of exponential type τ, that is, that

$$|f(z)| \leq A(\epsilon)e^{(\tau+\epsilon)|z|}, \tag{6.8.4}$$

can be replaced by the apparently more restrictive condition

$$|f(z)| \leq Ae^{\tau|z|} \tag{6.8.5}$$

or by the apparently still more restrictive condition

$$e^{-\tau|z|}|f(z)| = o(1); \tag{6.8.6}$$

and the condition that $f(x) \to 0$ as $|x| \to \infty$ can be added if desired because by 6.7.1 it is satisfied for any entire function of exponential type belonging to L^2 on the real axis, while it is satisfied for functions (6.8.2) either by the Riemann-Lebesgue lemma or by another appeal to 6.7.1.

In fact, on the one hand (6.8.6) is satisfied if $f(z)$ is defined by (6.8.2), since for $|y| > 0$

$$|f(x+iy)|^2 \leq \left\{\int_{-\tau}^{\tau} e^{-yt}|\phi(t)|\,dt\right\}^2$$

$$\leq \int_{-\tau}^{\tau} e^{-2yt}\,dt \int_{-\tau}^{\tau} |\phi(t)|^2\,dt$$

$$\leq A|y|^{-1}|\sinh 2\tau y|$$

(cf. 6.7.17). On the other hand, if we have proved that $f(z)$, belonging to L^2 on the real axis, has the representation (6.8.2) if it satisfies the most restrictive condition (6.8.6), and we are given a function satisfying (6.8.4) instead, we have the representation (6.8.2) with τ replaced by $\tau + \epsilon$ and $\phi(t)$ ostensibly depending on ϵ; but the uniqueness theorem for Fourier transforms shows that $\phi(t)$ is actually independent of ϵ and vanishes almost everywhere outside $(-\tau, \tau)$.

We can give added precision to 6.8.1 as follows.

6.8.7. *Theorem. If $f(z)$ is given by* (6.8.2) *and $\phi(t)$ does not vanish almost everywhere in any neighborhood of τ (or $-\tau$) then $f(z)$ is of order* 1 *and type τ (that is, not of exponential type less than τ).*

For, if $f(z)$ were of type $\sigma < \tau$, it would have the representation (6.8.2) with σ in place of τ, and since the representation is unique, $\phi(t)$ would have to vanish almost everywhere outside $(-\sigma, \sigma)$.

We now turn to the proof of the more difficult part of 6.8.1, that an entire function of exponential type has the form 6.8.2 if it belongs to L^2 on the real axis. We shall give two proofs.

First proof.[b] Let $\phi(t)$ be the Fourier transform of $f(x)$; then

$$f(x) = \int_{-\infty}^{\infty} \phi(t)e^{ixt}\,dt, \qquad -\infty < x < \infty, \tag{6.8.8}$$

where $\int_{-\infty}^{\infty}$ denotes the limit in mean square of $\int_{-T}^{T}$. It is necessary only show that $\phi(t) = 0$ for almost all t such that $|t| > \tau$, since this will establish (6.8.2) for real z and both sides of (6.8.2) are entire.

We have

$$2\pi\phi(t) = \int_{-\infty}^{\infty} e^{-itx}f(x)\,dx,$$

where the integral is again to be interpreted as a limit in the mean. Consider the contour integral

$$I = \int_{\Gamma} e^{-itz}f(z)\,dz,$$

where Γ consists of the upper three sides of the rectangle with vertices at $\pm T$, $\pm(T + iT)$. Since the integrand is an entire function, this is the negative of

$$\int_{-T}^{T} e^{-itx}f(x)\,dx. \tag{6.8.9}$$

We proceed to show that $I \to 0$ as $T \to \infty$ if $t < -\tau$ (we can use a similar argument with Γ in the lower half plane if $t > \tau$). We have

$$I = i\int_0^T e^{-itT+ty}f(T + iy)\,dy - \int_{-T}^{T} e^{-itx+tT}f(x + iT)\,dx - i\int_0^T e^{itT+ty}f(-T + iy)\,dy,$$

$$|I| \le \int_0^T e^{ty}\,|f(T + iy)|\,dy + \int_0^T e^{ty}\,|f(-T + iy)|\,dy + e^{tT}\int_{-T}^{T} |f(x + iT)|\,dx. \tag{6.8.10}$$

Now take the hypothesis in the form (6.8.5); since $f(x)$ belongs to L^2, $f(x) \to 0$ as $x \to \pm\infty$, by 6.7.1. By 6.2.4, $|f(x + iT)| \le Me^{\tau T}$ uniformly in x, so the third integral in (6.8.10) does not exceed $2TMe^{(t+\tau)T}$, and this approaches zero as $T \to \infty$ if $t < -\tau$. For the first integral in (6.8.10) (the second is handled in the same way) we consider separately $\int_0^R$ and $\int_R^T$,

$R > 0$. We have, again by 6.2.4,

$$\int_R^T e^{ty} \mid f(T + iy) \mid dy \leq M \int_R^T e^{(t+\tau)y} \, dy = M(t + \tau)^{-1}\{e^{(t+\tau)T} - e^{(t+\tau)R}\},$$

and if $t < -\tau$ this can be made small by taking R large and then fixing R and taking T large. Moreover, with a fixed R we have, by 6.2.8,

$$f(T + iy) \to 0$$

as $T \to \infty$, uniformly in $0 \leq y \leq R$; hence

$$\int_0^R e^{ty} \mid f(T + iy) \mid dy \to 0, \qquad T \to \infty.$$

We have now shown that $I \to 0$ for $\mid t \mid > \tau$, and so (6.8.9) converges to zero for $\mid t \mid > \tau$, while it converges in mean square to $\phi(t)$; so $\phi(t) = 0$ almost everywhere for $\mid t \mid > \tau$, as required.

The preceding proof is perhaps the most natural one. We shall give a second proof which is shorter and makes the smallest possible use of Fourier theory (requiring only the Riesz-Fischer theorem for Fourier series), but uses a uniqueness theorem from Chapter 9 (which we shall prove independently of 6.8.1).

Second proof. By 6.7.15, $\sum \mid f(2\pi n/\tau) \mid^2$ converges. According to the Riesz-Fischer theorem there is a function $\phi(t)$ of L^2 such that

$$f(2\pi n/\tau) = \int_{-\tau}^{\tau} e^{-2\pi i n t/\tau} \phi(t) \, dt, \quad n = 0, \pm 1, \pm 2, \cdots;$$

in other words the function

$$g(z) = \int_{-\tau}^{\tau} e^{-izt} \phi(t) \, dt$$

coincides with $f(z)$ at the points $2\pi n/\tau$. Now $g(z)$ is an entire function of exponential type satisfying $g(z) = o(e^{\tau|z|})$. By 9.4.4, $f(z) \equiv g(z)$ and (6.8.2) is established.

It is interesting to notice the essential difference between the two proofs: in the first one we have $f(z)$ represented by an integral over too large a range, and it is necessary to show that $\phi(t)$ vanishes outside $(-\tau, \tau)$; in the second proof, the integral is over the correct range but must be identified with $f(z)$ by using the fact that it coincides with $f(z)$ at the points of an arithmetic progression.

Theorem 6.8.1 has suggested a number of generalizations to classes other than L^2.

6.8.11.* *Theorem.*[c] *The entire function $f(z)$ is of exponential type and belongs to $L(-\infty, \infty)$ if and only if*

$$f(z) = \int_{-\tau}^{\tau} e^{izt}\phi(t)\,dt,$$

where $\phi(\tau) = \phi(-\tau) = 0$ *and the function obtained by extending* $\phi(t)$ *to be* 0 *outside* $(-\tau, \tau)$ *has an absolutely convergent Fourier series on the interval* $(-\tau - \delta, \tau + \delta)$, $\delta > 0$.

6.8.12.* *Theorem.*[d] *The entire function* $f(z)$ *is of exponential type and belongs to* $L^p(-\infty, \infty)$, $p > 1$, *if and only if*

$$f(z) = z\int_{-\tau}^{\tau} [\phi(t) - \phi(-t)]e^{izt}\,dt - 2\phi(\tau)z^{-1}\sin \tau z,$$

where $\phi(t)$ *is a continuous function of period* 2τ *whose Fourier coefficients* c_n *satisfy* $\sum |nc_n|^p < \infty$.

6.8.13.* *Theorem.*[e] *If* $f(z)$ *is an entire function of exponential type belonging to* $L^p(-\infty, \infty)$, $1 < p < 2$, *then* $f(z)$ *has the representation* (6.8.2) *with* $\phi(t) \in L^q$, $q = p/(p-1)$; *if* $\phi(t) \in L^p$, $1 < p < 2$, *and* $f(z)$ *has the form* (6.8.2), *then* $f(z)$ *is an entire function of exponential type belonging to* $L^q(-\infty, \infty)$.

6.8.14. *Theorem. If* $f(z)$ *is an entire function of exponential type* τ *and is bounded on the real axis, there are functions* $\phi_n(t)$, *each of bounded variation on* $(-\tau, \tau)$, *such that for each* z

$$f(z) = \lim_{n\to\infty} \int_{-\tau}^{\tau} e^{izt}\,d\phi_n(t). \tag{6.8.15}$$

To prove 6.8.14, consider $g(z) = z^{-1}\{f(z) - f(0)\}$. This is an entire function of exponential type τ and belongs to L^2 on the real axis. Then by 6.8.1,

$$g(z) = \int_{-\tau}^{\tau} e^{izt}\phi(t)\,dt, \qquad \phi(t)\in L^2,$$

$$= \lim_{n\to\infty} \int_{-\tau}^{\tau} e^{izt}\psi_n(t)\,dt,$$

where $\psi_n(t)$ are functions of bounded variation (step-functions, for example), with $\psi_n(\tau) = \psi_n(-\tau) = 0$. Integration by parts gives

$$g(z) = \lim_{n\to\infty} (i/z)\int_{-\tau}^{\tau} e^{izt}\,d\psi_n(t),$$

and so

$$f(z) = zg(z) + f(0) = f(0) + \lim_{n\to\infty} i\int_{-\tau}^{\tau} e^{izt}\,d\psi_n(t)$$

$$= \lim_{n\to\infty} \int_{-\tau}^{\tau} e^{izt}\,d\phi_n(t),$$

where $\phi_n(t) = i\psi_n(t) + f(0)\operatorname{sgn} t$.

A little further consideration shows that the limit in (6.8.15) is uniform on every bounded set.

6.9. The indicator diagram of a finite Fourier transform. As an application of 6.8.1 we prove the following theorem, which is unexpectedly difficult.

6.9.1. *Theorem. Let* $f(z) = \int_a^b e^{izt} g(t)\, dt$, *where* $g(t)$ *is integrable,* $|a| < b$, *and* $g(t)$ *does not vanish almost everywhere in any neighborhood of* b, *then* $f(z)$ *is an entire function of order* 1 *and type* b *(and not of smaller type).*

The main part of the theorem is the verification that $f(z)$ is not of type less than b. If $g(t)$ belongs to L^2, this follows immediately from 6.8.7. To establish it for the general case, put $G(t) = \int_a^t g(u)\, du$; then by integration by parts,

$$f(z) = f(0)e^{izb} - iz \int_a^b e^{izt} G(t)\, dt. \tag{6.9.2}$$

Suppose that $f(z)$ is of type c, $0 < c < b$; then $z^{-1}\{f(z) - f(0)e^{izc}\}$ is of type c and belongs to L^2 on the real axis. By 6.8.1,

$$f(z) = f(0)e^{izc} - iz \int_{-c}^{c} e^{izt} H(t)\, dt, \qquad H(t) \epsilon L^2. \tag{6.9.3}$$

Comparing (6.9.2) and (6.9.3) we have

$$f(0)(e^{izb} - e^{izc}) = if(0)z \int_c^b e^{izt}\, dt$$
$$= iz \int_{-c}^{c} e^{izt} H(t)\, dt - iz \int_a^b e^{izt} G(t)\, dt.$$

By the uniqueness theorem for Fourier transforms, we therefore have $f(0) = -G(t)$ almost everywhere in (c, b), hence $g(t) = 0$ almost everywhere in (c, b).

6.9.4. *Corollary. The indicator diagram of* $f(z)$ *in* 6.9.1 *is the segment* (ia, ib) *of the imaginary axis provided that* $g(t)$ *does not vanish almost everywhere in the neighborhood of* a.

That the indicator diagram is a segment of the imaginary axis with its upper end at ib follows from the fact that $f(z)$ is bounded on the real axis and of type b (and not less), by 6.2.4. The function

$$F(z) = e^{-izb} f(z) = \int_a^b e^{iz(t-b)} g(t)\, dt = \int_{a-b}^{c} e^{izu} g(u + b)\, du$$

has its indicator diagram shifted downward by b, and by the theorem (applied to $F(-z)$) $F(z)$ has the other end of its indicator diagram at $i(a - b)$; the conclusion follows.

A similar result is

6.9.5.* *Theorem.*[b] *If $F(x)$ is of bounded variation on $(-\infty, \infty)$, then $F(x)$ is constant except on some finite interval if and only if $f(z) = \int_{-\infty}^{\infty} e^{izt} dF(t)$ is an entire function of exponential type; and if $(-a, b)$ is the smallest interval outside which $F(x)$ is constant, $a = h_f(-\pi/2)$ and $b = h_f(\pi/2)$.*

6.10. Periodic and almost periodic functions. We shall show that if $f(z)$ is an entire function of exponential type which is periodic on the real axis, it is necessarily a trigonometric polynomial; if it is almost periodic, even in a very general sense, then its Fourier exponents are bounded.

6.10.1. *Theorem.*[a] *Let $f(z)$ be an entire function of exponential type τ, periodic on the real axis with period 2π. Then $f(z)$ has the form*

$$f(z) = \sum_{k=-n}^{n} a_k e^{ikz}, \qquad n \leq \tau. \tag{6.10.2}$$

6.10.3. *Theorem.*[b] *Let $f(z)$ be an entire function of exponential type τ, uniformly almost periodic on the real axis. Then the Fourier series of $f(z)$ is of the form*

$$f(z) \sim \sum a_k \exp(i\lambda_k z), \qquad |\lambda_k| \leq \tau. \tag{6.10.4}$$

This still holds for (e.g.) Stepanoff or Besicovitch almost periodic functions, since we have

6.10.5.* *Theorem.*[c] *Let $f(z)$ be an entire function of exponential type τ, such that the mean value*

$$\lim_{T\to\infty} (2T)^{-1} \int_{-T}^{T} |f(x)|\, dx$$

exists. Then for $|\lambda| > \tau$, the mean value

$$\lim_{T\to\infty} (2T)^{-1} \int_{-T}^{T} f(x) e^{i\lambda x}\, dx$$

exists and is zero.

That 6.10.5 is more general than 6.10.3 follows from the fact that there are Besicovitch almost periodic functions, of exponential type, which are not bounded on the real axis.

We shall give three proofs of 6.10.1. The shortest proof depends on a result from Chapter 11. We have

$$2\pi a_k = \int_{-\pi}^{\pi} e^{-ikt} f(t)\, dt,$$

and by repeated integration by parts this becomes

$$2\pi a_k = (ik)^{-m} \int_{-\pi}^{\pi} e^{-ikt} f^{(m)}(t)\, dt.$$

Since $|f^{(m)}(t)| \leq M\tau^m$, where $M = \max |f(x)|$ (11.1.2), we have

$$| 2\pi a_k | \leq M(\tau/k)^m,$$

and letting $m \to \infty$ we have $a_k = 0$ if $k > \tau$. Similarly $a_k = 0$ if $k < -\tau$.

By a different proof we can establish 6.10.1 and 6.10.3 together. We have for the a_k in (6.10.4)

$$a_k = \lim_{T\to\infty} (2T)^{-1} \int_{-T}^{T} f(x)e^{-i\lambda_k x}\, dx,$$

and this holds equally well for (6.10.2) with $\lambda_k = k$. We can now proceed along the lines of the first proof of 6.8.1 to show that $a_k = 0$ for $|\lambda_k| > \tau$. The details may be left to the reader.

We now give another proof of 6.10.1. This depends on the following lemma.

6.10.6. *Lemma.*[d] *If $\phi(w)$ is regular on the contour C and* $\int_C e^{zw}\phi(w)dw \equiv 0$, *then $\phi(w)$ is regular inside C.*

To prove the lemma, take contours C_1 inside C and C_2 outside C, on which $\phi(w)$ is still regular; then for w on C,

$$2\pi i\phi(w) = \int_{C_2} (t - w)^{-1}\phi(t)\, dt - \int_{C_1} (t - w)^{-1}\phi(t)\, dt = \phi_2(w) - \phi_1(w).$$

Then $\phi_2(w)$ is regular inside C_2 and $\phi_1(w)$ is regular outside C_1. We have

$$0 = 2\pi i \int_C e^{zw}\phi(w)\, dw = \int_C e^{zw}\{\phi_2(w) - \phi_1(w)\}\, dw = -\int_C e^{zw}\phi_1(w)\, dw,$$

since $e^{zw}\phi_2(w)$ is regular inside C_2. In the last integral replace C by a large circle and $\phi_1(w)$ by its Laurent series $\sum_1^\infty b_n w^{-n}$, and integrate term by term. We find

$$\sum_{n=1}^{\infty} b_n z^{n-1}/(n - 1)! = 0,$$

and so all $b_n = 0$ and $2\pi i\phi(w) \equiv \phi_2(w)$, which is indeed regular inside C.

Now suppose that $f(z)$ satisfies the hypotheses of 6.10.1; since

$$f(x + 2\pi) = f(x)$$

on the real axis implies $f(z + 2\pi) = f(z)$ for all z, we have

$$0 = f(z + 2\pi) - f(z) = \int_C \{e^{(z+2\pi)w} - e^{zw}\}\phi(w)\, dw$$

$$= \int_C e^{zw}(e^{2\pi w} - 1)\phi(w)\, dw,$$

where C surrounds the conjugate indicator diagram of $f(z)$. By the lemma, $\phi(w)(e^{2\pi w} - 1)$ must be regular inside C, so $\phi(w)$ must be regular except for simple poles at the zeros ik of $e^{2\pi w} - 1$ which are inside C. Thus $\phi(z)$ is a finite sum of terms of the form $a_k/(w - ik)$, and such a sum is the Laplace transform of a trigonometric polynomial.

The following result is a generalization of 6.10.1.

6.10.7.* *Theorem.*[e] *If $f(z)$ is an entire function of exponential type τ and $f(z + 2\pi) - f(z) = g(z)$ with $g(z)$ of zero exponential type then $f(z)$ is the sum of an entire function of zero exponential type and a trigonometric polynomial of degree not exceeding τ.*

Notes for Chapter 6

6.2a. Pólya and Szegö [1], vol. 2, p. 36; Duffin and Schaeffer [2]; Plancherel and Pólya [1].

6.2b. Duffin and Schaeffer [2]; another proof is given by Redheffer [3].

6.2c. Harvey [1].

6.2d. Boas [13] constructs an unbounded function of exponential type for which the mean value in (6.2.11) is finite and exists as a limit.

6.2e. S. Bernstein [1], [2]; cf. §§3.1 and 11.3.

6.3a. Theorems 6.3.6, 6.3.9, 6.3.10, 6.3.11, 6.3.12, 6.3.13, 6.3.14, in more or less general form, have been discovered by almost everybody who has had occasion to work with Carleman's formula. It would be difficult to trace them to their earliest occurrences, since they usually appear as lemmas in just enough generality to suffice for an immediate application. See e.g. Pólya and Szegö [1], vol. 1, p. 242, Valiron [8], Cartwright [4], [6], Ganapathy Iyer [9], Levinson [4], p. 28, Levin [6], Ahiezer [4], Chebotarëv and Meiman [1], p. 158.

6.3b. This is an integral form of the O-Tauberian theorem for $(C, 1)$ summability. Boas [3] proves this and some related theorems and gives further references.

6.4a. The case $\phi(x) = 1$ is due to Levin [6].

6.4b. Levin [6].

6.4c. Agmon [1]; cf. 6.6.9.

6.5a. The formula is due to R. Nevanlinna [1], who establishes it under hypothesis (6.5.1).

6.5b. The inequality (6.5.9) holds without any hypothesis about the zeros of $f(z)$.

6.6a. See e.g. Levinson [4], pp. 26 ff.; Levin [5].

6.6b. See e.g. Levinson [4], pp. 26 ff. Cartwright [4], [6] proves a more general result in which $f(z)$ only has to be of exponential type in an angle which contains the upper half plane in its interior.

6.6c. See also R. Nevanlinna [2], p. 43.

6.7a. Plancherel and Pólya [1].

6.7b. Boas [25], using the method of Plancherel and Pólya.

6.7c. Various special cases were discovered by various authors before the development of the general theory. Boas [4] gives references, to which Srivastava [2] should be added.

6.7d. Jain [1] for $p = 1$.

6.7e. The argument is due to Plancherel and Pólya [1].

6.7f. Plancherel and Pólya [1]. Boas [6] obtains a similar result by another method.

6.7g. Korevaar [1].

6.7h. Nikolskii [2], with a less favorable constant.

6.7i. Gabriel [1]; Ganapathy Iyer [1] and Levinson [6] obtained less precise results by other methods.

6.7j. Harvey [1].

6.8a. Paley and Wiener [1], p. 13. Many proofs have been given. There are analogous Fourier representations for functions $f(z)$ which are regular and of exponential type in the upper half plane and belong to L^2 on the boundary; $f(z)$ is represented as the Fourier transform of a function which vanishes over a half line. See Paley and Wiener [1], Hille and Tamarkin [1], Mikusiński [3]; Theorem 6.7.7 is relevant here.

6.8b. This proof resembles one given by G. H. Hardy in lectures, but is simplified by a different choice of the contour.

6.8c. The "only if" part is due to N. Wiener; the statement of the text is given by Boas [4], who states the condition on $\phi(t)$ in the incorrect form that $\phi(t)$ has an absolutely convergent Fourier series on $(-\tau, \tau)$. That this is incorrect follows from the fact that there are functions $\phi(t)$ such that the Fourier series of $\phi(t)$ over $(-\tau, \tau)$ converges absolutely while the Fourier series of the extended $\phi(t)$ does not.

6.8d. Plancherel and Pólya [1], Kawata [1].

6.8e. Plancherel and Pólya [1], Boas [4].

6.9a. Cf. Titchmarsh [2], pp. 323 ff.

6.9b. Pólya [4]; cf. Boas [5].

6.10a. The theorem seems to have been discovered by everybody who has happened to think about the problem. Some references are given by Boas [13]. See also Giaccardi [1].

6.10b. The converse was proved by Bohr, and he doubtless was aware of this theorem, although I have not found it earlier than Boas [13]. For Bohr's theorem see Szegö [1].

6.10c. Boas [13].

6.10d. Pólya: see Muggli [1].

6.10e. This theorem brings us into contact with Whittaker's theory of asymptotic periods: δ is an asymptotic period of $f(z)$ if $f(z+\delta) - f(z)$ is of lower order than $f(z)$. The theory is of interest chiefly for functions of order greater than 1. The principal facts are that the asymptotic periods of an entire function are either (i) an arithmetic progression or (ii) a dense set of points on a line, with measure zero and with the ratio of any two asymptotic periods either rational or transcendental; only (i) is possible for functions of order 1. See Whittaker [2], S. S. Macintyre [1], Gelfond [4], Sikkema [1].

CHAPTER 7

FUNCTIONS OF EXPONENTIAL TYPE, RESTRICTED ON A LINE. II. ASYMPTOTIC BEHAVIOR IN A HALF PLANE

7.1. Introduction. If $f(z)$ is an entire function of exponential type we know a number of quantities which are bounded, at least on one side; for example,

$$r^{-1} \log | f(re^{i\theta}) | \text{ is bounded above} \tag{7.1.1}$$

(for each θ);

$$r^{-1}n(r) \text{ is bounded;} \tag{7.1.2}$$

$$\sum_{|z_n| \leq r} 1/z_n \text{ is bounded.} \tag{7.1.3}$$

In this chapter and the next we shall be concerned with additional conditions which can be imposed on $f(z)$ to refine such "O-statements" into "o-statements," that is, to convert boundedness into approach to a limit; and with applications of the refined theorems. This chapter deals with those refinements of (7.1.1) which depend on having $f(z)$ bounded, or "almost bounded," on a line; results involving the distribution of the zeros in a more detailed way will be discussed in Chapter 8. The basic results of this chapter hold for functions which instead of being entire are merely regular and of exponential type in a half plane and satisfy appropriate conditions in the neighborhood of the boundary. The applications are mostly to properties of entire functions.

The most obvious refinement of (7.1.1) would be

$$\lim_{r \to \infty} r^{-1} \log | f(re^{i\theta}) | = h(\theta) \tag{7.1.4}$$

for each θ; but this is not a reasonable requirement, since the presence of zeros on or near the ray $\arg z = \theta$ prevents (7.1.4) from holding; and moreover it can be shown that (7.1.4) need not hold even in the interior of a zero-free angle.[a] However, it turns out that appropriate restrictions on the rate of growth of the function on the boundary of the half plane do suffice to make (7.1.4) hold except for a "small" set of θ's, for every θ if r is excluded from a "small" set, and without exception in the interior of a zero-free angle. Boundedness on the boundary is sufficient, but may be replaced by more general conditions. Suppose for definiteness that the half plane in question is the upper half plane; then it is sufficient to have either

$$\int_{-\infty}^{\infty} \frac{\log^+ |f(x)|}{1 + x^2}\,dx < \infty, \tag{7.1.5}$$

or more generally either

$$\int_1^{\to\infty} x^{-2} \log |f(\pm x)|\,dx \text{ exist} \tag{7.1.6}$$

or

$$\int_1^{\to\infty} x^{-2} \log |f(x)f(-x)|\,dx \text{ exists and } h(0) + h(\pi) = 0 \tag{7.1.7}$$

(that (7.1.6) is actually more general than (7.1.5) follows from 6.3.6; that (7.1.7) is a consequence of (7.1.6) for entire functions will be shown later (7.2.8)).

Conditions (7.1.5), (7.1.6), (7.1.7), and even the weaker condition

$$\int_1^{R} x^{-2} \log |f(x)f(-x)|\,dx \text{ is bounded above,} \tag{7.1.8}$$

imply (6.3.9) that

$$\sum \Im(1/z_n) \text{ converges} \tag{7.1.9}$$

(the sum is extended over the zeros z_n which are in the upper half plane). Thus we are dealing only with functions most of whose zeros lie near the real axis. When (7.1.9) holds and $f(z)$ is entire, it is true conversely that, when (7.1.4) holds with sufficiently few exceptions, (7.1.7) holds, so that we cannot expect an appreciably more general condition than (7.1.7) to work in this case. Results for *entire* functions which do not necessarily satisfy (7.1.9) will be considered briefly in §8.1.

7.2. The asymptotic behavior of a Blaschke product. Suppose that $f(z)$ is regular and of exponential type in the upper half plane, and satisfies (7.1.6), which is true in particular if $f(x)$ is bounded or if (7.1.5) holds. Then according to 6.5.4, $f(z)$ has the representation

$$\log |f(z)| = \log |B(z)| + y\pi^{-1} \int_{-\infty}^{\infty} \frac{\log |f(t)|\,dt}{(t - x)^2 + y^2} + cy, \tag{7.2.1}$$

where

$$B(z) = \prod_{n=1}^{\infty} \frac{1 - z/z_n}{1 - z/\bar{z}_n}, \tag{7.2.2}$$

a convergent Blaschke product (the name is by analogy with the more familiar product for the unit circle). By 6.6.2,

$$(\pi r)^{-1} y \int_{-\infty}^{\infty} \frac{\log |f(t)|}{(t-x)^2 + y^2}\, dt + cr^{-1} y$$

approaches $c \sin \theta$ along $\arg z = \theta$, $0 < \theta < \pi$. Hence the behavior of $r^{-1} \log |B(z)|$ determines what happens to $r^{-1} \log |f(z)|$. (By using the Riesz decomposition theorem for a subharmonic function,[a] we could consider a more general subharmonic function than $\log |f(z)|$: (7.2.1) represents $\log |f(z)|$ as the sum of a potential and a harmonic function of known asymptotic behavior.)

Concerning $B(z)$ we prove the following theorem.

7.2.3. *Theorem.*[b] *If $B(z)$ is defined by* (7.2.2), *with $\mathfrak{J}(z_n) > 0$ and $\sum \mathfrak{J}(1/z_n)$ convergent, then for all θ in $(0, \pi)$, except at most a set of outer capacity zero,*

(7.2.4)
$$\lim_{r \to \infty} r^{-1} \log |B(re^{i\theta})| = 0;$$

for a fixed θ_0, $0 < |\theta_0| < \pi/2$, there is an open set Δ_0 of values of r, having finite logarithmic length, such that (7.2.4) *holds uniformly in $\theta_0 \le \theta \le \pi - \theta_0$ provided that $r \to \infty$ outside Δ_0; and* (7.2.4) *holds without exception in any angle $0 < \alpha < \theta < \beta < \pi$ which contains no z_n.*

We recall the definitions of some terms used in the statement of 7.2.3. For logarithmic length see §1.5.

If E is a Borel set in $(0, \pi)$, and $\mu(\theta)$ is a nondecreasing function of θ whose points of increase are in E and for which $\int_0^\pi d\mu(\theta) = 1$, the capacity of E is defined as

(7.2.5)
$$\exp\left\{ - \inf_{\mu} \sup_{|\zeta| \le 1} \int_0^\pi \log \frac{1}{|e^{i\theta} - \zeta|}\, d\mu(\theta) \right\}.$$

A countable union of sets of capacity zero is of outer capacity zero. The property of a set of outer capacity zero which is relevant in most applications of Theorem 7.2.3 is that it is of measure zero, so that in particular the nonexceptional values of θ are everywhere dense. To see this, if E has positive measure, take $d\mu(\theta) = E(\theta)\, d\theta$, where $E(\theta) = 1/\text{meas } E$ on E and 0 otherwise; then

$$\sup_{\zeta} \int \log \frac{1}{|e^{i\theta} - \delta|}\, d\mu(\theta) < \infty,$$

and so the capacity of E cannot be zero. Hence the measure of a set of zero outer capacity is also zero, and the complement of such a set is everywhere dense.

The last clause of 7.2.3 follows because there are θ's arbitrarily near

α and β for which 7.2.4 holds, and then it holds uniformly between them (1.4.8) because $z^{-1} \log B(z)$ is regular and bounded there (by the estimates used in proving 6.4.8).

We shall abbreviate the conclusion of 7.2.3 by saying that $r^{-1} \log | B(z) |$ tends effectively to zero in $0 < \theta < \pi$. Using this terminology we now state a more general theorem which follows as soon as 7.2.3 has been proved.

7.2.6. *Theorem.*[c] *If $f(z)$ is regular and of exponential type in the upper half plane, continuous in the closed half plane, and satisfies*

$$\int_1^{\to\infty} x^{-2} \log | f(\pm x) | \, dx \text{ exist} \tag{7.2.7}$$

(or the more restrictive condition (7.1.5)), then $r^{-1} \log | f(re^{i\theta}) |$ tends effectively to $c \sin \theta$ in $0 < \theta < \pi$.

This is an extensive generalization of 1.4.3.

The following corollary justifies the remark made after (7.1.7).

7.2.8. *Corollary. If $f(z)$ is an entire function of exponential type satisfying (7.2.7) then $h(0) = h(\pi) = 0$.*

In fact, we have $h(\theta) = O(\sin \theta)$ as $\theta \to 0$ or π in either the upper or lower half plane, and hence $h(0) = h(\pi) = 0$, by the continuity of $h(\theta)$.

The following theorem is therefore a generalization of 7.2.6, at least for entire functions.

7.2.9. *Theorem.*[d] *If $f(z)$ is regular and of exponential type in the upper half plane, continuous in the closed half plane, and satisfies*

$$\int_1^{\to\infty} x^{-2} \log | f(x)f(-x) | \, dx \text{ exists and } h(0) = h(\pi) = 0, \tag{7.2.10}$$

then $r^{-1} \log | f(re^{i\theta}) |$ tends effectively to $c \sin \theta$ for $0 < \theta < \pi$.

The condition $h(0) = h(\pi) = 0$ can be weakened to

$$\int_r^{2r} t^{-1} \log | f(\pm t) | \, dt \le o(r)$$

at the expense of further complication in the proof.[e]

Since we do not have the representation (7.2.1) under the hypotheses of 7.2.9, the proof of 7.2.9 is more difficult than that of 7.2.6; we can, however, still restrict our attention to $f(z)/B(z)$ and deal with $B(z)$ by 7.2.3.

We shall refer to 7.2.3, 7.2.6, and 7.2.9 collectively as "the Ahlfors-Heins theorem."

We note the following theorem, which states a little more than is directly deducible from the corresponding special case of 7.2.9.

7.2.11.* *Theorem.*[f] *If $f(z)$ is an entire function of exponential type satisfy-*

ing (7.2.10), *then for all* θ

$$\lim_{r\to\infty} \frac{1}{\log r} \int_1^r t^{-2} \log | f(te^{i\theta})f(-te^{i\theta}) | \, dt = 2c \sin \theta.$$

This is trivial for the nonexceptional values of θ in 7.2.9, and indicates in another way that the presence of the zeros does not do too much to destroy the asymptotic regularity asserted in 7.2.9.

The proof of 7.2.3 is rather long. It would be considerably simpler to prove weaker theorems which would suffice for some applications; but in many applications the full force of the theorem is required or at least leads to great simplification in the proofs. It turns out to be slightly simpler notationally to work in a right-hand half plane, so for the rest of this section we assume that

$$B(z) = \prod_{n=1}^{\infty} \frac{1 - z/a_n}{1 + z/\bar{a}_n}, \qquad \Re(a_n) > 0, \qquad \sum \Re(1/a_n) \text{ convergent.}$$

To prove 7.2.3 we require several facts about the Green's function $g(z, \zeta)$ for the right-hand half plane with pole at ζ, namely

$$g(z, \zeta) = \log | (z + \bar{\zeta})/(z - \zeta) |,$$

whose relevance is clear from the fact that

$$\log | B(z) | = -\sum_{n=1}^{\infty} g(z, a_n).$$

The following notation will be preserved throughout the proof: $z = x + iy = re^{i\theta}$, $z_0 = x_0 + iy_0 = r_0e^{i\theta_0}$, $\zeta = \xi + i\eta = \rho e^{i\phi}$. Then

(7.2.12) $$g(\lambda z, \lambda\zeta) = g(z, \zeta) = g(1/z, 1/\zeta);$$

(7.2.13) $$g(z, \zeta) \le g(r, \rho);$$

(7.2.14) $$g(z, \zeta) \le g(e^{i\theta}, e^{i\phi});$$

and, for $r > r_0$, $x > \delta r$, $0 < \delta < 1$,

(7.2.15) $$\frac{g(z, \zeta)}{g(z_0, \zeta)} \le C(\delta) \frac{x}{x_0} \{g(e^{i\theta}, e^{i\phi}) + 1\},$$

(7.2.16) $$\frac{g(z, \zeta)}{g(z_0, \zeta)} \le C(\delta) \frac{x}{x_0} \{g(r, \rho) + 1\},$$

where $C(\delta)$ depends only on δ.

Of these, (7.2.12) is immediate and (7.2.13), (7.2.14) follow from

$$\left| \frac{re^{i\theta} + \rho e^{i\phi}}{re^{i\theta} - \rho e^{i\phi}} \right|^2 = \frac{r^2 + 2r\rho \cos (\theta + \phi) + \rho^2}{r^2 - 2r\rho \cos (\theta - \phi) + \rho^2}.$$

In fact, replacing $\theta \pm \phi$ by 0 makes the numerator as large and the denominator as small as possible, and thus we have (7.2.13). Also, if we write $r/\rho = t$, the right-hand side is

$$\frac{t + t^{-1} + 2 \cos (\theta + \phi)}{t + t^{-1} - 2 \cos (\theta - \phi)} = 1 + \frac{4 \cos \theta \cos \phi}{t + t^{-1} - 2 \cos (\theta - \phi)}$$

which is largest when $t + t^{-1}$ is smallest, i.e. at $t = 1$, $r = \rho$, and (7.2.14) then follows from (7.2.12).

For (7.2.15) it is enough to consider the case when $|\zeta| \geq |z_0|$, since when $|\zeta| \leq |z_0| \leq |z|$ we can apply (7.2.15) to $g(z^{-1}, \zeta^{-1})$ and $g(z_0^{-1}, \zeta^{-1})$. When $|\zeta| \geq |z_0|$, we have

$$(7.2.17) \quad g(z_0, \zeta) = -\frac{1}{2} \log \left\{ 1 - \frac{4x_0\zeta}{|z_0 + \bar{\zeta}|^2} \right\} \geq \frac{2x_0\zeta}{|z_0 + \bar{\zeta}^2|} \geq \frac{x_0\xi}{2|\zeta|^2},$$

$$(7.2.18) \quad g(z, \zeta) = \frac{1}{2} \log \left\{ 1 + \frac{4x\xi}{|z - \zeta|^2} \right\} \leq \frac{2x\xi}{|z - \zeta|^2}.$$

If $|z - \zeta| \geq \delta |\zeta|/2$, (7.2.18) implies

$$g(z, \zeta) \leq 8\delta^{-2}x\xi/|\zeta|^2,$$

and so

$$(7.2.19) \quad g(z, \zeta) \leq 8\delta^{-2}x\xi |\zeta|^{-2}[g(e^{i\theta}, e^{i\phi}) + 1].$$

On the other hand, if $|z - \zeta| \leq \delta|\zeta|/2$, or $|z^{-1} - \zeta^{-1}| \leq \frac{1}{2}\delta/|z|$, we have

$$\frac{\xi}{|\zeta|^2} \geq \frac{x}{|z|^2} - \frac{\delta}{2|z|} \geq \frac{\delta |z|}{|z|^2} - \frac{\delta}{2|z|} \geq \frac{\delta^2}{2x},$$

and by (7.2.14) we have (7.2.19) in this case also. Using (7.2.17), we obtain (7.2.15). By using (7.2.13) instead of (7.2.14), we prove (7.2.16) in the same way.

To prove the first part of the theorem, take a positive ϵ and consider the open set Ω where $\log |f(z)| \leq -\epsilon x$. Let E_R denote the radial projection, on the arc $|\theta| < \pi/2$ of the unit circle, of the part of Ω where $|z| > R$, and let $E_R(\delta)$ denote the part of E_R in the sector $x > \delta r$. Take any z_0, and $R \geq r_0$. To a θ in $E_R(\delta)$, there corresponds (at least one) z in Ω with $\arg z = \theta$, $r > r_0$. Then for this z we have

$$-\log |B(z)| = \sum_{n=1}^{\infty} g(z, a_n) \geq \epsilon x,$$

and hence by (7.2.15), for θ in $E_R(\delta)$,

$$(7.2.20) \quad \epsilon x_0 \leq C(\delta) \sum_{n=1}^{\infty} g(z_0, a_n) \log \frac{2e}{|e^{i\theta} - e^{i\phi_n}|}.$$

Let $\nu(\theta)$ be a nondecreasing function whose points of increase are in all the $E_R(\delta)$, $R > 0$, and for which $\int_{-\pi/2}^{\pi/2} d\nu(\theta) = 1$. If we multiply (7.2.17) by $d\nu(\theta)$ and integrate, we obtain

$$\epsilon x_0 \leq C(\delta) \sum_{n=1}^{\infty} g(z_0, a_n) \left\{ \log (2e) + \int_{-\pi/2}^{\pi/2} \log \frac{1}{| e^{i\theta} - e^{i\phi_n} |} d\nu(\theta) \right\}$$

and hence

$$\epsilon x_0 \leq C(\delta)\{ - \log | B(z_0) | \} \left\{ \log (2e) + \sup_{\phi} \int_{-\pi/2}^{\pi/2} \log \frac{1}{| e^{i\theta} - e^{i\phi} |} d\nu(\theta) \right\}.$$

Now we have $\log | B(z) | \leq 0$, $\log | B(iy) | = 0$, and so we can find values $z_0 \to \infty$, $| \theta_0 | < \pi/2$, such that $x_0^{-1} \log | B(z_0) | \to 0$. Letting $z_0 \to \infty$ in this way, we see that

$$\sup_{\phi} \int_{-\pi/2}^{\pi/2} \log \frac{1}{| e^{i\theta} - e^{i\phi} |} d\nu(\theta) \to \infty, \tag{7.2.21}$$

unless of course the intersection of all $E_R(\delta)$, $R > 0$, is empty, in which case Ω is empty and we may assign it 0 as capacity. From (7.2.21) we see that the capacity of $\bigcap_R E_R(\delta)$ is zero, and the union of these sets for $\delta = 1/n$, $n = 1, 2, \cdots$, is of outer capacity zero.

To prove the second part of the theorem, consider the set Ω as before, and let $P_R(\delta)$ be the set of r's for which $| z | = r$ intersects the part of Ω where $| z | > R$ and $x > \delta | z |$. With $R > | z_0 |$, we have, by (7.2.16), for r in $P_R(\delta)$,

$$\epsilon x_0 \leq C(\delta) \sum_{n=1}^{\infty} g(z_0, a_n)\{g(r, \rho) + 1\}.$$

Let $\mu(r)$ be a nondecreasing function whose points of increase are in $P_R(\delta)$, with $\int_0^{\infty} d\mu(r) = 1$, multiply by $d\mu(r)$, and integrate. We obtain

$$\epsilon x_0 \leq C(\delta)\{ - \log | f(z_0) | \} \left\{ 1 + \sup_{\xi > 0} \int_0^{\infty} \log \left| \frac{r + \zeta}{r - \zeta} \right| d\mu(r) \right\}.$$

Just as before, we see that

$$\sup_{\xi > 0} \int_0^{\infty} \log \left| \frac{r + \zeta}{r - \zeta} \right| d\mu(r) \to \infty \tag{7.2.22}$$

as $R \to \infty$. We now define a functional $V(E)$ as the infimum of the left-hand side of (7.2.22) for $\mu(r)$ with its points of increase in E and

$$\int_0^\infty d\mu(r) = 1,$$

and establish some properties of $V(E)$ from which we can conclude that for large enough R the logarithmic length of $P_R(\delta)$ is finite.

The necessary properties of $V(E)$ are as follows: (a) if $E_1 \subset E_2$ then $V(E_1) \geq V(E_2)$; (b) if E_2 is obtained from E_1 by the transformation $x' = kx$, then $V(E_1) = V(E_2)$; (c) if $E_1, \cdots, E_n$ are disjoint intervals and E is an interval whose logarithmic length is the sum of the logarithmic lengths of $E_1, \cdots, E_n$, then $V(\bigcup E_k) \leq V(E)$; (d) $V((0, \infty)) = 0$.

Of these, (a) is evident because any function μ which can be used for E_1 can also be used for E_2; and (b) is evident since the only effect of the transformation on the left-hand side of (7.2.22) is to change ξ. For (c) it is enough to consider two sets, say (a_1, b_1) and (a_2, b_2) with $b_1 < a_2$; we have to show that

$$\inf_\mu \sup_\xi \left\{ \int_{a_1}^{b_1} g(r, -\xi)\, d\mu(r) + \int_{a_2}^{b_2} g(r, -\xi)\, d\mu(r) \right\} \leq \inf_\nu \sup_\xi \int_{a_1a_2}^{b_1b_2} g(r, -\xi)\, d\nu(r). \tag{7.2.23}$$

The left-hand side is, by (7.2.12),

$$\inf_\mu \sup_\xi \left\{ \int_{a_1}^{b_1} g(r, -\xi)\, d\mu(r) + \int_{b_1}^{b_1b_2/a_2} g(r, -\xi)\, d\mu(b_1r/a_2) \right\} = \inf_\mu \sup_\xi \int_{a_1a_2}^{b_1b_2} g(r, -\xi)\, d\nu(r),$$

where $\nu(r) = \mu(a_2r)$ in (a_1, b_1), $\mu(b_1r)$ in $(b_1, b_1b_2/a_2)$, and (7.2.23) follows.

For (d) we have to show that there is a $\mu(r)$ with $\int_0^\infty d\mu(r) = 1$ for which

$$\sup_\xi \int_0^\infty g(r, -\xi)\, d\mu(r)$$

is arbitrarily small. There are functions $\mu(r)$ satisfying our hypotheses with the additional property that they are absolutely continuous and $\sup r\mu'(r)$ is arbitrarily small; and we have

$$\int_0^\infty g(r, -\xi)\mu'(r)\, dr \leq \sup \{r\mu'(r)\} \int_0^\infty r^{-1} g(r, -\xi)\, dr$$

$$= \sup \{r\mu'(r)\} \int_0^\infty t^{-1} g(t, -1)\, dt,$$

so that we can make the left-hand side arbitrarily small by choosing $\mu(r)$ appropriately.

It now follows from (a), (c) and (d) that $V(E) = 0$ if E has infinite logarithmic length. Since (7.2.22) shows that $V(P_R(\delta)) > 0$ for sufficiently large R, $P_R(\delta)$ must, for large R, have finite logarithmic length; in other words,

$$\log |f(re^{i\theta})| \geq -\epsilon r, \qquad r > 1, |\theta| < \theta_0,$$

except for a set E_ϵ of finite logarithmic length. Let $\epsilon_j \downarrow 0$; for each j, there is a λ_j such that $H_j = E_{\epsilon_j} \cap (r > \lambda_k)$ has logarithmic length less than 2^{-j}. The union of all the H_j has finite logarithmic length and

$$r^{-1} \inf_{|\theta| \leq \theta_0} \log |f(re^{i\theta})| \to 0$$

as $r \to \infty$ outside this union.

7.3. Asymptotic behavior with other boundary conditions. Before turning to the applications of the Ahlfors-Heins theorem we prove theorem 7.2.9. The same proof gives a weaker result under a weaker hypothesis; we state the result but leave the necessary modifications in the proof to the reader.

7.3.1.* *Theorem.*[a] *If $f(z)$ is regular and of exponential type in $y > 0$, continuous in $y \geq 0$, and satisfies*

$$\int_1^R x^{-2} \log |f(x)f(-x)| \, dx \textit{ is bounded} \tag{7.3.2}$$

then $r^{-1} \log |f(re^{i\theta})|$ is bounded, with the same exceptions as in 7.2.9.

If $f(z)$ is an entire function of exponential type, Theorem 7.3.1 has a more general hypothesis than 7.2.9, and the hypothesis (7.3.2) may be replaced by the equivalent one that $\sum |\Im(1/z_n)| < \infty$ (6.3.14).

To prove 7.2.9, we start again from (6.5.10), which states that

$$\log |f(re^{i\theta})| = \sum_{|z_n|<R} \log \frac{z - z_n}{z - \bar{z}_n} + \frac{y}{\pi} \int_{-R}^{R} P_1(R, z, t) \log |f(t)| \, dt$$

$$+ \frac{2Ry}{\pi} \int_0^\pi P_2(R, z, \phi) \log |f(Re^{i\phi})| \, d\phi = S + yI_1 + yI_2 + yI_3,$$

where

$$P_1 = \frac{1}{t^2 - 2tx + r^2} - \frac{R^2}{R^4 - 2tR^2x + r^2t^2},$$

$$P_2 = \frac{(R^2 - r^2) \sin \phi}{|R^2e^{2i\phi} - 2Rxe^{i\phi} + r^2|^2}.$$

Now we know (6.4.8) that S approaches a limit, which has the required asymptotic behavior (7.2.3). Step (iii) of the proof of 6.5.4 still applies, and shows that I_3 approaches a constant limit. It is therefore enough to show that

$$\lim_{r\to\infty} \limsup_{R\to\infty} (|I_1| + |I_2|) = 0, \qquad 0 < \theta < \pi. \tag{7.3.3}$$

We can write πI_1 in either of the two forms

$$\pi I_1 = \int_0^R \frac{\log|f(t)\,f(-t)|}{(t+x)^2+y^2}\,dt + \int_0^R \frac{4tx\log|f(t)|}{(t^2+r^2)^2-4t^2x^2}\,dt, \tag{7.3.4}$$

$$\pi I_1 = \int_0^R \frac{\log|f(t)\,f(-t)|}{(t-x)^2+y^2}\,dt - \int_0^R \frac{4tx\log|f(-t)|}{(t^2+r^2)^2-4t^2x^2}\,dt. \tag{7.3.5}$$

Suppose that $x > 0$, and consider (7.3.4). Then as in step (ii) of the proof of 6.5.4, an application of the second mean value theorem shows that

$$\lim_{R\to\infty} \int_0^R \frac{\log|f(t)\,f(-t)|\,dt}{(t+x)^2+y^2} = \int_0^\infty \frac{\log|f(t)\,f(-t)|\,dt}{(t+x)^2+y^2},$$

and 6.6.2 shows that the limit of the right-hand side is zero as $r \to \infty$, $0 < \theta < \pi$. This disposes of the first part of πI_1.

For the second part of πI_1, we have $\log|f(t)| \le o(t)$; then if $\epsilon > 0$ and M is large but fixed, while $R > M$, and A is a certain constant,

$$x\int_0^R \frac{t\log|f(t)|}{(t^2+r^2)^2-4t^2x^2}\,dt \le Ax\int_0^M \frac{t^2\,dt}{(t^2+r^2)-4t^2x^2} + \epsilon x\int_M^R \frac{t^2\,dt}{(t^2+r^2)^2-4t^2x^2}.$$

If $y/x \ge \delta > 0$, the right hand side is at most

$$A\int_0^{M/x} \frac{u^2\,du}{(1+u^2+\delta^2)^2-4u^2} + \epsilon\int_0^\infty \frac{u^2\,du}{(1+u^2+\delta^2)^2-4u^2}, \tag{7.3.6}$$

and this can be made arbitrarily small (independently of R) by taking ϵ small and then r large. Hence

$$\limsup_{r\to\infty} \{\limsup_{R\to\infty} I_1\} \le 0.$$

Using (7.3.5) instead of (7.3.4), we have similarly, if $x \ge 0$,

$$\liminf_{r\to\infty} \{\liminf_{R\to\infty} I_1\} \ge 0,$$

and so (7.3.3) holds for I_1 and $x > 0$. For $x < 0$ the same conclusion follows if we interchange the roles of (7.3.4) and (7.3.5).

Finally we estimate I_2 in a similar way. We have, instead of (7.3.4),

(7.3.5),

$$(7.3.7) \quad -\pi I_2 =$$

$$R^2 \int_0^R \frac{\log | f(t)f(-t) | \, dt}{R^4 \pm 2tR^2x + r^2t^2} \pm 4R^4x \int_0^R \frac{t \log | f(\pm t) | \, dt}{(R^4 + r^2t^2)^2 - 4t^2R^4x^2},$$

where the three upper or three lower signs are to be taken. Taking the upper signs and $x \geq 0$, we obtain an upper estimate for $-\pi I_2$; the corresponding lower estimate, and estimates for $x < 0$, are obtained similarly.

The first integral in (7.3.7) has the limit 0 as $R \to \infty$, by an application of the second mean value theorem similar to that used in dealing with I_2 in step (ii) of the proof of 6.5.4. For the second integral we have, when $y/x \geq \epsilon > 0$, the upper estimate

$$4ARx \int_0^R \frac{t^2 \, dt}{(R^4 + r^2t^2)^2 - 4t^2R^4x^2}$$

$$\leq 4xAR^3 \int_0^\infty \frac{u^2 \, du}{[R^2 + x^2u^2(1 + \epsilon^2)]^2 - 4R^2x^2u^2},$$

and the limit superior of this, as $R \to \infty$, is zero.

7.4. The limit of $r^{-1} \log M(r)$. As a first application of the Ahlfors-Heins theorem we shall prove the result (stated in §6.6) that $\lim_{r\to\infty} r^{-1} \log M(r)$ always exists for a function of exponential type in the upper half plane which is bounded on the real axis, and is not negative (unless $f(z) \equiv 0$). We shall, in fact, replace boundedness on the real axis by a less restrictive condition.

7.4.1. *Theorem.*[a] *If $f(z)$ is regular and of exponential type in the upper half plane and continuous in the closed half plane, $f(z) \not\equiv 0$,*

$$(7.4.2) \qquad \int_1^{\to\infty} x^{-2} \log | f(x)f(-x) | \, dx \text{ exists,}$$

$$(7.4.3) \qquad h(0) = h(\pi) = 0,$$

and

$$(7.4.4) \qquad \liminf_{r\to\infty} r^{-1} \log M(r) = \alpha,$$

then $\alpha \geq 0$ and

$$(7.4.5) \qquad \lim_{r\to\infty} r^{-1} \log M(r) = \alpha.$$

According to 6.6.6, when $f(z)$ is bounded on the real axis we may drop the hypothesis that $f(z)$ is of exponential type and assume only that $\alpha < \infty$, since the hypotheses then imply that $f(z)$ is in fact of exponential type.

By 6.3.6,

(7.4.6)
$$\int_{-\infty}^{\infty} \frac{\log^{+} |f(x)|}{1+x^2}\, dx$$

or

(7.4.7)
$$\int_{1}^{\infty} x^{-2} \log |f(\pm x)|\, dx \text{ exist}$$

implies (7.4.2); but these conditions imply (7.4.3) only if $f(z)$ is an entire function of exponential type. We could therefore substitute (7.4.6) or (7.4.7) for (7.4.2) in the theorem; of course if $f(z)$ is bounded on the real axis both (7.4.2) and (7.4.3) are satisfied.

The idea of the proof is as follows. Our hypotheses imply $h(\theta) = c \sin \theta$ for $0 < \theta < \pi$ and that $r^{-1} \log |f(re^{i\theta})|$ tends effectively to $h(\theta)$, while $\log M(r) \geq \log |f(re^{i\theta})|$; therefore taking θ near 0 shows that $\alpha \geq 0$; on the other hand, taking θ near $\pi/2$ shows, if $c > 0$, that $|f(re^{i\theta})|$ is generally largest near $\theta = \pi/2$ and hence that $\lim r^{-1} \log M(r) = c$ (so that $c = \alpha$).

To carry out the second step in detail, we consider first the case when $c > 0$. If δ_1 and δ_2 are small positive numbers such that δ_1 and $\pi - \delta_2$ are nonexceptional values of θ in 7.2.9, and $r \to \infty$ outside an open set of finite logarithmic length, then $r^{-1} \log |f(re^{i\theta})| \to c \sin \theta$ uniformly for

$$\delta_1 \leq \theta \leq \pi - \delta_2 .$$

Furthermore, by the Phragmén-Lindelöf theorem (1.4.2), $r^{-1} \log |f(re^{i\theta})| \leq O(c \sin \delta)$, uniformly for $0 \leq \theta \leq \delta_1$ and $\pi - \delta_2 \leq \theta < \pi$, where δ is the larger of δ_1 and δ_2. Therefore $r^{-1} \log M(r) \to c$ except perhaps for an open set of finite logarithmic length. If (a, b) is an interval of this set, and $a < r < b$, we have

$$\begin{aligned}(a/b)a^{-1} \log M(a) &= b^{-1} \log M(a) \leq r^{-1} \log M(r) \leq a^{-1} \log M(b)\\ &= (b/a)b^{-1} \log M(b),\end{aligned}$$

and since $a/b \to 1$ as $a \to \infty$, it follows that the exceptional intervals are not really exceptional.

If $c \leq 0$, on the other hand, we have $h(\theta) < 0$ for $0 \leq \theta \leq \pi$, so by 5.1.9 we have $\limsup r^{-1} \log M(r) \leq 0$. Since $\alpha \geq 0$, this means that

$$r^{-1} \log M(r) \to 0.$$

7.5. Factorization of a positive function. A positive trigonometric polynomial of degree n can be written[a] as the square of the absolute value of a trigonometric polynomial of order $n/2$. The following theorem extends this to more general entire functions; it follows easily from the Ahlfors-Heins theorem.

7.5.1. *Theorem.*[b] *Let $f(z)$ be an entire function of exponential type τ, with $f(x) \geq 0$ for real x. Then a necessary and sufficient condition for the existence of an entire function $F(z)$, of exponential type $\tau/2$, having all its zeros in the (closed) upper half plane, such that $f(z) = F(z)\bar{F}(z)$, is that*

$$\sum |\Im(1/z_n)| < \infty, \tag{7.5.2}$$

where z_n are the zeros of $f(z)$.

Theorem 7.5.1 does in fact contain the theorem about trigonometric polynomials which it is supposed to generalize. For, a trigonometric polynomial $T_n(z)$ of degree n is an entire function of exponential type n with period 2π; then the $F(z)$ of 7.5.1 is an entire function of exponential type $n/2$ with the same period, and by 6.10.1 reduces to a trigonometric polynomial of degree (at most) $n/2$.

Suppose first that $f(z) = F(z)\bar{F}(z)$, with $F(z)$ as specified in 7.5.1. Let w_n be the zeros of $F(z)$; then $\Im(w_n) \geq 0$ and every z_n is either a w_n or a $\bar{w}_n$. Since both $f(z)$ and $F(z)$ are of exponential type, by Lindelöf's theorem (2.10.1) the sums

$$\sum_{|z_n|<r} 1/z_n = \sum_{|w_n|<r} 2\Re(1/w_n)$$

and $\sum_{|w_n|<r} 1/w_n$ are bounded, hence so are $\sum_{|w_n|<r} \Im(1/w_n)$, which (since $\Im(1/w_n) \leq 0$) means that (7.5.2) is satisfied for $F(z)$ and therefore for $f(z)$.

For the converse, let the Hadamard factorization of $f(z)$ be

$$f(z) = Az^k e^{cz} \prod_{n=1}^{\infty} (1 - z/z_n) \exp(z/z_n).$$

Since k must be even we may consider $A^{-1}f(z)z^{-k}$ instead of $f(z)$, and so we may suppose that $f(0) = 1$. In addition, c must be real (so that $(z^{-k}f(z))'$ will be real at $z = 0$).

Let the zeros of $f(z)$ in the closed upper half plane be $a_n = \alpha_n + i\beta_n$ and put

$$F(z) = e^{cz/2} \prod_{n=1}^{\infty} (1 - z/a_n) \exp\{\alpha_n z/|a_n|^2\},$$

where a real zero (necessarily of even multiplicity) appears with half its multiplicity. The product converges because (7.5.2) is satisfied, and we have $f(z) = F(z)\bar{F}(z)$. It remains to show that the type of $F(z)$ is at most

$\tau/2$. We have

$$F(z)/\bar{F}(z) = \prod_{n=1}^{\infty} \frac{1 - z/a_n}{1 - z/\bar{a}_n}$$

a function of the form (7.2.2) and so, by 7.2.3, for almost all θ,

$$\lim_{r\to\infty} r^{-1}\{\log | F(re^{i\theta}) | - \log | F(re^{-i\theta}) | \} = 0.$$

On the other hand,

$$\limsup_{r\to\infty} r^{-1}\{\log | F(re^{i\theta}) | + \log | F(re^{-i\theta}) | \}$$

$$= \limsup_{r\to\infty} r^{-1} \log | f(re^{i\theta}) | \le \tau,$$

and the conclusion follows.

7.6. The type of a product. Another direct application of the Ahlfors-Heins theorem is to the problem of the type of the product of two entire functions of exponential type. If $f_1(z)$ and $f_2(z)$ are of order 1 and types τ_1, τ_2, and $f_{12}(z) = f_1(z)f_2(z)$ is of exponential type τ_{12}, we have

$$\tau_{12} \le \tau_1 + \tau_2 .$$

In general we cannot expect equality here, and we cannot even give a universal lower bound (other than the trivial 0) for τ_{12}. However, if for one of the functions involved, say $f_1(z)$, we have $r^{-1} \log | f(re^{i\theta}) |$ tending effectively (in the sense of §7.2) to $h(\theta)$, we have (with self-explanatory notation) $h_{12}(\theta) = h_1(\theta) + h_2(\theta)$ for a dense set of θ's, and so

$$\tau_{12} - \max h_{12}(\theta) \ge \max \{h_1(\theta) + h_2(\theta)\}.$$

Furthermore, if $h_j(\phi_j) = \tau_j$ the indicator diagram of $f_j(z)$ contains at least the point $\tau_j \exp (i\phi_j)$ and so $h_j(\theta) \ge \tau_j \cos (\phi_j - \theta)$. Suppose for definiteness that $h_1(\pi/2) = \tau_1$ and $h_2(\phi) = \tau_2$, with $f_1(z)$ satisfying the hypothesis mentioned above. Then

$$\tau_{12} \ge \max_{\theta} \{\tau_1 \sin \theta + \tau_2 \cos (\phi - \theta)\},$$

and so for a dense set of θ's,

$$\begin{aligned} h_{12}(\theta) &\ge (\tau_1 + \tau_2 \sin \phi) \sin \theta + \tau_2 \cos \phi \cos \theta \\ &= \cos (\theta - \omega)\cdot(\tau_1^2 + \tau_2^2 + 2\tau_1\tau_2 \sin \phi)^{1/2}, \end{aligned} \tag{7.6.1}$$

where

$$\sin \omega = \frac{\tau_1 + \tau_2 \sin \phi}{(\tau_1^2 + \tau_2^2 + 2\tau_1\tau_2 \sin \phi)^{1/2}}, \quad \cos \omega = \frac{\tau_2 \cos \phi}{(\tau_1^2 + \tau_2^2 + 2\tau_1\tau_2 \sin \phi)^{1/2}}.$$

Since we may let $\theta \to \omega$ we obtain

$$\tau_{12} \geq (\tau_1^2 + \tau_2^2 + 2\tau_1\tau_2 \sin\phi)^{1/2} = | i\tau_1 + \tau_2 e^{i\phi} | .$$

We now apply these considerations to the case when $f_1(z)$ satisfies the hypotheses of 7.2.9.

7.6.2. *Theorem.*[a] *If $f_1(z)$ and $f_2(z)$ are entire functions of order* 1 *and types* τ_1, τ_2, *if*

$$\int_1^{\to\infty} x^{-2} \log | f_1(x)f_1(-x) | \, dx$$

exists, $h_1(0) = h_1(\pi) = 0$, *and* $h_1(\pi/2) = \tau_1$, *while* $h_2(\phi) = \tau_2$, *then the type* τ_{12} *of* $f_1(z)f_2(z)$ *satisfies*

$$\tau_{12} \geq (\tau_1^2 + \tau_2^2 + 2\tau_1\tau_2 \sin\phi)^{1/2}. \tag{7.6.3}$$

We have $h_1(\theta) = c\,|\sin\theta| + b\sin\theta$, $b + c = \tau_1$, $c - b \leq \tau_1$, *and the result (sometimes more precise than* (7.6.3))

$$\tau_{12} \geq \{(b - c)^2 + \tau_2^2 + 2(b - c)\tau_2 \sin\phi\}^{1/2}. \tag{7.6.4}$$

The estimate (7.6.3) follows from our previous discussion. However, when $-\pi < \theta < 0$ the explicit form of $h_1(\theta)$ is sometimes more favorable than the estimate $h_1(\theta) \geq \tau_1 \sin\theta$, and (7.6.4) results from maximizing $c\,|\sin\theta| + b\sin\theta + \tau_2\cos(\phi - \theta)$ for $-\pi < \theta < 0$.

By applying 7.3.1 we can obtain a similar, but weaker, result.

7.6.5.* *Theorem.*[b] *If $f_1(z)$ and $f_2(z)$ are entire functions of order* 1 *and types* τ_1, τ_2, *and if*

$$\int_1^{R} x^{-2} \log | f_1(x)f_1(-x) | \, dx$$

is bounded, then there is a finite number c (depending only on $f_1(z)$) such that the type τ_{12} of $f_1(z)f_2(z)$ satisfies $\tau_{12} \geq \tau_2 - c$.

The point of the theorem is that c is independent of $f_2(z)$.

We may complete 7.6.2, in a special case, as follows.

7.6.6. *Theorem. If $f_1(z)$ and $f_2(z)$ are entire functions of order* 1 *and types* τ_1, τ_2, *if*

$$\int_1^{\to\infty} x^{-2} \log | f_j(x)f_j(-x) | \, dx$$

exists and $h_j(0) = h_j(\pi) = 0$ $(j = 1, 2)$, *then the indicator diagram of* f_1f_2 *is the sum of the indicator diagrams of* f_1 *and* f_2.

In this case $r^{-1} \log | f_j(re^{i\theta}) |$ tends effectively to $h_j(\theta)$ $(j = 1, 2)$, and so $h_{12}(\theta) = h_1(\theta) + h_2(\theta)$ for almost all θ and hence for all θ (by continuity). In particular, 7.6.6 applies when $f_1(z)$ and $f_2(z)$ are finite Fourier transforms (cf. 6.9.4).

7.7. A symmetry property of the indicator diagram. The condition

$$\sum |\Im(1/z_n)| < \infty \tag{7.7.1}$$

requires that, in a certain sense, the zeros of $f(z)$ are concentrated around the real axis. It is then to be expected that an entire function satisfying (7.7.1) should behave in some ways like a function with real zeros only. A function $f(z)$ of exponential type which has all its zeros on the real axis is of the form $Az^m e^{(a+ib)z}P(z)$, where $P(z)$ is the canonical product formed with the zeros and so is real on the real axis. The indicator diagram of $P(z)$ is symmetric about the real axis, and therefore that of $f(z)$ is symmetric about $y = -b$. This property of symmetry is preserved for entire functions satisfying (7.7.1).

7.7.2. *Theorem.*[a] *If $f(z)$ is an entire function of exponential type such that the series $\sum \Im(1/z_n)$, extended over its zeros z_n in the lower half plane, converges, then its indicator diagram has a horizontal axis of symmetry.*

The hypothesis is equivalent to (7.7.1), since the sums

$$\sum_{|z_n|<r} \Im(1/z_n)$$

are bounded (2.10.1).

The conclusion may be written

$$h(\theta) - h(-\theta) = 2a \sin \theta \tag{7.7.3}$$

for some a and all θ.

As in the proof of 6.4.5, let w_n be the zeros of $f(z)$ in $y < 0$, ζ_n the rest of the zeros together with $\bar{w}_n$. Let

$$\phi(z) = \prod_{n=1}^{\infty} \frac{1 - z/\bar{w}_n}{1 - z/w_n}, \tag{7.7.4}$$

and put $\omega(z) = \phi(z)f(z)$. Now $\omega(z)$ has zeros ζ_n, none of which are in the lower half plane, and as in (6.4.4) we can write

$$\omega(z) = Az^m e^{cz} \prod_{n=1}^{\infty} (1 - z/\zeta_n) \exp \{z\Re(1/\zeta_n)\}, \tag{7.7.5}$$

so that

$$\left|\frac{f(z)}{f(\bar{z})}\right| = \left|\frac{\omega(z)}{\omega(\bar{z})}\frac{\phi(\bar{z})}{\phi(z)}\right| = |e^{2iky}| \prod_{n=1}^{\infty} \left|\frac{1 - z/\zeta_n}{1 - \bar{z}/\zeta_n}\right| \left|\frac{\phi(\bar{z})}{\phi(z)}\right|$$

$$= \exp \{-2y\Im(c)\} \, |\mu(z)| \, |\phi(z)|^{-2},$$

where

$$\mu(z) = \prod_{n=1}^{\infty} \frac{1 - z/\zeta_n}{1 - z/\bar{\zeta}_n}.$$

Since $\phi(z)$ and $\mu(z)$ are Blaschke products of the form (7.2.2), by the Ahlfors-Heins theorem we have

$$h(\theta) = h(-\theta) - 2\Im(c) \sin \theta \tag{7.7.6}$$

for a dense set of values of θ, and hence, since $h(\theta)$ is continuous, for all θ. This establishes (7.7.3).

7.8. Functions with a zero-free half plane. Here we shall obtain some properties of entire functions of exponential type with no zeros in the upper half plane. These, aside from their intrinsic interest, will be useful later.

7.8.1. *Theorem.*[a] *Let $\omega(z)$ be an entire function of exponential type having no zeros for $y > 0$ and having $h(\alpha) \geq h(-\alpha)$ for some α, $0 < \alpha < \pi$. Then $|\omega(z)| \geq |\omega(\bar{z})|$ for $y > 0$.*

In other words, if the conclusion (even in a weakened form) holds for z on one ray, it holds on all rays.

In Theorem 6.4.2, take $f(z) = \omega(z)$, $g(z) = \bar{\omega}(z)$, $\phi(x) = 1$, where $\bar{\omega}(z)$ is the conjugate of $\omega(\bar{z})$. Then $\psi(z) = \bar{\omega}(z)/\omega(z)$ has $|\psi(x)| = 1$ for real x and is of exponential type in the upper half plane. By the Ahlfors-Heins theorem, for a dense set of θ's in $(0, \pi)$

$$\lim_{r\to\infty} r^{-1} \log |\psi(re^{i\theta})| = h_\psi(\theta) = c \sin \theta.$$

Since

$$\log |\bar{\omega}(re^{i\theta})| = \log |\psi(re^{i\theta})| + \log |\omega(re^{i\theta})|$$

the indicator of $\bar{\omega}(\theta)$ is, for a dense set of θ's and so for all θ's, equal to $h_\psi(\theta) + h_\omega(\theta)$. Since $|\bar{\omega}(re^{i\theta})| = |\omega(re^{-i\theta})|$, this means, in particular, for $\theta = \alpha$, that

$$h_\omega(-\alpha) = c \sin \alpha + h_\omega(\alpha),$$

and so $c \leq 0$. By 6.2.4, applied to $\psi(z)$, $|\psi(z)| \leq e^{cy}$ for $y > 0$, or

$$|\omega(\bar{z})| = |\bar{\omega}(z)| \leq e^{cy} |\omega(z)|, \qquad y > 0,$$

which is the desired conclusion (since $c \leq 0$).

We now introduce a definition.

7.8.2. *Definition.*[b] *An entire function $\omega(z)$ of exponential type having no zeros for $y < 0$ and satisfying one of the conditions (equivalent, by 7.8.1) $h(-\alpha) \geq h(\alpha)$ for some α, $0 < \alpha < \pi$, or $|\omega(z)| \geq |\omega(\bar{z})|$ for $y < 0$ is said to belong to class P.*

(Note the interchange of half planes from 7.8.1.)

The class P is important in a number of applications; cf. §11.7. We next give another characterization of the class.

7.8.3. *Theorem.*[c] *A function $\omega(z)$ of exponential type belongs to P if and only if it has the form*

$$\omega(z) = Az^m e^{cz} \prod_{n=1} (1 - z/z_n) \exp \{z\Re(1/z_n)\} \tag{7.8.4}$$

with $\Im(z_n) \geq 0$ *and* $2\Im(c) = h(-\pi/2) - h(\pi/2) \geq 0$.

If $\omega(z)$ has the specified form, it belongs to P because

$$\left| \frac{\omega(z)}{\omega(\bar{z})} \right| = \exp\{-2y\Im(c)\} \prod_{n=1}^{\infty} \left| \frac{1 - z/z_n}{1 - \bar{z}/z_n} \right| \tag{7.8.5}$$

$$= \exp\{-2y\Im(c)\} \prod_{n=1}^{\infty} \left| \frac{1 - z/z_n}{1 - z/\bar{z}_n} \right|,$$

and for $y < 0$ this is not less than 1.

If $\omega(z)$ belongs to P, it can be written in the form (7.8.4), where (since the current $\omega(z)$ corresponds to $f(z)$ in 7.7.2), $h(\theta) = h(-\theta) - 2\Im(c) \sin \theta$. Taking $\theta = \pi/2$ we have $h(\pi/2) - h(-\pi/2) = -2\Im(c)$, and this is not positive, by the definition of class P.

In 6.4.5 we found that an entire function $f(z)$ of exponential type satisfies $|f(x)| \leq |\omega(x)|$, $-\infty < x < \infty$, for some entire $\omega(z)$ of exponential type having no zeros for $y < 0$, if and only if $\sum |\Im(1/z_n)|$ converges, and that if this condition holds then $\omega(z)$ can be taken so that $|f(x)| = |\omega(x)|$. We see now that in addition $\omega(z)$ can be required to belong to class P, since we can write it in the form (7.8.4), and if $\Im(c)$ does not have the right value we can adjust it by replacing $\omega(z)$ by $\omega(z)e^{ipz}$ with a suitable value for p.

We now prove a companion theorem which gives conditions for a given $f(z)$ to satisfy $|f(x)| \leq |\omega(x)|$ with a specific $\omega(z)$.

7.8.6. *Theorem.*[c] *If $\omega(z)$ is an entire function of order* 1 *and type σ, and $f(z)$ is an entire function of exponential type $\tau \leq \sigma$, then*

$$|f(x)| \leq |\omega(x)|, \qquad -\infty < x < \infty,$$

and $\omega(z)$ belongs to P, if and only if $\phi_u(z) = f(z) - u\omega(z)$ belongs to P for every complex u of modulus at least 1.

First we show that the condition of the conclusion implies that

$$|f(x)| \leq |\omega(x)|$$

and that $\omega(z)$ belongs to P. It follows from the definition of class P that the limit, uniformly in every bounded region, of a sequence of elements of P belongs to P. Since $-u^{-1}\phi_u(z)$ belongs to P and approaches $\omega(z)$ as $u \to \infty$,

$\omega(z)$ belongs to P. The function $\phi_u(z)/\omega(z) = \{f(z)/\omega(z)\} - u$ is regular and has no zeros in $y < 0$, if $|u| \geq 1$, since both $\phi_u(z)$ and $\omega(z)$ belong to P. Since this holds for every u such that $|u| \geq 1$, $|f(z)/\omega(z)| < 1$ for $y < 0$, and hence $|f(x)/\omega(x)| \leq 1$ for real x.

To establish the converse, we suppose that $|f(x)| \leq |\omega(x)|$, where $\omega(z)$ belongs to P. Putting $\psi(z) = f(z)/\omega(z)$, we have $\psi(z)$ of exponential type in the lower half plane (6.4.2) and bounded on the real axis. As in the proof of 7.8.3, we have, from the Ahlfors-Heins theorem applied to $\psi(z)$,

$$h_f(\theta) = h_\psi(\theta) + h_\omega(\theta), \qquad -\pi < \theta < 0;$$

and $h_\psi(\theta) = k|\sin\theta|$, $-\pi < \theta < 0$, by 6.2.4. By the definition of class P (7.8.2), $h_\omega(\theta) \geq h_\omega(-\theta)$ for $-\pi < \theta < 0$, and so the type σ of $\omega(z)$, which is the maximum of $h_\omega(\theta)$, must be the maximum of $h_\omega(\theta)$ for $-\pi \leq \theta < 0$. If $k > 0$, then, we should have $h_f(\theta) > h_\omega(\theta)$ for $-\pi < \theta < 0$, and so $\tau > \sigma$, contrary to hypothesis. Therefore $k \leq 0$ and $|f(z)/\omega(z)| \leq 1$ for $y < 0$.

If $|f(z_0)/\omega(z_0)| = 1$ for some z_0 with $y_0 < 0$, $f(z) \equiv e^{i\gamma}\omega(z)$ (real γ) and $\phi_u(z)$ certainly belongs to P, so we may assume that $|f(z)/\omega(z)| < 1$ for $y < 0$. In this case $\phi_u(z)$ has no zeros in the lower half plane if $|u| \geq 1$, and to show that $\phi_u(z) \in P$ we have to show that $h_\phi(\pi/2) \leq h_\phi(-\pi/2)$, where h_ϕ is the indicator of $\phi_u(z)$.

Suppose first that $|u| > 1$. Then

$$|f(z)/\omega(z) - u| \geq |u| - 1$$

for $y \leq 0$, so

$$|f(z) - u\omega(z)| = |\omega(z)|\,|f(z)/\omega(z) - u| \geq (|u| - 1)\,|\omega(z)|,$$

whence $h_\phi(\theta) \geq h_\omega(\theta)$ for $-\pi < \theta < 0$. Since $h_\omega(\theta) = \sigma$ for some θ in this range, and $\tau \leq \sigma$, we also have $h_\phi(\theta) = \sigma$ for some θ in $-\pi < \theta < 0$. On the other hand, since $\tau \leq \sigma$, $h_\phi(-\theta) \leq \sigma$ for $-\pi < \theta < 0$. Thus for some θ in $(-\pi, 0)$, $h_\phi(-\theta) \leq h_\phi(\theta)$. By 7.8.1 (restated for a lower half plane) this implies that $h_\phi(-\theta) \leq h_\phi(\theta)$ for all θ in $(-\pi, 0)$, so that $\phi_u(z)$ does belong to P.

If $|u| = 1$, we let $|u| \to 1$ through values such that $|u| > 1$ and use again the fact that the limit of functions of class P is of class P.

Notes for Chapter 7

7.1a. See §8.1.

7.2a. For subharmonic functions in general and this theorem in particular see Radó [1].

7.2b. The more general result for a subharmonic function with bounded boundary values is proved by Ahlfors and Heins [1] and by Lelong-Ferrand [1]; I follow the exposition given by Ahlfors and Heins. Lelong-Ferrand, here and in later papers, extends the theorem to n dimensions and generalizes it in other ways.

7.2c. For functions which are bounded on the boundary see the preceding note. With regular asymptotic behavior instead of effective convergence, Theorem 7.2.6 was given by Cartwright [4], [6] under condition (7.1.5), and by Levin [5]. Krein [9] establishes regular asymptotic behavior if $f(z)$ satisfies (7.1.5) and is the quotient of two regular bounded functions (beschränktartig). A necessary and sufficient condition for the latter property is (Wishard [1], Franck [1]) that

$$\int_{-\infty}^{\infty} \{x^2 + (1 + y)^2\}^{-1} \log^+ |f(z)| \, dx$$

is bounded ($0 < y < \infty$); when $f(z)$ is of exponential type, (7.1.5) implies this (6.7.11). Krein [2] shows conversely that if $f(z)$ is entire, and is the quotient of two regular bounded functions in both the upper and lower half planes, it is of exponential type and (7.1.5) is satisfied.

7.2d. With regular asymptotic behavior, this is proved by Pfluger [5] for entire functions, and stated by Levin [5]. By applying 7.2.3 we can deduce effective convergence from regular asymptotic behavior.

7.2e. Boas [30].

7.2f. Ganapathy Iyer [9], with condition (7.1.5).

7.3a. Boas [30].

7.4a. Heins [1], with boundedness on the real axis. A (different) deduction from 7.2.6 is given (in this special case) by Lelong-Ferrand [1].

7.5a. Fejér and Riesz [1].

7.5b. Ahiezer [4], Chebotarëv and Meiman [1], pp. 167 ff. In the second reference the theorem is generalized to functions of finite order in general. Condition (7.5.2) is equivalent to the condition that $\int_1^r x^{-2} \log |f(x)f(-x)| \, dx$ is bounded (6.3.14). Under more restrictive conditions the representation $f(x) = |F(x)|^2$ has been known for a long time (cf. S. Bernstein [2]), and has been repeatedly rediscovered.

7.6a. A result similar to the first part is given by S. Bernstein [13].

7.6b. A theorem of Cartwright [7] shows that the hypothesis on $f_1(z)$ can be replaced by the hypothesis that $h_1(0) = h_1(\pi) = 0$; neither hypothesis implies the other.

7.7a. Ahiezer [4], Chebotarëv and Meiman [1], p. 174, Levin [6].

7.8a. Levin [6].

7.8b. The terminology was introduced by Levin [6].

7.8c. Levin [6].

CHAPTER 8

FUNCTIONS OF EXPONENTIAL TYPE: CONNECTIONS BETWEEN GROWTH AND DISTRIBUTION OF ZEROS

8.1. Introduction. In Chapter 4 we saw that, for entire functions of order less than 1, with zeros on or close to a half line, there is a close connection between the asymptotic properties of the zeros and of the function itself. In this chapter we study in more detail the corresponding results for entire functions of order 1. The simplest case is that in which the function $f(z)$ is even, with real zeros; then $f(-z^{1/2})$ is an entire function of order $\frac{1}{2}$ with real negative zeros, and some facts about $f(z)$ can be read off from the corresponding properties of functions of order $\frac{1}{2}$. These facts suggest generalizations to other functions of order 1, but there are additional phenomena of importance which have no counterparts for functions of order $\frac{1}{2}$.

For entire functions of order less than 1 (and indeed for any nonintegral order) with real negative zeros, the conditions $n(r) \sim \lambda r^{\rho}$ and

$$\log | f(re^{i\theta}) | \sim \mu(\theta) r^{\rho}$$

are equivalent (with appropriate relations between λ and μ). For functions of order 1 we shall need a number of different asymptotic conditions, and we begin by collecting these, giving them abbreviations for convenience in reference. We shall need not only the condition

$$\text{(D)} \qquad \lim_{r\to\infty} r^{-1} n(r) = D,$$

that the zeros have a density, but also the condition that they have equal densities in the right and left half planes,

$$(\text{D}^{+}) \qquad \lim_{r\to\infty} r^{-1} n_{+}(r) = \lim_{r\to\infty} r^{-1} n_{-}(r) = D/2,$$

where $n_{+}(r)$ and $n_{-}(r)$ count, respectively, the zeros in $x > 0$ and $x < 0$. We shall also need the condition

$$\text{(S)} \qquad S(r) = \sum_{|z_n| \le r} 1/z_n \to S, \qquad r \to \infty.$$

This is suggested for consideration by the fact that $S(r)$ is bounded for entire functions of exponential type.

For functions of fractional order with real negative zeros the asymptotic behavior of the function on any ray is governed by its behavior on the positive real axis; we might therefore expect that the asymptotic behavior of an entire function of order 1 with real zeros (or zeros close to the real axis) would be determined by its behavior on the imaginary axis. This, however, is not always the case, and we shall introduce not only the condition

(A) $$\lim_{y \to +\infty} y^{-1} \log | f(iy) | = h(\pi/2),$$

but also conditions describing the asymptotic behavior in general. We have already had occasion to consider the condition that $r^{-1} \log | f(re^{i\theta}) |$ tends effectively (in the sense of §7.2) to $h(\theta)$ for $-\pi \le \theta \le \pi$. We may also consider the condition of regular asymptotic behavior (4.3.2), which requires that $r^{-1} \log | f(re^{i\theta}) | \to h(\theta)$ for each θ if r is excluded from a set of zero linear density, and is therefore a less restrictive hypothesis or a weaker conclusion. We shall refer to both of these as "Condition A*", with the understanding that (unless the contrary is specified) the weaker hypothesis or the stronger conclusion is to be asserted. Condition A does not usually imply A*, and A* (even in the stronger form) implies A only when the ray $\theta = \pi/2$ is in a zero-free angle.

By 6.3.14, if $f(z)$ is an entire function of exponential type the conditions

(C) $$\sum_{n=1}^{\infty} | \Im(1/z_n) | < \infty$$

and

(B) $$\int_1^R x^{-2} \log | f(x)f(-x) | \, dx \text{ bounded (or bounded above)}$$

are equivalent; C says that the zeros are close to the real axis, and B, that $| f(z) |$ is not too often either very large or very small on the real axis. We introduce the more refined condition (already used in Chapter 7)

(I) $$\int_1^{\to\infty} x^{-2} \log | f(x)f(-x) | \, dx \text{ converges;}$$

we recall that (6.3.6) the condition

$$\int_{-\infty}^{\infty} (1 + x^2)^{-1} \log^+ | f(x) | \, dx < \infty$$

implies this.

Finally, we let V stand for the statement that the indicator diagram of $f(z)$ is a vertical line segment, i.e.

(V) $$h(0) + h(\pi) = 0.$$

We can now summarize the principal results of this chapter rather compactly. Suppose that the zeros are either real or at least satisfy C. Then

I (which implies C) and D are equivalent, and A implies D; D implies A if the positive imaginary axis is in a zero-free angle, and implies "almost" A in any case. However, D alone does not imply A*, even if strengthened to D^+. To obtain A* we must impose an additional condition. If D (or I or A) holds, then V and S are equivalent; and if V (or S) holds in addition to D, then D^+ and A* hold. Conversely, A* implies I and V; in particular, then, (C + A*) is equivalent to (I + V). This shows incidentally that A* (with C) is possible only if $h(\theta)$ has the special form imposed by V.

The role of S may be seen quite clearly from the following condition which is implied by D^+ (with C). Let $S(r) = S'(r) + iS''(r)$. Then there are constants a and b such that

$$(\mathrm{A}^{**}) \quad r^{-1} \log |f(re^{i\theta})| = a \cos\theta + b \sin\theta + S'(r)\cos\theta - S''(r)\sin\theta + \pi D|\sin\theta|/2 + \epsilon(r,\theta),$$

where $\epsilon(r,\theta)$ tends effectively (§7.2) to zero. It is clear from this why condition A can be expected to occur even when A* does not, since A** reduces (almost) to A for $\theta = \pi/2$ irrespective of what $S'(r)$ does ($S''(r)$ always approaches a limit when C holds). On the other hand, A** prevents A* from holding unless S holds; and if S does hold, A** implies V. If D holds but S does not, we can produce extreme irregularity in the behavior of $r^{-1} \log |f(re^{i\theta})|$, even in a zero-free angle, by making $S(r)$ oscillate.

The most striking of these results, and the one with the most interesting applications, is that conditions I and V, which appear to bear only on the growth of the function, imply D^+; we can replace I and V by the condition $\int_{-\infty}^{\infty} (1 + x^2)^{-1} \log^+ |f(x)|\, dx < \infty$. This is satisfied for any function which is bounded on the real axis, and so for example for functions of the form[a] $f(z) = \int_{-c}^{c} e^{izt}\phi(t)\, dt$. The theory developed in this chapter is, in fact, chiefly motivated by a desire to infer the distribution of the zeros from more accessible properties of the function.

If we do not even require that C is true, the behavior of the function is no longer governed by its behavior on the real axis; however, there are still connections between regularity of asymptotic behavior and regularity of the distribution of the zeros.[b] We may first ask for conditions under which A* holds; a necessary and sufficient condition is that S holds and the zeros are measurable (according to definition 4.3.3), i.e. the zeros have a density in each fixed sector, or more precisely

$$n(r, \phi_1, \phi_2) = \{N(\phi_1) - N(\phi_2)\}r + o(r)$$

for every two points of continuity ϕ_1, ϕ_2 of a monotone nondecreasing function $N(\phi)$, where $n(r, \phi_1, \phi_2)$ is the number of zeros of modulus not exceeding r in $\phi_1 \leq \arg z < \phi_2$. If we drop S but retain the measurability of the zeros, we have the following generalization of A**:

$$r^{-1} \log f(re^{i\theta}) = q(\theta) + \{\bar{C}(\pi i - 1) + \overline{q(0)}\}e^{i\theta} + \epsilon(r, \theta),$$

where

$$q(\theta) = -i \int_0^{2\pi} \phi e^{-i\phi}\, dN(\theta + \phi), \qquad C = \int_0^{2\pi} e^{i\phi}\, dN(\phi),$$

$\Re\epsilon(r, \theta) \to 0$ on a set of unit linear density,

$$| \Im\epsilon(r,\theta) | < \pi\{N(\theta^+) - N(\theta^-)\} + o(1),$$

and $\epsilon(r, \theta) \to 0$ in any zero-free angle.

8.2. Entire functions with real zeros. In this section we prove the equivalence of conditions I, D and A for entire functions $f(z)$ of order 1 with real zeros (even if they are not of finite type). Since these conditions are unaffected by the value of $f(0)$, we shall suppose that $f(0) = 1$. Conditions I and D are unaffected if $f(z)$ is multiplied by $e^{\alpha z}$, and this changes A only by changing the value of $h(\pi/2)$. We may therefore suppose that $f(z)$ is a canonical product with real zeros. The theorem to be proved is now as follows.

8.2.1. *Theorem.*[a] *If $f(z)$ is a canonical product of order* 1 *with real zeros, the three conditions*

(8.2.2) $$\lim_{R \to \infty} \int_{-R}^{R} x^{-2} \log | f(x) |\, dx = -\pi^2 B,$$

(8.2.3) $$\lim_{r \to \infty} r^{-1} n(r) = 2B,$$

(8.2.4) $$\lim_{y \to \pm\infty} | y |^{-1} \log | f(iy) | = \pi B$$

are equivalent.

In (8.2.2) the integral is a principal value at 0, existing because $f(0) = 1$; then (8.2.2) is equivalent to condition I. Condition (8.2.3) is D and (8.2.4) is A (since $| f(iy) | = | f(-iy) |$).

It is enough to prove the theorem when $f(z)$ is even and so can be written in the form

(8.2.5) $$f(z) = \prod (1 - z^2/\lambda_n^2);$$

for, if $f(z)f(-z)$ satisfies one of (8.2.2), (8.2.3), (8.2.4), $f(z)$ satisfies the corresponding condition with $B/2$ instead of B.

We assume then that $f(z)$ has the form (8.2.5). The sums $S(r)$ for such a function vanish identically and so $f(z)$ is of exponential type by Lindelöf's theorem (2.10.1) (even though the original $f(z)$ in 8.2.1 need not be of exponential type). The regularity of behavior expressed by 8.2.1 may now be traced to the fact that a function (8.2.5) behaves like a function of order ½, since if

$$g(z) = \prod_{n=1}^{\infty} (1 + z/\lambda_n^2) \tag{8.2.6}$$

we have $g(-z^2) = f(z)$. Now $g(z)$ is a function of order ½ with real negative zeros, and from Chapter 4 we know that

$$n_g(r) \sim Br^{1/2} \tag{8.2.7}$$

is equivalent to

$$\log g(r) \sim \pi Br^{1/2} \tag{8.2.8}$$

Condition (8.2.7) is the same as $n_g(r) \sim 2Br$ (since $-\lambda_n$ is a zero of the even function $f(z)$ when λ_n is a zero); and (8.2.8) is

$$\log f(ir^{1/2}) \sim \pi Br^{1/2},$$

or

$$\log f(iy) \sim \pi B \mid y \mid.$$

Thus we have the equivalence of (8.2.3) and (8.2.4).

Next, we have from 4.4.1 that

$$\int_0^\infty x^{-3/2} \log \mid g(-x) \mid dx = -\pi \lim_{x\to\infty} x^{-1/2} \log g(x),$$

which is the same as

$$2 \int_0^\infty t^{-2} \log \mid f(t) \mid dt = -\pi \lim_{y\to\infty} y^{-1} \log \mid f(iy) \mid.$$

Thus (8.2.2) and (8.2.4) are equivalent.

We note also that the results of Chapter 4 show that more than 8.2.1 is true for even functions: I, D or A implies A* (except perhaps for the directions 0, π).

8.2.9. *Theorem. If $f(z)$ is an even entire function of exponential type, given by* (8.2.5), *and* (8.2.2) *or* (8.2.3) *or* (8.2.4) *is satisfied, then*

$$\lim_{r\to\infty} r^{-1} \log \mid f(re^{i\theta}) \mid = \pi B \mid \sin \theta \mid, \qquad \theta \neq 0, \pi. \tag{8.2.10}$$

In particular, the indicator diagram of $f(z)$ is a segment of the imaginary axis.

In fact, if $g(z)$ is defined again by (8.2.6), we have from 4.1.1 that (8.2.7) implies

$$\log | g(re^{i\theta}) | \sim \pi B r^{1/2} \cos \theta/2, \qquad -\pi < \theta < \pi,$$

and (8.2.10) follows.

8.3 Entire functions with real zeros, continued. We now prove that, for an entire function with real zeros, S and V are equivalent if I, D or A is satisfied, and imply D^+ and A*. We can no longer reduce the problem to one for even functions, since S is trivial for even functions, while for even functions D implies D^+ trivially and implies V by 8.2.9.

We can simplify the formulas slightly by assuming again that $f(z)$ is a canonical product, since a factor $e^{\alpha z}$ does not affect any of our conditions except the form of $k(\theta)$ in A and A*.

8.3.1. *Theorem.*[a] *If $f(z)$ is a canonical product of exponential type, with real zeros, and satisfies one of the equivalent conditions* I, D, A (8.2.2), (8.2.3), (8.2.4), *then* (V) *the indicator diagram of $f(z)$ is a vertical line segment if and only if*

(8.3.2)(S) $$S(r) = \sum_{|z_n|<r} 1/z_n \text{ converges,}$$

and (8.3.2) *implies that* (D^+) *the zeros have equal densities in the right-hand and left-hand half planes. Moreover, if* I, D *or* A *is satisfied, and* D^+ *is also satisfied, then for each* θ, $0 < | \theta | < \pi$,

(8.3.3) $$r^{-1} \log | f(re^{i\theta}) | = S(r) \cos \theta + \pi B | \sin \theta | + o(1).$$

Conclusion (8.3.3) shows in particular that I, D or A, together with S or V, implies A*. That (I + V) implies A* was shown (in a more general setting) in 7.2.9. It is also true that A* and D (or I) implies V (and S); cf. the discussion at the end of §8.4. A somewhat simpler proof can be given when the zeros are real.

Under (8.3.3) the indicator diagram is a rectangle with sides parallel to the axes.

8.3.4. *Corollary.*[b] *If $f(z)$ is a canonical product of exponential type with real zeros, and* $n(r) = o(r)$, *then for* $0 < | \theta | < \pi$,

$$\lim_{r \to \infty} \{ r^{-1} \log | f(re^{i\theta}) | - S(r) \cos \theta \} = 0.$$

Here D and D^+ are satisfied (both densities being zero).

The proof of 8.3.1 will be broken into three steps.

First,

8.3.5.[c] D *and* V *imply* S.

Next,

8.3.6.[d] *When the zeros are real,* D *and* S *imply* D^+.

Finally,

8.3.7.[e] *When the zeros are real,* D *and* D^+ *imply* (8.3.3), *so that in particular* D *and* S *imply* V.

Proof of 8.3.5. If we write Carleman's formula for the right and left hand half planes and subtract one from the other we obtain

$$\Re f'(0) + \sum_{r_n < r} \frac{\cos \theta_n}{r_n} \left(1 - \frac{r_n^2}{r^2}\right) = \frac{1}{\pi r} \int_{-\pi}^{\pi} \log | f(re^{i\theta}) | \cos \theta \, d\theta. \tag{8.3.8}$$

Now condition D, by Jensen's formula, implies

$$\lim_{r \to \infty} (2\pi r)^{-1} \int_{-\pi}^{\pi} \log | f(re^{i\theta}) | \, d\theta = 2B, \tag{8.3.9}$$

and since $h(\theta) = a \cos \theta + b \sin \theta + \pi B \mid \sin \theta \mid$ when the indicator diagram is a vertical line segment of length $2\pi B$, V implies

$$(2\pi)^{-1} \int_{-\pi}^{\pi} h(\theta) \, d\theta = 2B,$$

so that we have, for any positive ϵ and $r > r_0(\epsilon)$,

$$-2\epsilon < \int_{-\pi}^{\pi} \{h(\theta) + \epsilon - r^{-1} \log | f(re^{i\theta}) | \} \, d\theta < 2\epsilon.$$

By property (5.4.14) of $h(\theta)$, the integrand is positive when r_0 is sufficiently large, uniformly in θ, and hence

$$-2\epsilon < \int_{-\pi}^{\pi} \{h(\theta) + \epsilon - r^{-1} \log | f(re^{i\theta}) | \} \cos \theta \, d\theta < 2\epsilon.$$

Therefore

$$\lim_{r \to \infty} r^{-1} \int_{-\pi}^{\pi} \log | f(re^{i\theta}) | \cos \theta \, d\theta = \int_{-\pi}^{\pi} h(\theta) \cos \theta \, d\theta = \pi H,$$

say. Hence (8.3.8) yields

$$\Re f'(0) + \lim_{r \to \infty} \sum_{r_n < r} \frac{\cos \theta_n}{r_n} \left(1 - \frac{r_n^2}{r^2}\right) = H.$$

Now let r take the values $m/(2B)$, where m is an integer, so that $n(r) - m = o(m)$, and

$$\lim_{m \to \infty} \sum_{n=1}^{m} \frac{\cos \theta_n}{r_n} \left(1 - \frac{r_n^2 \cdot 4B^2}{m^2}\right) = H - \Re f'(0).$$

If σ_m denotes the sum on the left,

$$(m + 1)^2\sigma_{m+1} - m^2\sigma_m = (2m + 1) \sum_{n<m} \frac{\cos \theta_n}{r_n}$$

$$+ \frac{\cos \theta_m}{r_m} \{(m + 1)^2 - 4B^2r_m^2\},$$

so that

$$\sum_{n<m} \frac{\cos \theta_n}{r_n} = \frac{(m + 1)^2\sigma_{m+1} - m^2\sigma_m}{2m + 1} + o(1),$$

$$p^{-1} \sum_{m=1}^{p} \sum_{n=1}^{m} \frac{\cos \theta_n}{r_n} = p^{-1} \sum_{m=1}^{p} \frac{(m + 1)^2\sigma_{m+1} - m^2\sigma_m}{2m + 1} + o(1)$$

$$= p^{-1} \sum_{m=1}^{p} \frac{2m^2\sigma_m}{4m^2 - 1} + \frac{(p + 1)^2}{p(2p + 1)} \sigma_{p+1} + o(1).$$

As $p \to \infty$, this tends to the same limit as σ_p, which means that

$$\sum r_n^{-1} \cos \theta_n$$

is summable $(C,1)$ to $H - \Re f'(0)$. Since the nth term is $O(1/n)$, this series converges to the same limit. Thus the real part of $S(r)$ converges.

By proceeding similarly with Carleman's formula for the upper and lower half planes (or considering $f(iz)$) we see that the imaginary part of $S(r)$ also converges. We have therefore proved 8.3.5.

This proof that S holds actually uses only the fact that

$$r^{-1} \int_{-\pi}^{\pi} \log | f(re^{i\theta}) | \, d\theta \to \int_{-\pi}^{\pi} h(\theta) \, d\theta,$$

and so is true whenever

$$2\pi r^{-1} n(r) \to \int_{-\pi}^{\pi} h(\theta) \, d\theta. \qquad (8.3.10)$$

This is, however, a much more restrictive condition than it looks. In fact, the following theorem can be proved.

8.3.11.* *Theorem.*[f] *An entire function of exponential type satisfies* (8.3.10) *if and only if it has regular asymptotic behavior.*

Proof of 8.3.6. This is a simple result about sequences of real numbers which has nothing in particular to do with entire functions. We assume that $\sum_{r_n<r} z_n^{-1}$ converges, with real z_n; we may write this as $\sum_{n=1}^{\infty} \epsilon_n/r_n = S$, where $\epsilon_n = \pm 1$. We have $r_n \sim n/(2B)$ and we have to show that $\sum_{n=1}^{m} \epsilon_n = o(m)$. Now since $r_n \to \infty$ and $\sum \epsilon_n/r_n$ converges, by a theorem

of Kronecker[g] (whose proof is straightforward)

$$r_n^{-1} \sum_{k=1}^{n} (\epsilon_k/r_k)\, r_k \to 0,$$

which means $\sum_{k=1}^{n} \epsilon_k = o(r_n) = o(n)$.

Proof of 8.3.7. Consider $F(z) = f(z)f(-z)$, $\Phi(z) = f(z)/f(-z)$; then $f(z)^2 = F(z)\Phi(z)$. The function $F(z)$ has the form (8.2.5), and by 4.1.1 applied to $g(z)$ of (8.2.6), we have

(8.3.12) $$\lim_{r\to\infty} r^{-1} \log |F(re^{i\theta})| = 2\pi B |\sin\theta| .$$

We can write $\Phi(z)$ in the form $\Phi(z) = \phi(z)/\phi(-z)$, where $\phi(z)$ is the canonical product formed with the zeros and poles of $\Phi(z)$ which are on the positive real axis, that is, it is of the form

$$\phi(z) = \prod_{n=1}^{\infty} \frac{(1 - z/\lambda_n)\exp(z/\lambda_n)}{(1 - z/\mu_n)\exp(z/\mu_n)},$$

where λ_n are the positive zeros and $-\mu_n$ are the negative zeros of $f(z)$ (duplicates cancelling). If we put $\Lambda(t) = n_+(t) - n_-(t)$, we can write

$$\log\phi(z) = \int_0^\infty \left\{\log\left(1 - \frac{z}{t}\right) + \frac{z}{t}\right\} d\Lambda(t)$$

$$= -\int_0^\infty \frac{z^2\Lambda(t)}{t^2(t-z)}\, dt.$$

Now we have $\Lambda(t) = o(t)$ and so there is a monotonic nonincreasing function $\eta(r)$ such that $\eta(r) \to 0$ and $\Lambda(r) < r\eta(r)$, $r \geq 0$ (for instance, $\eta(r) = \sup\{t^{-1}\Lambda(t)\}, t \geq r$). We have

$$S(r) = \int_0^r t^{-1} d\Lambda(t) = r^{-1}\Lambda(r) + \int_0^r t^{-2}\Lambda(t)\, dt$$

$$= r^{-1}\Lambda(r) + U(r).$$

Then $U(r)$ is bounded, $S(r) - U(r) \leq \eta(r)$, and

$$|U(r_2) - U(r_1)| = \left|\int_{r_1}^{r_2} t^{-1}\Lambda(t)t^{-1}\, dt\right| \leq \eta(r)\log k$$

if $k > 1$, $r \leq r_1 \leq r_2 \leq kr$.

Now write

$$\int_0^\infty \frac{z^2\Lambda(t)}{t^2(t-z)}\, dt = \int_0^{2r} + \int_{2r}^{kr} + \int_{kr}^\infty = J_1 + J_2 + J_3, \qquad k > 2.$$

We have

$$J_3 = U(kr)\,\frac{z^2}{kr - z} + \int_{kr}^{\infty} U(t)\,\frac{z^2 dt}{(t - z)^2},$$

so that $|J_3| \le Ck^{-1}r$ with some constant C.

Let $R(t)$ and $J(t)$ denote the real and imaginary parts of $(t - z)^{-1}$; we need the properties that $R(t)$ and $J(t)$ are positive and decreasing for $2r \le t < \infty$, and that $R(2r) \le 3/r$, $J(2r) \le 1/r$. Then

$$\begin{aligned} J_2 &= z^2 \int_{2r}^{kr} t^{-2}\Lambda(t)\{R(t) + iJ(t)\}\,dt \\ &= z^2\left\{R(2r)\int_{2r}^{r_1} t^{-2}\Lambda(t)dt + iJ(2r)\int_{2r}^{r_2} t^{-2}\Lambda(t)dt\right\}, \end{aligned}$$

$$2r < r_i < kr,$$

by the second mean value theorem. Hence, because $\Lambda(r) < r\eta(r)$,

$$|J_2| \le 4r\eta(r)\log k.$$

Write J_1 in the form

$$\begin{aligned} J_1 &= \int_0^{r/k} \frac{z\Lambda(t)\,dt}{t(t - z)} + \int_{r/k}^{2r} \frac{z\Lambda(t)}{t(t - z)}\,dt - z\int_0^{2r} t^{-2}\Lambda(t)\,dt \\ &= J_4 + J_5 - zU(2r). \end{aligned}$$

Now $|J_4| \le -\log(1 - k^{-1})\eta(0)r < \eta(0)r/k$, and, for $|\arg z| \ge \alpha > 0$,

$$|t - z| \ge \sin\alpha\,(t + r)/2,$$

so that

$$|J_5| \le \frac{\log 3}{\sin\alpha/2}\,\eta(r/k)r < K\alpha^{-1}\eta(r/k)r.$$

Combining inequalities, we have

$$\begin{aligned} &|\log\phi(z) - zS(r)| \\ &\qquad \le r\{(C + \eta(0))k^{-1} + K\alpha^{-1}\eta(r/k) + (5\log k + 1)\eta(r)\}. \end{aligned}$$

Now let k be a function $k(r)$ such that $k(r) \to \infty$, $r/k(r) \to \infty$, and $\eta(r)\log k(r) \to 0$. Thus we have a function $\eta_1(r)$ such that $\eta_1(r) \to 0$ and

$$|\log\phi(z) - z\,S(r)| \le r\eta_1(r)$$

for $\alpha \le |\arg z| \le \pi - \alpha$, and hence 8.3.7 follows.

8.4. Entire functions with zeros close to the real axis. The greatest interest attaches to the case where the zeros are not assumed to be real, since a function arising in applications is likely to be known to satisfy, say, condition I, although it may be difficult to decide whether its zeros are all real. We shall assume condition C, as representing a reasonable requirement of closeness to the real axis, noting that it is equivalent to B, which appears to bear only on the growth of the function. We first carry over as much as possible of Theorem 8.2.1. In that theorem we proved that when the zeros are real, conditions A, D and I are equivalent. Here we shall prove that D (with C) and I (which implies C) are equivalent in any case, while A (with C) implies D (and hence I). Conversely, D (with C) implies as much in the direction of A as it can, namely that $y^{-1} \log | f(iy) |$ approaches a limit if a y-set of finite logarithmic length is omitted, and without exception if the positive imaginary axis is in a zero-free angle.

8.4.1. *Theorem.*[a] *Let*

$$f(z) = e^{cz} \prod_{n=1}^{\infty} (1 - z/z_n) \exp(z/z_n) \tag{8.4.2}$$

be an entire function of exponential type, with $f(0) = 1$. *Then the conditions*

$$\lim_{r\to\infty} r^{-1}n(r) = D, \qquad \sum_{n=1}^{\infty} r_n^{-1} | \sin \theta_n | = \pi C < \infty \tag{8.4.3}$$

and

$$\lim_{r\to\infty} \int_{-r}^{r} x^{-2} \log | f(x) | \, dx = -\pi^2 I \neq \pm\infty \tag{8.4.4}$$

are equivalent (that is, D + C *is equivalent to* I), *and* $D = 2C + 2I$.

We prove 8.4.1 by comparing $f(z)$ with another function which has real zeros. Let

$$F(z) = \prod_{n=1}^{\infty} (1 - z^2/r_n^2);$$

since

$$f(z) f(-z) = \prod_{n=1}^{\infty} (1 - z^2/z_n^2),$$

we have

$$\log \left| \frac{f(x) f(-x)}{F(x)} \right| = \sum_{n=1}^{\infty} \log \left| \frac{z_n^2 - x^2}{r_n^2 - x^2} \right|.$$

The right-hand side is uniformly convergent except in neighborhoods of

$x = r_n$, and dominatedly convergent there, so

$$(8.4.5)\quad 2\int_{-R}^{R} x^{-2}\log|f(x)|\,dx = \int_{-R}^{R} x^{-2}\log|F(x)|\,dx$$
$$+ \sum_{n=1}^{\infty}\int_{-R}^{R} x^{-2}\log\left|\frac{z_n^2 - x^2}{r_n^2 - x^2}\right| dx.$$

We now show that, as $R \to \infty$, the limit (finite or infinite) of the sum on the right is

$$(8.4.6)\qquad 2\pi\sum_{n=1}^{\infty} r_n^{-1}|\sin\theta_n|.$$

To do this, take a fixed n for which z_n is not real, take $R > r_n$, put

$$\phi(z) = \log(z_n^2 - z^2)/(r_n^2 - z^2),$$

and integrate $z^{-2}\phi(z)$ around the contour consisting of the semicircle $|z| = R$ from R to $-R$ and the real axis from $-R$ to R, with indentations at $\pm r_n$ and 0. Then $\phi(z)$ increases by $2\pi i$ as we traverse the contour starting at $z = R$, and by integration by parts

$$\int z^{-2}\phi(z)\,dz = -2\pi i/R + \int \phi'(z)z^{-1}\,dz$$
$$= -2\pi i/R + \frac{2\pi i}{r_n(\pm\cos\theta_n + i|\sin\theta_n|)}.$$

As $R \to \infty$, the integral along the semicircle approaches zero, so we have

$$(8.4.7)\qquad \int_{-\infty}^{\infty} x^{-2}\log\left|\frac{z_n^2 - x^2}{r_n^2 - x^2}\right| dx = 2\pi r_n^{-1}|\sin\theta_n|.$$

Since the integrand in the sum on the right of (8.4.5) is nonnegative, the terms of the series are nonnegative and increase as R increases, approaching the limit $2\pi r_n^{-1}|\sin\theta_n|$. Hence the sum in (8.4.5) tends to the limit (8.4.6).

Now $F(z)$ has real zeros and by 8.2.1, since $F(z)$ has twice as many zeros as $f(z)$,

$$(8.4.8)\qquad \lim_{r\to\infty} n(r)/r = -\pi^{-2}\lim_{r\to\infty}\int_{-r}^{r} x^{-2}\log|F(x)|\,dx,$$

if either limit exists. Suppose first that (8.4.3) holds; then (8.4.6) is finite; the limit on the left of (8.4.8) exists, so that the limit on the right exists and hence the limit of the left-hand side of (8.4.5) exists and has the stated value. Suppose next that (8.4.4) holds; then (8.4.6) is finite by 6.3.9 and so the first term on the right in (8.4.5) approaches a finite limit. Then by

(8.4.8), $r^{-1}n(r)$ approaches a finite limit and (8.4.3) is satisfied. This completes the proof of 8.4.1.

To complete the extension of 8.2.1 to the case when reality of the zeros is replaced by condition C, we prove the following theorem.

8.4.9. *Theorem.*[b] *Let*

$$f(z) = e^{cz} \prod_{n=1}^{\infty} (1 - z/z_n) \exp (z/z_n) \tag{8.4.10}$$

be an entire function of exponential type with $f(0) = 1$. *If*

$$\sum_{n=1}^{\infty} r_n^{-1} \mid \sin \theta_n \mid < \infty \tag{8.4.11}$$

then

$$\lim_{r\to\infty} r^{-1}n(r) \text{ exists} \tag{8.4.12}$$

if and only if

$$\lim_{y\to\infty} y^{-1} \log \mid f(iy) \mid \text{ exists if a set of finite logarithmic length is excluded;} \tag{8.4.13}$$

and when (8.4.12) *holds, the exceptional set in* (8.4.13) *is empty if there is an angle* $\mid \arg z - \pi/2 \mid < \epsilon$, $\epsilon > 0$, *containing no* z_n.

This means in particular that conditions A and C together imply D, while D and C together imply A as nearly as they can.

Since (8.4.11) holds, we can write $f(z)B(z) = g(z)$, where $B(z)$ is the Blaschke product (6.4.7) with poles at the zeros z_n of $f(z)$ which are in the upper half plane, so that $g(z)$ is an entire function of exponential type with no zeros in the upper half plane and with $\mid g(x) \mid = \mid f(x) \mid$ for real x. By 7.2.3, $B(z)$ satisfies (8.4.13) (and the sharper statement made at the end of the theorem), so that it is enough to prove 8.4.9 for $g(z)$. For, on the one hand, if (8.4.12) holds for $f(z)$, then $f(z)$ satisfies I, so does $g(z)$ (because $\mid g(x) \mid = \mid f(x) \mid$), and hence $g(z)$ satisfies (8.4.12). On the other hand, if (8.4.13) holds for $f(z)$, it holds for $g(z)$. Therefore we may suppose from now on that $f(z)$ has $\Im(z_n) \leq 0$. Because of (8.4.11) we can write $f(z)$ in the form

$$f(z) = e^{bz} \prod_{n=1}^{\infty} (1 - z/z_n) \exp \{z\Re(1/z_n)\},$$

and then

$$\mid f(iy) \mid^2 = \exp \{-2\Im(b)y\} \prod_{n=1}^{\infty} \mid 1 - iy/z_n \mid^2.$$

We shall compare $f(z)$ with the function

$$F(z) = \prod_{n=1}^{\infty} (1 - z^2/r_n^2) \exp \{2i\Im(b)z\},$$

which has real zeros. We have

$$\log \{ | f(iy) |^2/F(iy)\} = \sum_{n=1}^{\infty} \log \left\{1 - \frac{2yr_n \sin \theta_n}{r_n^2 + y^2}\right\}, \tag{8.4.14}$$

and $\sin \theta_n \leq 0$, so that

$$0 \leq \log \{ | f(iy) |^2/F(iy)\} \leq y \sum_{n=1}^{\infty} \frac{r_n | \sin \theta_n |}{r_n^2 + y^2} = o(y), \tag{8.4.15}$$

because $\sum | \sin \theta_n | /r_n$ converges.

Now $n(r)$ is the same for $f(z)^2$ as for $F(z)$, and so (8.4.12) implies that $F(z)$ satisfies (8.4.13) (actually, in the stronger form); therefore by (8.4.15), $f(z)$ satisfies (8.4.13) also. On the other hand, if $f(z)$ satisfies (8.4.13) so does $F(z)$, and then (8.4.12) follows (since $F(z)$ has real zeros) provided we show that the exceptional set in (8.4.13) is empty for $F(z)$. This follows from 7.4.1, according to which $r^{-1} \log M_F(r)$ approaches a limit, since $M_F(r) = F(ir)$.

Theorem 8.3.1 can also be carried over to the more general case when $f(z)$ satisfies C instead of having all its zeros real. This requires two things: that C, D and S imply D$^+$, and that C, D and D$^+$ imply A** and, in particular, the equivalence of S and V. The first part is the more important in applications, and we formulate it separately.

8.4.16. *Theorem.*[c] *If $f(z)$ is an entire function of exponential type k such that*

$$\lim_{r\to\infty} \int_{-r}^{r} x^{-2} \log | f(x) | \, dx$$

exists and is finite, and if $h(0) + h(\pi) \leq 0$ then the zeros of $f(z)$ have density $D \leq k/\pi$ in the right-hand and left-hand half planes.

In other words, I and V together imply D$^+$. That $D \leq k/\pi$ follows from Jensen's theorem combined with the fact that $h(\theta) \leq k | \sin \theta | + c \cos \theta$.

It is desirable to notice just which of our previous results are required in proving 8.4.16. We use 8.4.1, which depends on 8.2.1, to prove that I implies C and D. Then we use 8.3.5 to show that D and V imply S. Finally, we have to prove, much as in 8.3.6, that C, D and S imply D$^+$.

For the last step, we have $\sum r_n^{-1} \exp (i\theta_n)$ and $\sum | \sin \theta_n | /r_n$ convergent, so as in the proof of 8.3.6,

$$\sum_{r_n \leq r} \cos \theta_n = o(r), \tag{8.4.17}$$

$$\sum_{r_n \leq r} | \sin \theta_n | = o(r); \tag{8.4.18}$$

we have to prove

$$\sum_{r_n \le r} \epsilon_n = o(r), \qquad \epsilon_n = \operatorname{sgn} \cos \theta_n . \tag{8.4.19}$$

(The zeros, if any, on the imaginary axis can be neglected since by (8.4.18) they have zero density.) By (8.4.17),

$$\sum_{r_n \le r} \epsilon_n = \sum_{r_n \le r} (\epsilon_n - \cos \theta_n) + o(r),$$

and

$$\epsilon_n - \cos \theta_n = 2 \sin^2 \tfrac{1}{2} \theta_n, \qquad \epsilon_n = +1;$$

$$\epsilon_n - \cos \theta_n = -2 \sin^2 \tfrac{1}{2} (\pi - \theta_n), \qquad \epsilon_n = -1.$$

Therefore $| \epsilon_n - \cos \theta_n | \le 2 | \sin \theta_n |$, and by (8.4.18) it follows that

$$\sum_{r_n \le r} | \epsilon_n - \cos \theta_n | = o(r),$$

which implies (8.4.19).

That C, D and D^+ imply A** is part of the content of the following theorem.

8.4.20. *Theorem.*[d] *If $f(z)$ is an entire function of exponential type whose zeros z_n have a density and satisfy*

$$\sum | \Im(1/z_n) | < \infty, \tag{8.4.21}$$

the indicator diagram of $f(z)$ is a vertical line segment if and only if $\sum 1/z_n$ converges. If in addition the zeros of $f(z)$ have equal densities in the right and left half planes, then if

$$S(r) = S'(r) + iS''(r) = \sum_{r_n \le r} 1/z_n ,$$

$$\begin{aligned} r^{-1} \log | f(re^{i\theta}) | &= a \cos \theta + b \sin \theta \\ &\quad + S'(r) \cos \theta - S''(r) \sin \theta + \pi A | \sin \theta | + \epsilon(r, \theta), \end{aligned} \tag{8.4.22}$$

where $\epsilon(r, \theta) \to 0$ as $r \to \infty$ in any zero-free angle, and on a set of r's of linear density 1 for every θ.

In fact, the conclusion can be strengthened to correspond to the strong form of A*: $\epsilon(r, \theta) \to 0$ except for a set of θ's of outer logarithmic capacity 0, and for each θ except for a set of r's of finite logarithmic length. The argument is the same as that used for the second part of 8.4.9.

In §8.1 we remarked that conditions A* and C are together equivalent to I and V. We can now justify this statement. That (I + V) implies (A* + C) was shown in Chapter 7. For the converse, by 8.3.11, and the remark made just before it, A* implies D and S; by 8.4.20, D, C and S imply V; and by 8.4.1, D and C imply I.

Theorem 8.4.20 is proved by comparing $f(z)$ with a function having only real zeros, along lines similar to those used for 8.3.7. The main part of the proof consists in proving the following lemma.

8.4.23. *Lemma. Let $f(z)$ be a canonical product of exponential type whose zeros z_n have a density. Put $\lambda_n = r_n$ for $-\pi/2 < \theta_n \leq \pi/2$, $\lambda_n = -r_n$ for $\pi/2 < \theta_n \leq 3\pi/2$,*

$$S(r) = \sum_{r_n \leq r} 1/z_n , \qquad S_0(r) = \sum_{|\lambda_n| \leq r} 1/\lambda_n ,$$

$$g(z) = \prod_{n=1}^{\infty} (1 - z/\lambda_n) \exp (z/\lambda_n).$$

For each positive α there are functions $\epsilon(r)$ and $\sigma(r)$ (depending on α), which are $o(1)$ as $r \to \infty$ and are such that

$$\log | f(z)/g(z) | - \Re[z\{S(2R) - S_0(2R)\}] < R\epsilon(R), \tag{8.4.24}$$

$$| z | \leq R, \ \alpha \leq | \arg z | \leq \pi - \alpha,$$

and

$$\log | f(z)/g(z) | - \Re[z\{S(2R) - S_0(2R)\}] > - R\epsilon(R) \tag{8.4.25}$$

in the same domain with the exception of a set of circles of total length less than $R\sigma(R)$.

First we show that the lemma implies 8.4.20. Since

$$| z_n^{-1} - \lambda_n^{-1} | = \left| \frac{1 - | \cos \theta_n | - i \sin \theta_n}{r_n} \right| \leq 3r_n^{-1} | \sin \theta_n |, \tag{8.4.26}$$

and (8.4.21) holds, the sums $S_0(r)$ are bounded if $S(r)$ is; so $g(z)$ is of exponential type. If $S(r) = S'(r) + iS''(r)$, (8.4.21) shows that $S''(r)$ approaches a limit, so $S(r)$ approaches a limit if and only if $S'(r)$ does, and by (8.4.26) $S'(r)$ approaches a limit if and only if $S_0(r)$ does. That is, $S(r)$ and $S_0(r)$ both approach limits or both fail to do so. Hence the first part of 8.4.20 follows from the first part of 8.3.1.

Next suppose that the zeros have equal densities in the right-hand and left-hand half planes (D^+). From the proof of 8.3.7 we see that $S_0(2R)$ and $S_0(r)$ differ by $o(1)$ if $R/k \leq r \leq R$ and $k \to \infty$ sufficiently slowly. The same is true for $S(2R)$ and $S(r)$, as we may see as follows:

$$\sum_{r \leq r_n \leq ar} z_n^{-1} = \sum r_n^{-1}(\cos \theta_n - i \sin \theta_n) = \sum r_n^{-1} \cos \theta_n + o(1).$$

By (D^+), $\sum_{r_n \le r}$ sgn $\cos \theta_n = o(r)$; as in the proof of (8.4.19),

$$\left| \sum_{r_n \le r} \cos \theta_n \right| \le \left| \sum \operatorname{sgn} \cos \theta_n \right| + 2 \left| \sum \sin \theta_n \right| = o(r),$$

$$\sum_{r \le r_n \le ar} r_n^{-1} \cos \theta_n \le r^{-1} o(ar) = o(1).$$

Hence the rest of 8.4.20 follows from the last part of 8.3.1.

We now prove 8.4.23. Take $\delta > 0$ and let W'' denote the part $\delta < |\arg z| < \pi - \delta$ of the plane, W' the rest; let z_n' and z_n'' be the zeros in W', W''; and let $n'(r)$, $n''(r)$ determine the numbers of z_n', z_n'' in the usual way. Since $\sum r_n^{-1} |\sin \theta_n|$ converges and $|\sin \theta_n''| \ge \sin \delta$, $\sum 1/r_n''$ converges and so $n/r_n'' \to 0$. Hence there is a decreasing function $\eta(r, \delta)$ such that $\eta = o(1)$ as $r \to \infty$ for each δ and $n''(r) \le \eta(r,\delta) r$.

Let $r \le R$, and put

$$\log (f/g) = z\{S(2R) - S_0(2R)\} + \log (P''/Q'') + \log (P'/Q') + \log H,$$

where

$$\log H(z) = \sum_{r_n \le 2R} \left\{ \log\left(1 - \frac{z}{z_n}\right) + \frac{z}{z_n} \right\} - \sum_{r_n > 2r} \left\{ \log\left(1 - \frac{z}{\lambda_n}\right) + \frac{z}{\lambda_n} \right\},$$

$$P'(z) = \prod_{r_n' \le 2R} (1 - z/z_n'), \qquad Q'(z) = \prod_{r_n' \le 2R} (1 - z/\lambda_n'),$$

and P'', Q'' are defined similarly.

We can write

$$\log H(z) = \sum_{r_n > 2R} \log \left\{ 1 - \frac{z(z_n^{-1} - \lambda_n^{-1})}{1 - z/\lambda_n} \right\} + z \sum_{r_n > 2R} (z_n^{-1} - \lambda_n^{-1}),$$

where the last series converges because of (8.4.26). Since $|z| \le R$ and $|R/\lambda_n| < \frac{1}{2}$, we have

$$\left| \log \left| 1 - \frac{z(z_n^{-1} - \lambda_n^{-1})}{1 - z/\lambda_n} \right| \right| \le 2r \left| z_n^{-1} - \lambda_n^{-1} \right|,$$

and so by (8.4.26)

$$\left| \log |H(z)| \right| \le 9r \sum_{r_n > 2R} r_n^{-1} |\sin \theta_n|;$$

hence

$$\left| \log |H(z)| \right| \le R\eta_1(R), \qquad |z| \le R, \qquad \eta_1(R) = o(1). \tag{8.4.27}$$

Now let C_ν denote the arc of $|z| = r_\nu'$ between z_ν' and λ_ν'. If $\alpha > 0$, $\delta <$

$\alpha/2$, $\alpha \leq |\arg z| \leq \pi - \alpha$, and ζ is on C_ν we have

$$|\zeta - z| > (r + r_\nu') \sin \alpha/4,$$

$$|\log (1 - z/z_\nu') - \log (1 - z/\lambda_\nu')| = \left| \int_{C_\nu} \frac{z d\zeta}{\zeta(z - \zeta)} \right| \leq \frac{\delta}{\sin \alpha/4} \frac{r}{r + r_\nu}.$$

Since $n'(t) = O(t)$ we have, as an estimate for $\log(P'/Q')$,

$$\begin{aligned} |\log |P'(z)/Q'(z)|| &\leq \frac{\delta}{\sin \alpha/4} \int_0^{2R} \frac{r}{r + t}\, dn'(t) \\ &\leq \frac{\delta r}{\sin \alpha/4} \left\{ \frac{n'(2R)}{2R} + \int_0^{2R} \frac{n'(t)\, dt}{(r + t)^2} \right\} \\ &\leq k \delta \alpha^{-1} A R, \qquad k^{-1} R \leq |z| \leq R, \\ &\qquad\qquad \alpha \leq |\arg z| \leq \pi - \alpha, \end{aligned}$$

where A is independent of k, δ and α.

Next, we have, for either $\log |P''(z)|$ or $\log |Q''(z)|$, the upper estimate

$$\begin{aligned} &\int_0^{2R} \log (1 + 2R/t)\, dn''(t) \\ &\qquad = n''(2R) \log 2 + 2R \left(\int_0^{R/k} + \int_{R/k}^{2R} \right) \frac{n''(t)}{(t + 2R)t}\, dt \\ &\qquad\qquad \leq AR\{\eta(2R, \sigma) + \eta(R/k, \sigma) + 1/k\}. \end{aligned}$$

For a lower estimate, take $H = 2hR$ in the Boutroux-Cartan lemma (3.4.1); then for either $\log |P''(z)|$ or $\log |Q''(z)|$ we have the lower estimate

$$\eta''(2R)\{\log (2hR) - \log R - 1\} \geq 2R\eta(2R, \delta)\{\log (2h) - 1\}$$

outside circles the sum of whose radii is at most $4hR$. Thus

$$|\log |P''(z)/Q''(z)|| \leq AR\{\eta(2R, \delta)[1 + |\log h|] + \eta(R/k, \delta) + 1/k\}$$

for $R/k \leq |z| \leq 2R$, outside the exceptional circles.

Combining our estimates for $\log |H|$, $\log |P'/Q'|$ and $\log |P''/Q''|$, let us take δ, k and h to be functions of R such that

$$\delta(R) \to 0, \qquad \eta(R, \delta(R)) \to 0, \qquad h(R) \to 0,$$

$|\log h(R)|\, \eta(2R, \delta(2R)) \to 0$, $k(R) \to \infty$, $k(R)\, \delta(R) \to 0$ and $R/k(R) \to \infty$. It follows that (8.4.24) and (8.4.25) hold except in circles the sum of whose radii is at most $\{4h(R) + 1/k(R)\}R = o(R)$. Since $f(z)/g(z)$ is regular in the upper and lower half planes, (8.4.24) holds without exception when R is large enough, by the maximum principle.

NOTES FOR CHAPTER 8

The exposition follows Pfluger [5] except as noted.

8.1a. For integrals $\int_{-c}^{c} e^{izt}\phi(t)dt$ under various hypotheses on $\phi(t)$ see Cartwright [1], [2]. A still more special case is that of an exponential polynomial; a survey is given by Langer [1].

8.1b. The results quoted here are proved by Pfluger [6] in a more general form applying to entire functions of integral order in general.

8.2a. Paley and Wiener [1] formulate and prove this, and some related results, for even functions, as a Tauberian theorem. Pfluger [5] observed that the general case is reducible to the even case.

8.3a. Pfluger [5].

8.3b. Cartwright [3].

8.3c. Cartwright [3] gives a more general result.

8.3d. Rademacher [1].

8.3e. Pfluger [5].

8.3f. Pfluger [3].

8.3g. Knopp [1], p. 129.

8.4a. Pfluger [5]. Pfluger also notes that the proof shows that $r^{-1}n(r) \to D$ implies that $\int_{-r}^{r} x^{-2} \log | f(x) | \, dx$ always approaches a limit, which is finite only if

$$\Sigma \, r_n^{-1} \, | \sin \theta_n |$$

converges. Compare 7.2.11, which gives the asymptotic behavior of the integral when the series diverges.

8.4b. Noble [1] shows that (8.4.11) together with (8.4.13) (with no exceptions and with the positive imaginary axis in a zero-free angle) imply (8.4.12); in his unpublished thesis he removed the condition about the location of the zeros. Cf. Boas [29].

8.4c. Equivalent results are due to Levinson and Cartwright; the present formulation is that given by Levinson [4].

8.4d. Pfluger [5].

CHAPTER 9

UNIQUENESS THEOREMS

9.1. Introduction. The typical theorem of this chapter states that a function with certain growth properties must vanish identically if it has, in some sense, too many zeros. Since the classes of functions with which we shall be concerned contain the linear combinations of their members, a theorem of this kind states that a function of a specified class is completely determined by its values on a certain set of points.

We are thus concerned with the problem of the determination of the zeros by the growth of the function; since some aspects of this problem were discussed in Chapter 8, from a different point of view, we can formulate some of the results of Chapter 8 as uniqueness theorems. Other uniqueness theorems involve growth restrictions of different kinds. We shall also consider theorems in which the zeros belong, not to the function itself, but to some transform of it. In Chapter 10 we shall consider theorems of a more general kind in which zeros are replaced by points at which the function is bounded or satisfies some other special condition, and something is deduced about the behavior of the function in the large.

When a set $\{a_n\}$ is such that there is at most one entire function $f(z)$, belonging to a class C, having the values $f(a_n)$, it is natural to ask whether an element of C can be found with prescribed values $f(a_n)$. The answer is complete, but trivial, when C is the whole class of entire functions (which of course does not have the uniqueness property)[a]: for any set $\{a_n\}$ whose only limit point is ∞ and for arbitrary numbers b_n there is an entire function $f(z)$ such that $f(a_n) = b_n$. The problem of characterizing the sequences $\{a_n\}$ and $\{b_n\}$ for which there is a function f with $f(a_n) = b_n$ and f belonging to a more interesting class, e.g. the class of entire functions of given exponential type, is more difficult and will not be considered here. To represent a given function $f(z)$ by an interpolation series in which the numbers $f(a_n)$ appear as coefficients is a much easier task, and although we shall not give a systematic theory we shall encounter several cases in which such a representation is possible (cf. §§9.10, 10.2, 11.5)[b].

The theorems of Chapter 2 which involve an upper bound for $n(r)$ can be interpreted as uniqueness theorems. For example, we have

9.1.1. *Theorem. If $f(z)$ is of growth (ρ, τ), if $\{a_n\}$ is a set of complex numbers for which either*

$$\tau < \rho^{-1} \liminf_{r\to\infty} r^{-\rho} n_a(r) \tag{9.1.2}$$

or

$$\tau < (e\rho)^{-1} \limsup_{r\to\infty} r^{-\rho} n_a(r), \tag{9.1.3}$$

and if $f(a_n) = 0$ *for all* n, *then* $f(z) \equiv 0$.

This is another way of stating 2.5.13. For $\rho = 1$ we have the corollary:

9.1.4. *Theorem.*[c] *If* $f(z)$ *is of exponential type* τ, $f(z) \equiv 0$ *if* $f(a_n) = 0$ *where either*

$$\tau < \liminf_{r\to\infty} r^{-1} n_a(r)$$

or

$$\tau < \limsup_{r\to\infty} (er)^{-1} n_a(r).$$

9.2. Zeros at the positive integers. We get a much sharper uniqueness theorem if we impose a strong condition of regularity on the set of points at which the function is assumed to vanish. Our first theorem is the best-known result of this kind.

9.2.1. *Carlson's theorem.*[a] *If* $f(z)$ *is regular and of exponential type in the half plane* $x \geq 0$ *and* $h(\pi/2) + h(-\pi/2) < 2\pi$ *then* $f(z) \equiv 0$ *if* $f(n) = 0$, $n = 0, 1, 2, \cdots$.

The example $f(z) = \sin \pi z$ shows that $h(\pi/2) + h(-\pi/2) = 2\pi$ is not admissible; however, the hypothesis $h(\pi/2) + h(-\pi/2) < 2\pi$ can be weakened somewhat (see the next theorem).

We shall give several proofs of 9.2.1; or, more accurately speaking, we shall exhibit it in several ways as a special case of more general theorems.

The hypothesis placed on $h(\theta)$ in (9.2.1) means that the indicator diagram (defined only for $-\pi/2 \leq \theta \leq \pi/2$) of $f(z)$ has width less than 2π in the direction of the imaginary axis. Carlson's theorem is then included in the following more general result.

9.2.2. *Theorem.*[b] *If* $f(z)$ *is regular and of exponential type in the half plane* $x \geq 0$ *and its indicator diagram is not bounded on the right by a vertical line segment of length* 2π *or more, then* $f(z) \equiv 0$ *if* $f(n) = 0$, $n = 0, 1, 2, \cdots$.

The proof of 9.2.1 is only slightly simpler than that of 9.2.2, so we prove only 9.2.2. The proof could be abbreviated by using geometrical properties of the indicator diagram when $f(z)$ is entire (or in the general case if we had discussed the geometry of the indicator diagram for functions which are not entire). The hypothesis about the indicator diagram means analytically that there are real numbers α and β such that

$$\lim_{\theta\to\pm 0} \frac{h(\theta) - \alpha\cos\theta - \beta\sin\theta}{\sin\theta} = \pm c, \qquad 0 \leq c \leq \pi, \tag{9.2.3}$$

and it is in this form that we shall use it.

We assume $f(z) \not\equiv 0$ and deduce a contradiction. Consider the function $g(z) = f(z)e^{-(\alpha - i\beta)z}$, with α and β from 9.2.3. Then

$$h_g(\theta) = h_f(\theta) - \alpha \cos \theta - \beta \sin \theta,$$

and so for some $c' < \pi$ and $\delta > 0$ we have

$$(9.2.4) \qquad |g(re^{\pm i\delta})| = O(e^{c'r \sin \delta}).$$

Now consider the function $G(z) = g(z)/\sin \pi z$; since the zeros of $f(z)$ cancel those of $\sin \pi z$, $G(z)$ is regular in $x \geq 0$. Since $|\sin \pi z|$ is bounded away from zero, independently of n, on the semicircles $r = n + \frac{1}{2}$, $|\theta| \leq \pi/2$, and $2|\sin \pi z| \sim e^{\pi r \sin \delta}$ for $\arg z = \pm\delta$, by (9.2.4) we have $G(z)$ of exponential type in $|\arg z| \leq \delta$ and $h_G(\pm\delta) \leq (c' - \pi) \sin \delta < 0$. Then $h_G(\theta) < 0$ for $|\theta| < \delta$ and so for $\theta = 0$. But for real positive x, $|f(x)| \leq e^{\alpha x}|g(x)| \leq e^{\alpha x}|G(x)|$, and so $h_f(0) < \alpha$, contradicting the fact that (9.2.3) implies $h_f(0) = \alpha$.

9.3. Applications of Carleman's theorem. A different line of proof leads to a theorem which generalizes Carlson's theorem by allowing the width of the indicator diagram to reach 2π if a supplementary hypothesis is satisfied.

9.3.1. *Theorem. If $f(z)$ is regular and of exponential type in $x \geq 0$, $f(n) = 0$ $(n = 0, 1, 2, \cdots)$, and*

$$(9.3.2) \qquad \log |f(iy)f(-iy)| \leq 2\pi|y| - \sigma(y),$$

where

$$(9.3.3) \qquad \int^{\infty} y^{-2}\sigma(y)\, dy \text{ diverges},$$

then $f(z) \equiv 0$.

According to Carleman's theorem (1.2.2), unless $f(z) \equiv 0$ we have (since the points $z = n$ are some of the zeros of $f(z)$)

$$\sum_{k \leq r} \left(\frac{1}{k} - \frac{k}{R^2}\right) \leq \frac{1}{\pi R} \int_{-\pi/2}^{\pi/2} \log |f(Re^{i\theta})| \cos \theta \, d\theta$$
$$+ \frac{1}{2\pi} \int_1^R \left(\frac{1}{y^2} - \frac{1}{R^2}\right) \log |f(iy)f(-iy)| \, dy + O(1).$$

Using the fact that $\log|f(z)| \leq cr$ with some positive c, uniformly in $|\theta| \leq \pi$, we have

$$\log R \leq (2\pi)^{-1} \int_1^R y^{-2} \log |f(iy)\, f(-iy)| \, dy + O(1).$$

If (9.3.2) and (9.3.3) hold, this leads to a contradiction.

The same proof can be used to establish a more general result in which the zeros of $f(z)$ do not have to be at the integers. In the following theorem $\Lambda(t)$ denotes the number of λ_n not exceeding t, where $\{\lambda_n\}$ is an increasing sequence of positive numbers.

9.3.4. *Theorem*[a]. *If $f(z)$ is regular and of exponential type in $x \geq 0$, $f(\lambda_n) = 0$,*

$$\Lambda(x) \geq x + \delta(x), \tag{9.3.5}$$

and

$$\log |f(iy)f(-iy)| \leq 2\pi\{|y| + \sigma(|y|)\}, \tag{9.3.6}$$

then $f(z) \equiv 0$ if

$$\limsup_{R\to\infty} \int_1^R y^{-2}\{\delta(y) - \sigma(y)\}\,dy = +\infty. \tag{9.3.7}$$

The divergence of the integral in (9.3.7) can result either from a surplus of zeros or a reduction in the rapidity of growth of $f(iy)$ (δ and σ are allowed to have negative values).

Unless $f(z) \equiv 0$, Carleman's theorem gives us

$$\sum_{\lambda_k \leq R} \left(\frac{1}{\lambda_k} - \frac{\lambda_k}{R^2}\right) \leq \frac{1}{\pi R} \int_{-\pi/2}^{\pi/2} \log |f(Re^{i\theta})| \cos\theta\, d\theta$$

$$+ \frac{1}{2\pi} \int_1^R \left(\frac{1}{y^2} - \frac{1}{R^2}\right) \log |f(iy)\,f(-iy)|\,dy + O(1);$$

or, as we see by writing the left-hand side as a Stieltjes integral and integrating by parts,

$$\int_0^R \Lambda(t)(t^{-2} + R^{-2})\,dt \leq \log R + (2\pi)^{-1} \int_1^R \sigma(y)(y^{-2} - R^{-2})\,dy + O(1).$$

Using (9.3.5) and (9.3.6), we then have

$$\int_0^R \delta(t)(t^{-2} + R^{-2})\,dt \leq \int_1^R \sigma(y)(y^{-2} - R^{-2})\,dy + O(1)$$

and hence

$$\int_1^R t^{-2}\{\delta(t) - \sigma(t)\}\,dt \leq O(1),$$

which contradicts (9.3.7).

A result allowing a still smaller $\delta(x)$ can be established[b] if $f(z)$ is entire by applying Carleman's theorem to $f(z^2)$ instead of $f(z)$; and in the same way the theorem appropriate to an angle of some other size can be established by applying Carleman's theorem to $f(z^\alpha)$.

Theorem 9.3.4 contains in particular the first part of the following result; this is also a consequence of 6.3.9.

9.3.8. *Theorem. If* $f(z)$ *is regular and of exponential type in* $x \geq 0$, $f(\lambda_n) = 0$ *and* $\sum 1/\lambda_n = \infty$, *then* $f(z) \equiv 0$ *if*

$$\int_1^r u^{-2} \log | f(iu) f(-iu) | \, du \tag{9.3.9}$$

is bounded above. The condition $\sum 1/\lambda_n = \infty$ *is also necessary, in the sense that if* $\sum 1/\lambda_n$ *converges there is a function* $f(z)$ *satisfying the hypotheses (and even having* $f(iy)$ *bounded), but not identically zero.*

Actually it is not necessary for the λ_n to be real if $\sum \Re(1/\lambda_n)$ diverges, and a corresponding sharpening of 9.3.4 is easily given.

The necessity of the divergence condition is shown by the function[c]

$$\phi(z) = \prod_{n=1}^{\infty} \frac{(z - \lambda_n)(z - \bar{\lambda}_n)}{(z + \lambda_n)(z + \bar{\lambda}_n)} = \prod_{n=1}^{\infty} \left(1 - \frac{2z}{z + \lambda_n}\right)\left(1 - \frac{2z}{z + \bar{\lambda}_n}\right).$$

The convergence of $\sum 1/|\lambda_n|$ makes this product converge to a regular function, not identically zero, for $x \geq 0$; and $|\phi(z)| \leq 1$ there, while $\phi(\lambda_n) = 0$.

Since Carleman's formula counts multiple zeros according to their multiplicities, we can require fewer zeros if we require them to be multiple. Thus, in particular, it follows from 9.3.4 that if $f(z)$ is regular and of exponential type in $x \geq 0$, $h(\pm \pi/2) < \pi$ and $f(z)$ has zeros of multiplicity at least 2 at $z = 2n$, $n = 1, 2, \cdots$, then $f(z) \equiv 0$.[d]

9.4. Zeros at the positive and negative integers. If $f(z)$ is an entire function of exponential type and has zeros at all the integers, we can conclude that $f(z) \equiv 0$ under a hypothesis which is less restrictive than that of Carlson's theorem; the theorem which we shall prove neither includes nor is included by 9.3.4.

9.4.1. *Theorem.*[a] *If* $f(z)$ *is an entire function of exponential type whose indicator diagram has width at most* 2π *in the direction of the imaginary axis, and does not contain two horizontal straight line segments whose perpendicular distance apart is* 2π, *then if* $f(z) = 0$ *for* $z = 0, \pm 1, \pm 2, \cdots$,

$$f(z) = e^{\alpha z} \phi(z) \sin \pi z,$$

where $\phi(z)$ *is an entire function of zero exponential type.*

9.4.2. *Corollary. If* $f(z)$ *is an entire function of exponential type,* $f(z) = 0$ *for* $z = 0, \pm 1, \pm 2, \cdots$, *and*

$$|f(z)| \leq \epsilon(r) e^{\pi r}, \qquad \epsilon(r) = O(r^p), \tag{9.4.3}$$

then $f(z) = P(z) \sin \pi z$, *where* $P(z)$ *is a polynomial of degree not exceeding* p.

9.4.4. *Corollary. If* $\epsilon(r) = o(1)$ *in (9.4.2),* $f(z) \equiv 0$.

9.4.5. *Corollary. If* (9.4.3) *is replaced by*

$$| f(z) | \leq \epsilon(r) e^{\pi r(|\sin\theta| + \delta(\theta)|\cos\theta|)}$$

where $\epsilon(r) = o(1)$ *and* $\delta(\theta) \to 0$ *either as* $\theta \to \pi/2$ *or as* $\theta \to -\pi/2$, *then* $f(z) \equiv 0$.

An important difference between 9.3.1 and 9.4.4 or 9.4.5 is that 9.3.1 is still true if a finite or even (by 9.3.4) a sparse infinite set of zeros is removed, while 9.4.5 fails to hold if even one zero is removed: for example, consider $z^{-1} \sin \pi z$. Another difference is that 9.4.4 or 9.4.5 demands (and necessarily so) more than a requirement of the form $f(iy) = o(e^{\pi|y|})$.

The corollaries follow from the theorem because (9.4.3) makes the indicator diagram of $f(z)$ lie inside the circle $| z | = \pi$, and because an entire function $\phi(z)$ of zero exponential type satisfying $| \phi(iy) | \leq | y |^p$ is a polynomial, and is identically zero if $\phi(iy) = o(1)$ as $| y | \to \infty$ (6.2.13). The condition in (9.4.5) implies the hypothesis of 9.4.1 about the form of the indicator diagram.

To prove Theorem 9.4.1, we may assume that the indicator diagram is of width 2π in the direction of the imaginary axis, since if its width is less the conclusion follows from Carlson's theorem (9.2.1). Then multiplying $f(z)$ by some $e^{\alpha z}$ shifts the indicator diagram so that it has $y = \pm\pi$ as supporting lines and so that $h(\pi/2) = \pi$. The hypothesis on the indicator diagram then states that (for the modified function)

$$h(\theta) \leq \pi | \sin \theta | + \epsilon(\theta) | \cos \theta |,$$

where $\epsilon(\theta) \to 0$ either as $\theta \to \pi/2$ or as $\theta \to -\pi/2$ (we shall suppose the former, for definiteness). Now consider $g(z) = f(z)/\sin \pi z$; then because $| \sin \pi z |$ is bounded away from zero on the circles $| z | = n + \frac{1}{2}$, $g(z)$ is an entire function of exponential type, and its indicator satisfies, first for $\theta \neq 0, \pi$, and then, by continuity, for all θ,

$$h_g(\theta) = h_f(\theta) - \pi | \sin \theta | \leq \epsilon(\theta) | \cos \theta |.$$

If $0 < \beta < \pi/2$, we have $h_g(\beta) \leq \epsilon(\beta) \cos \beta$, $h_g(-\pi/2) \leq 0$, and so (since $\beta < \pi/2$), $h_g(\theta) \leq \epsilon(\beta) \cos \theta$, $-\pi/2 \leq \theta \leq \beta$. As $\beta \to \pi/2$, $\epsilon(\beta) \to 0$, and so $h_g(\theta) \leq 0$ for $| \theta | \leq \pi/2$. Similarly $h_g(\theta) \leq 0$ for $\pi/2 \leq \theta \leq 3\pi/2$. Hence $h_g(\theta) \equiv 0$, so that $g(z)$ is of zero exponential type.

9.5. General distribution of zeros on a half line. For functions of exponential type in a half plane there is a very general uniqueness theorem which is particularly interesting because it is the best possible result of its kind.

9.5.1. *Fuchs's theorem.*[a] *If* $f(z)$ *is regular for* $x \geq 0$ *and* $O(e^{c|z|})$

and if $f(\lambda_n) = 0$, *with* $\lambda_n > 0$, $\lambda_{n+1} - \lambda_n > \delta > 0$, *then* $f(z) \equiv 0$ *if*

(9.5.2) $$\limsup_{r\to\infty} r^{-2c/\pi}\psi(r) = \infty$$

where

(9.5.3) $$\psi(r) = \exp\left\{2\sum_{\lambda_n<r} \lambda_n^{-1}\right\};$$

and if (9.5.2) *is not true there is an* $f(z)$ *satisfying all the hypotheses of the theorem and not identically zero.*

9.5.4. *Corollary. If* $f(z)$ *satisfies the hypotheses of* 9.5.1, $c < \pi$, *and* (9.5.2) *is replaced by*

(9.5.5) $$\liminf_{n\to\infty} n/\lambda_n \geq 1$$

(*i.e.*, $\{\lambda_n\}$ *has lower density at least* 1), *then* $f(z) \equiv 0$.

Condition (9.5.5) is the same as $\liminf_{r\to\infty} n(r)/r \geq 1$; we have

$$\sum_{\lambda_n<r} \lambda_n^{-1} = \int_0^r t^{-1}\,dn(t) = r^{-1}n(r) + \int_0^r t^{-2}n(t)\,dt;$$

if $n(r) \geq (1 - \epsilon)r$, $r > r_0$, $\epsilon > 0$, we have

$$\sum_{\lambda_n<r} \lambda_n^{-1} \geq (1 - \epsilon)\log r + O(1),$$

$$\psi(r) \geq Ar^{2(1-\epsilon)},$$

and so (9.5.2) holds.

We could replace (9.5.5) by $\lim_{n\to\infty} n/\lambda_n = 1$, since omitting some of the λ_n only strengthens the theorem.

In §9.7 we shall prove a uniqueness theorem in which (9.5.5) is replaced by a much weaker condition, under more stringent restrictions on the growth of $f(z)$.[b]

To prove 9.5.1 we require two lemmas.

9.5.6. *Lemma. Let* $W = \log|(1 - z)(1 + z)^{-1}e^{2z}|$. *Then*

(9.5.7) $$W \leq 2x, \qquad x \geq 0;$$

(9.5.8) $$|W| = O(xr^2), \qquad r \leq \tfrac{1}{2}, \quad x > 0.$$

For $x \geq 0$, $|(1 - z)/(1 + z)| \leq 1$, and (9.5.7) follows at once. For (9.5.8) we note that

$$W = 2x + \frac{1}{2}\log\left|\frac{1 - z}{1 + z}\right|^2$$

$$= 2x + \frac{1}{2}\log\left\{1 - \frac{4x}{(1 + x)^2 + y^2}\right\};$$

expanding the last logarithm in a power series, we have

$$W = 2x\{1 - [(1+x)^2 + y^2]^{-1} - 2x[(1+x)^2 + y^2]^{-2} + O(x^2)\}$$
$$= 2x[(1+x)^2 + y^2]^{-2}\{(1+x)^4 + 2y^2(1+x)^2 + y^4 - (1+x)^2 - y^2 - 2x\} + O(xr^2)$$
$$= O(xr^2).$$

9.5.9. *Lemma. For $x \geq 0$, the function*

$$H(z) = \prod_{n=1}^{\infty} \frac{\lambda_n - z}{\lambda_n + z} \exp(2z/\lambda_n) \tag{9.5.10}$$

is regular and satisfies

$$|H(z)| \leq \{A\psi(r)\}^x, \tag{9.5.11}$$

and

$$|H(z)| \geq \{B\psi(r)\}^x \tag{9.5.12}$$

outside circles of radius $\delta/3$ with centers at the λ_n.

Here $\psi(r)$ and δ are defined in 9.5.1 and A and B are constants.

Since the separation condition $\lambda_{n+1} - \lambda_n > \delta$ makes $\sum \lambda_n^{-1-\epsilon}$ converge ($\epsilon > 0$), $H(z)$ is the quotient of two convergent canonical products and so is regular when $z \neq -\lambda_n$. To prove (9.5.11) and (9.5.12), write $H(z) = \Pi_1\Pi_2$, where Π_1 contains the terms with $\lambda_n < 2r$. Apply (9.5.7) to the factors in Π_1, and (9.5.8) to those in Π_2. Then, with various constants A,

$$\log|\Pi_1| \leq 2x \sum_{\lambda_n < 2x} \lambda_n^{-1} = x \log \psi(2x) < x \log \psi(x) + Ax, \tag{9.5.13}$$

and

$$|\log|\Pi_2|| \leq A \sum_{\lambda_n \geq 2r} xr^2/\lambda_n^3 \leq Ax; \tag{9.5.14}$$

we used the inequality $\lambda_n \geq An$ in the last part of (9.5.13) and (9.5.14). We have now established (9.5.11).

For (9.5.12), we can again use (9.5.14) for Π_2. For Π_1, suppose first that $r \geq 3x$. There are $O(r)$ numbers λ_n occurring in Π_1, and if λ is one of them,

$$\left|\frac{\lambda - z}{\lambda + z}\right|^2 = 1 - \frac{4\lambda x}{(\lambda + x)^2 + y^2} > 1 - 4\lambda x/r^2 \geq 1 - 2x/r \geq 1/3.$$

Hence, again with A denoting various constants,

$$|\Pi_1| > (1 - 2x/r)^{Ar} \exp\left\{2x \sum_{\lambda_n < 2r} \lambda_n^{-1}\right\} > \{A\psi(2r)\}^x > \{A\psi(r)\}^x,$$

since $(1 - 2x/r) > \exp(Ax/r)$ for $x/r \leq \frac{1}{3}$. This establishes (9.5.12) for

$r \geq 3x$. If $r < 3x$, let N be the number of λ_n occurring in Π_1. Then if N is a positive even number and z is outside the specified circles, $\lambda_{n+1} - \lambda_n > \delta$ gives

$$\prod_1 |\lambda_n - z|$$
$$\geq (N/A)^N (\delta/3)^2 (\delta + \delta/3)^2 (2\delta + \delta/3)^2 \cdots ((N-2)\delta/2 + \delta/3)^2$$

by Stirling's formula. The same estimate holds for odd N (with a new A). Also

$$\prod_1 |\lambda_n + z| \leq (Ar)^N < (3Ax)^N,$$

and so

$$\left| \prod_1 \frac{\lambda_n - z}{\lambda_n + z} e^{2z} \right| \geq \left(\frac{N}{Ax} \right)^N \{\psi(r)\}^x \geq \{A\psi(r)\}^x,$$

since $(N/a)^N \geq e^{-a/e}$, $a > 0$. Thus (9.5.12) follows.

To prove 9.5.1, we suppose first that $f(\lambda_n) = 0$ and $f(z) \not\equiv 0$, with (9.5.2) satisfied. We suppose for convenience that $|f(z)| < e^{cr}$, and consider

$$g(z) = f(z) z^{2cz/\pi} / H(z),$$

where $H(z)$ is defined in (9.5.10). This function is regular for $x \geq 0$ and continuous for $x \geq 0$, and by (9.5.9),

$$|g(z)| = |f(z)| \exp \{2c\pi^{-1} \Re(z \log z)\} / |H(z)|$$
$$\leq \exp \{cr - 2c\pi^{-1} r\theta \sin \theta + 2c\pi^{-1} x \log r\} \{A\psi(r)\}^{-x}.$$

This holds initially except perhaps in small circles around the λ_n, but since $g(z)$ is regular in the circles it holds inside them also, by the maximum principle. Now for $|\theta| \leq \pi/2$ we have $1 - 2\pi^{-1}\theta \sin \theta \leq 2 \cos \theta$, since (for $0 < \theta < \pi/2$)

$$\frac{1 - 2\pi^{-1}\theta \sin \theta}{\cos \theta} = \frac{1 - \sin \theta}{\cos \theta} + \frac{2(\pi/2 - \theta) \sin \theta}{\pi \cos \theta}$$
$$= \tan \frac{\pi - 2\theta}{4} + \frac{2 \sin \theta}{\pi} \frac{\pi/2 - \theta}{\sin (\pi/2 - \theta)}$$
$$< 1 + (2/\pi)(\pi/2) = 2.$$

Thus

(9.5.15) $$|g(z)| \leq \{A\psi(r) r^{-2c/\pi}\}^{-x},$$

where A is a new constant.

We now have, by Carleman's theorem applied to $g(z)$,

$$(9.5.16) \qquad 0 \le (\pi R)^{-1} \int_{-\pi/2}^{\pi/2} \log | g(Re^{i\theta}) | \cos\theta \, d\theta + O(1),$$

since the contribution from $g(iy)$ is negative by (9.5.15). Furthermore, (9.5.15) shows that the integral in (9.5.16) is less than

$$-\pi^{-1} \int_{-\pi/2}^{\pi/2} \log \{A\psi(r) r^{-2c/\pi}\} \cos^2\theta \, d\theta = -\tfrac{1}{2} \log \{\psi(r) r^{-2c/\pi}\} + O(1),$$

and thus (9.5.16) contradicts (9.5.2). This shows that $f(z) \equiv 0$ if $f(\lambda_n) \equiv 0$ and (9.5.2) holds.

Now suppose that (9.5.2) fails, so that $\psi(r) r^{-2c/\pi} < K$, $r > 1$. With $L > 0$, the function

$$f(z) = (1 + z)^{-2cz/\pi} L^z H(z) = (1 + z^{-1})^{-2cz/\pi} z^{-2cz/\pi} L^z H(z)$$

is regular for $x \ge 0$ and has zeros at λ_n. By 9.5.9, for $r > 2$ and $x \ge 0$,

$$\begin{aligned} | f(z) | &\le | (1 + z^{-1})^{-2cz/\pi} | \, r^{-2cx/\pi} \exp \{2c\pi^{-1} r\theta \sin\theta\} \{AL\psi(r)\}^x \\ &\le | (1 + z^{-1})^{-2cz/\pi} | \, (ALK)^x e^{cr}. \end{aligned}$$

Since $\arg (1 + z^{-1}) = O(1/r)$ and $| 1 + z^{-1} | > \tfrac{1}{2}$ for $r > 2$ and $x > 0$,

$$\begin{aligned} | (1 + z^{-1})^{-2cz/\pi} | &= \exp \{2c\pi^{-1} [- x \log | 1 + z^{-1} | + y \arg (1 + z^{-1})]\} \\ &= O(e^{cx/\pi}). \end{aligned}$$

Hence by making L small enough we can make $| f(z) | = O(e^{cr})$, and so we have constructed the function required in the second part of 9.5.1.

9.6. Zeros near the integers. Although 9.5.1 is the best result of its kind, there are more precise results when more restrictive hypotheses are imposed on the λ_n.

9.6.1. *Valiron's theorem.*[a] *If $f(z)$ is an entire function such that*

$$(9.6.2) \qquad f(z) = o(r^{-1} e^{\pi r}),$$

uniformly in θ, and if $f(z)$ vanishes at least once in every interval $n \le x < n + 1$, $n = 0, \pm 1, \pm 2, \cdots$, then $f(z) \equiv 0$.

We can prove this along the lines of the proof of 9.4.1 provided that we have suitable estimates for an entire function having zeros at the points where $f(z)$ is assumed to vanish. The following lemma provides the necessary information.

9.6.3. *Lemma. Let $n \le \lambda_n < n + 1$, $-n - 1 \le \mu_n < -n$, $n = 0, 1, 2, \cdots$, and let*

$$(9.6.4) \qquad \phi(z) = (z - \lambda_0)(z - \mu_0) \prod_{n=1}^{\infty} (1 - z/\lambda_n)(1 - z/\mu_n).$$

Then $\phi(z)$ *is an entire function of exponential type, and*

$$|\phi(iy)| \geq Ae^{\pi|y|}/|y|, \qquad |y| \geq 1, \tag{9.6.5}$$

where A *is a positive constant;*

$$\phi(iy) = O(|y| e^{\pi|y|}), \qquad |y| \to \infty; \tag{9.6.6}$$

$$\phi(x) = O(|x|), \qquad |x| \to \infty; \tag{9.6.7}$$

$$\phi(z) = O(|z| e^{\pi|z|}), \qquad |z| \to \infty. \tag{9.6.8}$$

The canonical product

$$\prod_{n=1}^{\infty} (1 - z/\lambda_n) \exp(z/\lambda_n)(1 + z/\mu_n) \exp(-z/\mu_n) \tag{9.6.9}$$

(where $1 - z/\lambda_0$ is to be interpreted as z if $\lambda_0 = 0$) is an entire function of exponential type by 2.6.5, and it can be rearranged to give

$$\prod_{n=1}^{\infty} (1 - z/\lambda_n)(1 + z/\mu_n) \prod_{n=1}^{\infty} \exp\left\{z \frac{\lambda_n - \mu_n}{\lambda_n \mu_n}\right\}. \tag{9.6.10}$$

Since $\sum |\lambda_n - \mu_n|/(\lambda_n\mu_n) \leq \sum 2/n^2 < \infty$, the second product is of the form e^{cz} and so $\phi(z)$ is of exponential type.

Since $n \leq \lambda_n < n + 1$, and $n < \mu_n \leq n + 1$, we have

$$|\phi(x + iy)| = \prod_{n=1}^{\infty} \left| 1 + z\left(\frac{1}{\mu_n} - \frac{1}{\lambda_n}\right) - \frac{z^2}{\lambda_n\mu_n} \right| |z - \lambda_0| \, |z - \mu_0|,$$

$$\frac{|\phi(iy)|}{|iy - \lambda_0| \, |iy - \mu_0|} \geq \prod_{n=1}^{\infty} \{1 + y^2/(\lambda_n\mu_n)\} \geq \prod_{n=1}^{\infty} \{1 + y^2/(n + 1)^2\}$$

$$= \left|\frac{\sin \pi iy}{\pi iy}\right| \div (1 + y^2),$$

and (9.6.5) follows. Also,

$$\frac{|\phi(iy)|}{|iy - \lambda_0| \, |iy - \mu_0|} \leq \prod_{n=1}^{\infty} (1 + y^2/n^2) = \left|\frac{\sin \pi iy}{\pi iy}\right| = O(|y|^{-1} e^{\pi|y|}),$$

and (9.6.6) follows.

For $x > 0$, if $k < x < k + 1$,

$$\frac{|\phi(x)|}{|(x - \lambda_0)(x + \mu_0)|} \leq \left|\left(1 - \frac{x}{\lambda_k}\right)\left(1 + \frac{x}{\mu_k}\right)\right| \prod_{n=1}^{k-1} \left|1 - \frac{x}{n}\right| \left(1 + \frac{x}{n}\right)$$

$$\times \prod_{n=k+1}^{\infty} \left(1 - \frac{x}{n + 1}\right)\left(1 + \frac{x}{n}\right)$$

$$= \frac{|1 - x/\lambda_k| \, (1 + x/\mu_k)}{|1 - x/(k + 1)| \, |1 - x^2/k^2|} \frac{|\sin \pi x|}{\pi x}$$

$$= O(x^{-1}),$$

and so (9.6.7) follows for $x > 0$ (similarly for $x < 0$). From this we infer (9.6.8) by using 6.2.4. (This of course includes (9.6.6) again.)

Now put $g(z) = f(z)/\phi(z)$, where the zeros of $\phi(z)$ are the zeros assumed for $f(z)$; then $g(z)$ is entire. It is also of exponential type by 2.10.1: in the first place it is of order 1 (at most) by an application of 2.6.18; and since the sums $\sum_{|z_n|\leq r} 1/z_n$ are bounded when z_n are all the zeros of $f(z)$, and the corresponding sums with the zeros of $\phi(z)$ are bounded, so are the sums with the zeros of $g(z)$. Reasoning as in 9.4.1, we find that $g(z)$ is a constant, and $f(z) = K\phi(z)$. But $yf(iy)e^{-\pi y} \to 0$ as $y \to \infty$, and $y\phi(iy)e^{-\pi|y|}$ does not, so K must be zero.

We can proceed similarly to obtain a uniqueness theorem associated with any function of exponential type whose behavior is known with sufficient precision. For example, the estimates needed in §10.5 in the next chapter lead to a uniqueness theorem whose formulation may be left to the reader.

As an example of a still more delicate theorem, we prove

9.6.11. *Theorem.*[b] *If* $\{\lambda_n\}$ *is an increasing sequence of real numbers,* $\lambda_n \neq 0$, *and*

(9.6.12)
$$\int_1^r t^{-1}n_\lambda(t)\,dt > 2r - \alpha \log r - B,$$

while $H(z)$ *is an entire function satisfying*

(9.6.13)
$$|y|^\beta H(x+iy)e^{-\pi|y|} \to 0, \qquad |y| \to \infty,$$

and $\alpha \leq \beta$, *then* $H(z) \equiv 0$ *if* $H(\lambda_n) = 0$.

9.6.14.* *Corollary. The conclusion of* 9.6.11 *holds if* $\lambda_{-n} = -\lambda_n$ *and, for large* n, $0 < \lambda_n \leq n + \alpha/2 - \frac{1}{2}$.

To prove 9.6.11, suppose that $H(z) \not\equiv 0$. Since (9.6.13) implies, in particular, that $H(z)$ is of exponential type, we have $n_\lambda(t) = O(t)$, since λ_n are some of the zeros of $H(z)$. More precisely, Jensen's theorem yields

$$\int_1^r t^{-1}n(t)\,dt \leq (2\pi)^{-1}\int_0^{2\pi} \log|H(re^{i\theta})|\,d\theta + O(1),$$

(9.6.15)
$$\int_1^r t^{-1}n(t)\,dt \leq 2r - \beta \log r + O(1).$$

Then the product

$$F(z) = \prod_{n=-\infty}^{\infty} (1 - z/\lambda_n) \exp(z/\lambda_n)$$

is an entire function of order 1 at most, and $\phi(z) = H(z)/F(z)$ is entire and also of order 1 at most. If $n_\phi(r)$ and $n_H(r)$ denote the number of zeros of ϕ and H of modulus not exceeding r, we have $n_\phi(r) = n_H(r) - n_\lambda(r)$, and so by (9.6.12) and (9.6.15),

$$\int_1^r u^{-1} n_\phi(u)\, du \le (\alpha - \beta) \log r + O(1).$$

Since $\alpha < \beta$, the right-hand side is bounded as $r \to \infty$ and so $n_\phi(r) \equiv 0$; that is, $\phi(z)$ has no zeros. Thus $\phi(z) = ae^{bz}F(z)$ and so, in particular,

$$\log | H(iy) | = cy + \log | F(iy) | + O(1), \tag{9.6.16}$$

with a real c.

On the other hand, using (9.6.12) we have

$$\begin{aligned} \log | F(iy) | &= \frac{1}{2} \int_0^\infty \log (1 + y^2/u^2)\, dn_\lambda(u) \\ &= \int_0^\infty \frac{2y^2 u}{(u^2 + y^2)^2}\, du \int_0^u t^{-1} n_\lambda(t)\, dt \\ &\ge \pi | y | - \alpha \log | y | - O(1), \end{aligned}$$

and so by (9.6.16)

$$\log | H(iy) | \ge cy + \pi | y | - \alpha \log | y | - O(1). \tag{9.6.17}$$

This, however, contradicts (9.6.13), as we see by letting $y \to \infty$ with its sign that of c if $c \ne 0$.

9.7. Zeros less regularly distributed. While the results of the preceding sections are quite sharp, they demand rather regular behavior of the sequence of assumed zeros: in most cases that it has a density and even rather more. Uniqueness theorems which allow much more irregular behavior of the sequence of zeros (but demand more restrictive hypotheses on the growth of the function) can be deduced from the results of Chapter 8 which assert that under certain circumstances the zeros of an entire function have a density, which is smaller when the function grows less rapidly. The assumed zeros are some of the zeros of the function, and if they are occasionally so dense that they force the density of the whole set of zeros to be too large for the growth of the function, it follows that the function vanishes identically. This idea is made precise in the following theorem.

9.7.1. *Theorem.*[a] *Let $f(z)$ be an entire function of exponential type k. If $h(0) = h(\pi) = 0$ and if*

$$\int_1^{\to\infty} x^{-2} \log | f(x) f(-x) |\, dx \textit{ converges,} \tag{9.7.2}$$

it is impossible to have $f(\lambda_n) = 0$ for an increasing positive sequence $\{\lambda_n\}$ having maximum density exceeding k/π.

In particular, then (cf. 6.3.6 and 7.2.8), if $f(z)$ is of exponential type less than π and $f(\lambda_n) = 0$ for a sequence of maximum density at least 1, and if

$$\int_{-\infty}^{\infty} \frac{\log^+ |f(x)|}{1 + x^2}\, dx \text{ converges,}$$

$f(z) \equiv 0$. This differs from 9.5.4 by allowing a smaller set of zeros but requiring that the function grows slowly both in the direction in which the zeros lie and in the opposite direction; it makes essential use of the fact that $f(z)$ is entire.

We recall that the maximum density D is defined by

$$D = \lim_{\xi \to 1-} \limsup_{u \to \infty} \frac{n_\lambda(u) - n_\lambda(\xi u)}{u - \xi u}.$$

Let $n_+(r)$ denote the number of zeros of $f(z)$ (assumed not identically zero) in the right-hand half plane. Then by 8.4.16, $n_+(r)/r \to B \leq k/\pi$. Therefore if $0 < \xi < 1$,

$$\lim_{r \to \infty} \frac{n_+(r) - n_+(\xi r)}{r - \xi r} = B.$$

Since the λ_n are among the zeros of $f(z)$ which are counted in $n_+(r)$,

$$n_+(r) - n_+(\xi r) \geq n_\lambda(r) - n_\lambda(\xi r),$$

and so the maximum density D of $\{\lambda_n\}$ satisfies $D \leq B \leq k/\pi$, as required.

9.8. Functions with a sequence of small values. Our uniqueness theorems state roughly that a function of exponential type cannot have too many zeros unless it vanishes identically. With some modifications of the hypotheses, there are more general theorems which state that a sequence of points at which the modulus of the function is very small has the same effect as a sequence of zeros. Of course the points must be subjected to some kind of a separation condition, since a function is small throughout a neighborhood of each of its zeros. We shall give some of the results which have been obtained in this direction; there is, however, no really systematic theory as yet.

There are at least two possible methods of proof: one can first infer the existence of a large number of zeros from the existence of a sequence of points at which the function is small, or infer that the function is small all along a line and so identically zero. In the first case the proof is completed by the theorems of this chapter; in the second case one has to use theorems which will be discussed in Chapter 10.

Our first two theorems deal with functions which are only of exponential type in an angle; in one there is little restriction on the function on the boundary, but the sequence of points at which the function is small has to be quite dense; in the other the sequence is much less restricted but the function is severely restricted on the boundary.

9.8.1. *Theorem. Let* $f(z)$ *be regular and of exponential type for* $|\arg z| \leq \alpha$, $\alpha > \pi/2$, *and let* $h(\pm\pi/2) < \pi$. *Let* $\{\lambda_n\}$ *be a sequence with* $n/\lambda_n \to 1$,

$$|\lambda_m - \lambda_n| \geq \delta |n - m|, \qquad \delta > 0.$$

If $\lambda_n^{-1} \log |f(\lambda_n)| \to -\infty$, $f(z) \equiv 0$.

This follows from 10.3.1, 5.1.4, and 5.1.13. (A sharper result, for entire functions, follows from 10.3.1 and 5.4.5.)

9.8.2.* *Theorem. Let* $f(z)$ *be regular and of exponential type for* $x \geq 0$, *and let*

$$\int_1^r y^{-2} \log |f(iy)f(-iy)| \, dy$$

be bounded. Let $\{\lambda_n\}$ *be a sequence with* $\arg \lambda_n \to 0$,

$$||\lambda_m| - |\lambda_n|| \geq \delta |n - m|, \qquad \delta > 0,$$

and $\sum 1/|\lambda_n|$ *divergent. If* $|\lambda_n|^{-1} \log |f(\lambda_n)| \to -\infty$, $f(z) \equiv 0$.

This is deducible from 7.3.1 along the lines of 10.7.4.

Next we consider some theorems for entire functions, in which progressively stronger supplementary conditions are imposed on the function along the real axis and the growth of $\{f(\lambda_n)\}$ is progressively less severely restricted.

9.8.3.* *Theorem.*[a] *Let* $f(z)$ *be an entire function of exponential type* k, *and* $\{\lambda_n\}$ *a positive increasing sequence of maximum density greater than* k/π, *such that* $\lambda_{n+1} - \lambda_n \geq \delta > 0$. *If* $h(0) = h(\pi) = 0$ *and* $f(\lambda_n) = O\{\exp(-\epsilon\lambda_n)\}$, $\epsilon > 0$, *then* $f(z) \equiv 0$.

The (weaker) theorem in which we assume only $h(\pi) \leq 0$ but require that $n/\lambda_n \to k$ is deducible from 10.3.1 and 5.1.12.

9.8.4. *Theorem.*[b] *Let* $f(z)$ *be an entire function of exponential type* k, $\{\lambda_n\}$ *a positive increasing sequence of density* $D > k/\pi$, *such that*

$$\lambda_{n+1} - \lambda_n \geq \delta > 0.$$

If

$$\int_{-\infty}^{\infty} \frac{\log^+ |f(x)|}{1 + x^2} \, dx < \infty \tag{9.8.5}$$

and

$$|f(\lambda_n)| = O\{\exp(-\eta(\lambda_n))\}, \tag{9.8.6}$$

where $\eta(t)$ *is a nondecreasing function such that* $\int^{\infty} u^{-2}\eta(u)\,du$ *diverges, then* $f(z) \equiv 0$.

For some similar theorems for entire functions of zero exponential type see §10.7.

If we had (9.8.6) for all positive x, or even for sufficiently long intervals, instead of just for $x = \lambda_n$, the conclusion of 9.8.4 would follow from 6.3.6, according to which (9.8.5) implies the convergence of

$$\int_{-\infty}^{\infty} \frac{\log^- |f(x)|}{1 + x^2}\,dx. \tag{9.8.7}$$

We shall now show that (9.8.6) implies the same inequality with λ_n replaced by x for intervals which are long enough to make (9.8.7) diverge.

Since $f(z)$ is an entire function of exponential type satisfying (9.8.5), its zeros in the right-hand half plane have a density not exceeding k/π (6.3.6), (6.6.5), (8.4.16). There is no loss of generality from assuming that $f(0) = 1$, so that

$$f(z) = e^{cz} \prod_{n=1}^{\infty} (1 - z/z_n) \exp(z/z_n).$$

Consider an auxiliary function $F(z)$ with real zeros:

$$F(z) = \exp\{z\Re(c)\} \prod_{n=1}^{\infty} (1 - z\Re(1/z_n)) \exp\{z\Re(1/z_n)\}.$$

Since

$$|1 - x\Re(1/z_n)| = |1 - xr_n^{-1}\cos\theta_n| \leq |1 - xr_n^{-1}e^{-i\theta_n}|$$

we have $|F(x)| \leq |f(x)|$, so that $F(z)$ satisfies (9.8.5) and (9.8.6). The zeros of $F(z)$ are at $r_n \sec\theta_n$ and have moduli at least as large as those of $f(z)$, so that we infer first that $F(z)$ is not of order greater than 1, and then, since the sums $\sum_{r_n \leq r} r_n^{-1}\cos\theta_n$ are bounded, that $F(z)$ is of exponential type (2.10.1). Then since $F(z)$ satisfies (9.8.5), its zeros in the right-hand half plane have a density B, which cannot exceed that for $f(z)$ (since $F(z)$ has no more zeros in any circle about 0 than $f(z)$ does). Therefore

$$D - B = 4\alpha > 0.$$

Since $\{\lambda_n\}$ has density D,

$$\lim_{r\to\infty} r^{-1}\{n_\lambda(2r) - n_\lambda(r)\} = D,$$

and so for large integers m there are at least $2^m(D - \alpha)$ of the λ_n in

$$(2^m, 2^{m+1}).$$

On the other hand, since the zeros of $F(z)$ have density B, for large m we have at most $2^m(B + \alpha)$ zeros in $(2^m, 2^{m+1})$. With each λ_n associate the interval $(\lambda_n - \delta, \lambda_n + \delta)$; these intervals do not overlap, there are at least $2^m(D - \alpha)$ of them in $(2^m, 2^{m+1})$, and so at least

$$2^m(D - \alpha) - 2^m(B + \alpha) = 2^{m+1}\alpha$$

of them are free of zeros of $F(z)$. Since $F(z)$ is real on the real axis, Laguerre's theorem (2.8.1) shows that $F'(z)$ vanishes just once between successive zeros of $F(z)$, and so at most once in any of the $2^{m+1}\alpha$ intervals which contain no zeros of $F(z)$. Therefore each of these intervals contains an interval of length c in which $| F(x) | \leq | F(\lambda_n) | \leq | f(\lambda_n) | \leq \exp\{- \eta(\lambda_n)\}$. Since $\lambda_n \geq 2^m$ in $(2^m, 2^{m+1})$ and $\eta(x)$ does not decrease,

$$\int_{2^m}^{2^{m+1}} \log^- | F(x) | \, dx \leq -2^{m+1}\alpha\{c\eta(2^m)\},$$

and so

$$\int_{2^m}^{2^{m+1}} x^{-2} \log^- | F(x) | \, dx \leq -2^{-m-1}\alpha c\eta(2^m) \leq -\frac{\alpha c}{2} \int_{2^{m-1}}^{2^m} x^{-2}\eta(x) \, dx.$$

If we add these inequalities for $m = m_0, m_0 + 1, \cdots$, starting from a large m_0, we see that the divergence of $\int^\infty x^{-2}\eta(x) \, dx$ implies that of (9.8.4), which is impossible unless $F(z) \equiv 0$, as we have already noted; but $F(z) \equiv 0$ means that $f(z) \equiv 0$.

9.9. Alternation theorems. The uniqueness theorems of the preceding sections can be used to establish some theorems of a different kind. The following result will serve as an example.

9.9.1. *Theorem.*[a] *Let $f(z)$ be an entire function which is real on the real axis and satisfies*

$$f(z) = o(r^{-1}e^{\pi r}) \qquad (9.9.2)$$

uniformly in θ. Then it is impossible to have

$$f(n)f(n + 1) < 0, \qquad n = 0, \pm 1, \pm 2, \cdots . \qquad (9.9.3)$$

In fact, if (9.9.3) holds, $f(z)$ has a zero in each interval $(n, n + 1)$, and this fact together with (9.9.2) implies that $f(z) \equiv 0$ (9.6.1), contradicting (9.9.3).

A theorem of the same kind can be written down for any of the uniqueness theorems that involve real zeros. Both the requirement that $f(z)$ is real for real z and the form of (9.9.3) can be weakened. For, the function

$$F(z) = f(z) + \overline{f(\bar{z})}$$

is entire, and of the same exponential type as $f(z)$; more precisely, if

$$h_f(\pm\pi/2) < \pi,$$

then

$$h_F(\pm\pi/2) < \pi$$

also. Hence if we assume that $(-1)^n \Re f(n) \geq 0$ and $h_f(\pm\pi/2) < \pi$, we infer that $F(z) \equiv 0$ by 9.3.4. Then $f(z)$ is pure imaginary on the real axis, and any hypothesis about $f(n)$ which makes this impossible leads to a contradiction. Thus we have, for example,

9.9.4. *Theorem.*[b] *Let $f(z)$ be an entire function of exponential type with $h(\pm\pi/2) < \pi$. It is impossible for the numbers $f(n)$ to lie alternately in the open half-planes $x > 0$, $x < 0$.*

9.9.5.* *Theorem.*[c] *If $f(z)$ is an entire function of exponential type with $h(\pm\pi/2) \leq c < \pi$, and* $\arg f(n) = \theta_n$ *with*

$$\liminf_{n\to\infty} \{\cos(\theta_n + n\alpha)\}^{1/n} > 0, \qquad c < \alpha \leq \pi,$$

then $f(z) \equiv 0$.

It is clear from 9.3.8 that the numbers $f(n)$, $f(n+1)$ can be replaced by $f(\lambda_n)$, $f(\lambda_n + 1)$ for a sufficiently dense sequence of integers λ_n.

9.10.[a] **Uniqueness theorems with operators.** Another kind of uniqueness theorem states that $f(z) \equiv 0$ if some set of functionals vanish. A general method by which many theorems of this kind can be proved is to exhibit the functionals as the coefficients in a series which represents the function. Suppose that we have an expansion

$$e^{zw} = \sum_{n=0}^{\infty} u_n(z) g_n(w), \tag{9.10.1}$$

where $g_n(w)$ are continuous and the convergence is uniform on a contour Γ; if the conjugate indicator diagram of $f(z)$ is inside Γ, we have[b]

$$\begin{aligned} f(z) &= \int_\Gamma e^{zw}\phi(w)\,dw = \sum_{n=0}^{\infty} u_n(z) \int_\Gamma \phi(w) g_n(w)\,dw \\ &= \sum_{n=0}^{\infty} u_n(z) T_n(f), \qquad T_n(f) = \int_\Gamma \phi(w) g_n(w)\,dw. \end{aligned} \tag{9.10.2}$$

Then if all $T_n(f) = 0$, it follows that $f(z) \equiv 0$. In most applications the functionals $T_n(f)$ are independent of Γ and can be expressed in other ways; for example, $g_n(w) = w^n \exp\{a_n w\}$ makes $T_n(f) = f^{(n)}(a_n)$.

More generally, if (9.10.1) is uniformly summable on Γ by a method which admits term-by-term integration, the same conclusion follows. We actually have more than a uniqueness theorem: we have an expansion theorem, since $f(z)$ has the expansion (9.10.2), convergent or summable according to what we know about (9.10.1). Kernels other than e^{zw} can also be used in (9.10.1) and (9.10.2).[c]

We illustrate the method by considering the case where $g_n(w) = \{\zeta(w)\}^n$, with $\zeta(w)$ regular and univalent in a neighborhood Ω_w of the origin, and $\zeta(0) = 0$. Let Ω_ζ be the image of Ω_w under $\zeta = \zeta(w)$; then e^{zw} is a regular function of ζ in Ω_ζ, and we have

$$e^{zw} = \sum c_n(z)\zeta^n,$$

(9.10.3)
$$c_n(z) = \frac{d^n}{d\zeta^n}(e^{zw})\bigg|_{\zeta=0},$$

with the series converging in the largest circle Δ_ζ, with center at the origin, contained in Ω_ζ, and Mittag-Leffler summable (§1.7) in the Mittag-Leffler star of Ω_ζ with respect to the origin. If Δ_w is the w-image of the circle and Ω_ζ^* the w-image of the star, we then have

(9.10.4)
$$e^{zw} = \sum c_n(z)\{\zeta(w)\}^n$$

with uniform convergence in any compact subset of Δ_w and uniform Mittag-Leffler summability in any compact subset of Ω_w^*.

We can substitute (9.10.4) into (9.10.2) provided that the conjugate indicator diagram of $f(z)$ is inside Δ_w or Ω_w^*; and if the convergence in (9.10.4) is uniform on part or all of the boundary and $f(z)$ satisfies conditions (5.5.1) making it possible to take Γ as the boundary of the conjugate indicator diagram, we may even have the conjugate indicator diagram touching the boundary of Δ_w or Ω_w^*.

We summarize the discussion in a formal theorem.

9.10.5. *Theorem.*[d] *Let*

$$T_n(f) = \int_C \phi(w)\{\zeta(w)\}^n\,dw,$$

where $\zeta(w)$ is regular and univalent at 0, $\zeta(0) = 0$, *and Δ_w, Ω_w^* are respectively the images under $w = w(\zeta)$ of a circle with center at the origin lying in the ζ-image of a region of univalence of $\zeta(w)$ and of the Mittag-Leffler star of the ζ-image of a region of univalence of $\zeta(w)$. Then if the conjugate indicator diagram D of $f(z)$ is interior to Δ_w, $f(z)$ admits the convergent expansion* (9.10.2) *with coefficients* (9.10.3), *while if D is interior to Ω_w^*, $f(z)$ admits the Mittag-Leffler summable expansion* (9.10.2). *In either case, $T_n(f) = 0$ for all n implies $f(z) \equiv 0$.*

We illustrate the discussion with some examples.

(i) *Newton series.*[e] Here

$$T_n(f) = \Delta^n f(0) = (-1)^n \sum_{k=0}^{n} \binom{n}{k} (-1)^k f(k).$$

These functionals are of the form just considered with $\zeta(w) = e^w - 1$, and the polynomials $c_n(z)$ turn out to be

$$\binom{z}{n} = z(z - 1)\cdots(z - n + 1)/n!.$$

Take Ω_w to be the strip $|v| < \pi$; then Ω_ζ is the ζ-plane with a cut along the negative real axis from $-\infty$ to -1. The circle Δ_ζ is $|\zeta| < 1$, and Δ_w is the part of Ω_w such that $|e^w - 1| < 1$; its boundary is defined by

$$u = \log(2 \cos v).$$

The Mittag-Leffler star of Ω_ζ is Ω_ζ itself, and so Ω_w^* is just Ω_w. Finally we note that the supporting function of Δ_w is

(9.10.6) $$k(\theta) = \cos\theta \log(2\cos\theta) + \theta \sin\theta,$$

with a minimum of log 2. Hence we have

9.10.7. *Theorem. If $f(z)$ is an entire function of exponential type such that $h(\theta) < k(\theta)$, where $k(\theta)$ is defined by* (9.10.6), *and in particular if $f(z)$ is of type less than* log 2, *then*

(9.10.8) $$f(z) = \sum_{n=0}^{\infty} \Delta^n f(0) z(z - 1)\cdots(z - n + 1)/n!,$$

the series converging uniformly to $f(z)$ in any bounded set. If the indicator diagram of $f(z)$ is inside the strip $|v| < \pi$ of the w-plane, the expansion (9.10.8) *is Mittag-Leffler summable to $f(z)$.*

9.10.9. *Corollary. If $f(z)$ is an entire function of exponential type with $h(\pm\pi/2) < \pi$, and $f(n) = 0$ for $n = 0, 1, 2, \cdots$, then $f(z) \equiv 0$.*

This follows from 9.10.7 since $f(n) = 0$ for all $n \geq 0$ implies $\Delta^n f(0) = 0$. This furnishes another proof of Carlson's theorem (9.2.1) for entire functions. The uniqueness theorem for Newton series is thus not particularly interesting. An application of the expansion theorem will be given in §9.12.

(ii) *Abel series.*[f] Here $T_n(f) = f^{(n)}(n)$, $g_n(w) = w^n e^{nw}$; $\zeta(w) = we^w$; $u_0(z) = 1$, $u_n(z) = z(z - n)^{n-1}/n!$. By computation we find that a region Ω_w is bounded by the curve with the polar equation $\rho = (\pi - |\phi|) \csc|\phi|$, and then that Ω_ζ is the ζ-plane cut from $-\infty$ to $-1/e$ along the real axis. The circle Δ_ζ is $|\zeta| < 1/e$, and Δ_w is the (convex) subset of Ω_w where $|we^{1+w}| < 1$, while $\Omega_w^* = \Omega_w$. The supporting function of Δ_w is

$$k(\theta, \Delta_w) = \begin{cases} -(\tfrac{1}{2}) \sin 2\psi \csc\theta, & 0 \leq |\theta| \leq 3\pi/4, \\ |\cos\theta|, & 3\pi/4 < |\theta| \leq \pi, \end{cases}$$

where $\log\{-\cos\psi \csc\theta\} + \sin\psi(\sin\psi - \cot\theta\cos\psi) = 0$. The supporting function of $\Omega_w = \Omega_w^*$ is

(9.10.10) $$k(\theta, \Omega_w) = -\beta(\sin\theta + \cos\theta\cot\beta), \qquad \pi/2 \leq \theta \leq \pi,$$
$$\tan\theta = -\cot\beta + \beta\csc^2\beta.$$

The minimum of $k(\theta, \Delta_w)$ is the positive root (approximately .278) of $xe^{1+x} = 1$, while the minimum of $k(\theta, \Omega_w)$ is 1. Hence the Abel series converges to $f(z)$ if $h(\theta) < k(\theta, \Delta_w)$, and in particular if $f(z)$ is of exponential type less than .278, while it is Mittag-Leffler summable if $h(\theta) < k(\theta, \Omega_w)$, and in particular if $f(z)$ is of type less than 1. The uniqueness theorem is

9.10.11. *Theorem. If $f(z)$ is an entire function of exponential type with $h(\theta) < k(\theta, \Omega_w)$* (9.10.10), *and in particular if $f(z)$ is of type less than* 1, *then $f^{(n)}(n) = 0$, $n = 0, 1, 2, \cdots$ implies $f(z) \equiv 0$.*

Generalized Abel series are discussed in §9.11.

(iii) *Lidstone series.*[g] This series arises from $g_{2n}(w) = w^{2n}$, $g_{2n+1}(w) = e^w w^{2n}$, so that $T_{2n}(f) = f^{(2n)}(0)$, $T_{2n+1}(f) = f^{(2n)}(1)$. These functionals are not of the form considered at the beginning of this section, but the same method applies to them. The result is as follows.

9.10.12.* *Theorem. The Lidstone series of an entire function $f(z)$ of exponential type converges if $f(z)$ is of type less than π, and is Mittag-Leffler summable if the conjugate indicator diagram does not contain any points of the imaginary axis with $|v| \geq \pi$ (and in particular if $h(\pm\pi/2) < \pi$). In the second case, $f(z) \equiv 0$ if $f^{(2n)}(0) = f^{(2n)}(1) = 0$ for $n = 0, 1, 2, \cdots$.*

9.10.13.* *Theorem.*[h] *If $f(z)$ is an entire function of exponential type τ and $f^{(2n)}(0) = f^{(2n)}(1) = 0$ for $n = 0, 1, 2, \cdots$, then $f(z)$ is a sine polynomial*

$$\sum_{k=1}^{m} a_k \sin k\pi z, \qquad m \leq \tau.$$

The expansion of a function in Lidstone series, and the associated uniqueness problem, have been generalized both by considering more general sequences of derivatives and by considering more than two points at which the derivatives are assigned.[i]

Further expansion theorems will be discussed in §12.9, where the emphasis is on the expansion properties rather than on the associated uniqueness theorems.

9.11. Generalized Abel series.[a] Another way of establishing the expansion 9.10.1 when the generating functions $g_n(w)$ of the functionals $T_n(f)$ are not powers of a single function is to study the functions $u_n(z)$. As an illustration of the possibilities we consider the case when $g_n(z) = w^n \exp(a_n w)$; then $T_n(f) = f^{(n)}(a_n)$. We prove the following uniqueness theorem.

9.11.1. *Theorem.*[b] *If $\{a_n\}$ is a sequence of complex numbers of modulus not exceeding* 1, *and if $f(z)$ is an entire function of exponential type less than* log 2, *then $f(z)$ vanishes identically if $f^{(n)}(a_n) = 0$, $n = 0, 1, 2, \cdots$.*

The example $f(z) = \sin \pi z/4 + \cos \pi z/4$ shows that no number greater than $\pi/4 = .785 \cdots$ can replace $\log 2 = .693 \cdots$. A more effective example is $f(\lambda z)$, where $f(z) = e^z + \omega^2 e^{\omega z} + e^{\omega^2 z}$, ω is an imaginary cube root of unity, and λ is the modulus of the zero of $f(z)$ which is closest to the origin; this reduces the upper bound to .748. If the largest number which can be used is W, the so-called Whittaker constant, it can be shown by improving the estimates used in proving 9.11.1 that $W \geq .7259$, and by considering more elaborate examples that $W < .7378$. The exact value of W is unknown. However, if the problem is varied somewhat, exact constants can be obtained. For example, $\pi/4$ is the exact constant if the a_n in 9.11.1 are required to be real, or even to have only real limit points; another case with an exact constant is given in 9.11.4. Among other generalizations we mention the following.

9.11.2.* *Theorem.*[c] *If $f(z)$ is an entire function of exponential type less than* log 2 *and at most a finite number of its derivatives are univalent in* $|z| < 1$ *then $f(z)$ is a polynomial.*

From our present point of view this appears as the uniqueness theorem associated with the functionals $T_0(f) = f(0)$, $T_n(f) = f^{(n-1)}(a_n) - f^{(n-1)}(b_n)$, $|a_n| < 1$, $|b_n| < 1$, $a_n \neq b_n$.

9.11.3.* *Theorem.*[d] *If $f(z)$ is an entire function of exponential type less than* $\log(2 + \sqrt{3}) = 1.31 \cdots$, $f^{(2n+1)}(0) = 0$, *and* $f^{(2n)}(a_n) = 0$, *where* $|a_n| < 1$, *then* $f(z) \equiv 0$.

In particular, this improves 9.11.1 when $f(z)$ is even or odd.

9.11.4.* *Theorem.*[e] *If $f(z)$ is an entire function of exponential type less than* $1/e$, *and* $\{a_n\}$ *is a sequence of complex numbers such that*

$$\limsup_{n\to\infty} n^{-1}(|a_1| + \cdots + |a_n|) \leq 1, \tag{9.11.5}$$

then $f(z)$ vanishes identically if $f^{(n)}(a_n) = 0$, $n = 0, 1, 2, \cdots$. *Furthermore,* $1/e$ *cannot be replaced by any larger number.*

Here the condition that all the a_n have modulus less than 1 is replaced by the condition that their moduli are less than 1 on the average.

In proving 9.11.1 we may assume that $a_0 = 0$ (otherwise consider $f(z) - f(0)$). We then try to find an expansion

$$e^{zw} = \sum_{n=0}^{\infty} c_n(z) w^n \exp(a_n w). \tag{9.11.6}$$

By formal rearrangement we have

$$\begin{aligned} e^{zw} &= \sum_{n=0}^{\infty} c_n(z) \sum_{k=n}^{\infty} a_n^{k-n} w^k/(k-n)! \\ &= \sum_{k=0}^{\infty} w^k \sum_{n=0}^{k} c_n(z) a_n^{k-n}/(k-n)! \end{aligned} \tag{9.11.7}$$

and hence

$$\sum_{n=0}^{k} c_n(z) a_n^{k-n}/(k-n)! = z^k/k!. \tag{9.11.8}$$

From this we can compute the $c_n(z)$ by recursion, starting from $c_0(z) = 1$. However, without calculating the $c_n(z)$ explicitly we can find a bound for them which justifies the rearrangement in (9.11.7) and hence verifies (9.11.6) for a certain w-domain. We shall show by induction that

$$|c_n(z)| \leq (1/\log 2)^n, \qquad |z| \leq 1; \tag{9.11.9}$$

this justifies (9.11.6) for $|w| < \log 2$ and $|z| \leq 1$, and hence, if $f(z)$ is of type less than log 2, the series

$$\sum_{n=0}^{\infty} c_n(z) f^{(n)}(a_n)$$

converges to $f(z)$ at least for $|z| \leq 1$. Therefore if $f^{(n)}(a_n) = 0$ for all n, $f(z)$ vanishes for $|z| \leq 1$ and so for all z. Actually it can be shown by more careful estimates that (9.11.6) converges for all z.

To prove (9.11.9), we note that it is true for $n = 0$; assuming it true for $n \leq k - 1$, we write (9.11.8) in the form

$$c_k(z) + \sum_{n=1}^{k-1} c_n(z) a_n^{k-n}/(k - n)! = z^k/k!,$$

recalling that $a_0 = 0$. Thus since $|a_n| \leq 1$, and $|z| \leq 1$,

$$|c_k(z)| \leq \sum_{n=1}^{k-1} (1/\log 2)^{k-n}/n! + 1/k!$$

$$= (\log 2)^{-k} \sum_{n=1}^{k} (\log 2)^n/n! < (\log 2)^{-k}(e^{\log 2} - 1) = (\log 2)^{-k},$$

so that (9.11.9) holds for $n = k$.

As a matter of fact, it is not hard to obtain the explicit form of $c_n(z)$. By differentiating (9.11.6) repeatedly with respect to w and setting $w = 0$, we see that $c_n(z)$ is a polynomial of degree n depending only on $a_0, a_1, a_2, \cdots, a_{n-1}$. If we exhibit this dependence by writing $c_n(z) = c_n(z; a_0, \cdots, a_{n-1})$, we obtain, by differentiating (9.11.6) with respect to z,

$$\sum_{n=1}^{\infty} c_n'(z) w^{n-1} \exp(a_n w) = \sum_{n=0}^{\infty} c_{n+1}'(z) w^n \exp(a_{n+1} w)$$

$$= \sum_{n=0}^{\infty} c_n(z) w^n \exp(a_n w);$$

this will be satisfied if

$$c_n'(z; a_0, a_1, \cdots, a_{n-1}) = c_{n-1}(z; a_1, a_2, \cdots, a_{n-1}),$$

and only if this is true, since the recursion (9.11.8) determines $c_n(z)$ uniquely.

We can then show by induction that

$$c_n(z) = \int_{a_0}^{z} dz' \int_{a_1}^{z'} dz'' \cdots \int_{a_{n-1}}^{z^{(n-1)}} dz^{(n)};$$

these are known as the Gontcharoff polynomials.

9.12. Integral-valued entire functions. Requiring that functionals $T_n(f)$ are zero (or small) is not the only condition which imposes a special form on an entire function of exponential type. We now prove a theorem which can be stated in the form that 2^z is the smallest transcendental entire function taking integral values at the positive integers.

9.12.1. *Theorem.*[a] *Let $f(z)$ be an entire function of exponential type less than* log 2. *If $f(n)$ is an integer for $n = 0, 1, 2, \cdots$, $f(z)$ is a polynomial.*

This is readily obtainable from 9.10.7, according to which $f(z)$ has the representation

$$f(z) = \sum_{n=0}^{\infty} \Delta^n f(0) z(z-1)\cdots(z-n+1)/n! = \sum_{n=0}^{\infty} \Delta^n f(0) c_n(z).$$

Here all the $\Delta^n f(0)$ are integers. Let $z = -1$; then all the $c_n(-1)$ are ± 1. Since a convergent infinite series of integers must terminate, $f(z)$ is a polynomial.

This theorem has inspired a great many further results, of which we quote a few. Although of course the number log 2 in 9.12.1 cannot be replaced by anything larger, a slightly larger type for $f(z)$ still allows only a very restricted class of integral-valued functions.

9.12.2.* *Theorem.*[b] *Let $f(z)$ be an entire function of exponential type τ, with $f(n)$ an integer for $n = 0, 1, 2, \cdots$. If*

$$\tau < \left| \log\left(\frac{3}{2} + \frac{i\sqrt{3}}{2}\right) \right| = .7588\cdots,$$

$f(z)$ is of the form $P_0(z) + P_1(z)2^z$, where P_0 and P_1 are polynomials. If $\tau < .80$,

$$f(z) = P_0(z) + P_1(z)2^z + P_2(z)\left(\frac{3}{2} + \frac{i\sqrt{3}}{2}\right)^z + P_3(z)\left(\frac{3}{2} - \frac{i\sqrt{3}}{2}\right)^z.$$

Results of this kind, showing that an integral-valued function of exponential type is the sum of finitely many functions of the form $P(z)a^z$, where a is an algebraic integer, cannot be extended as far as $\tau = \pi$, since $\sin \pi z$ is integral-valued. A more recondite example has been constructed to show that $\tau = \log 5 < 1.61$ is not admissible, but just how far one can go is not known. Additional restrictions on the growth of the function lead to further results, for example

9.12.3.* *Theorem.*[c] *If $f(z)$ is an entire function of exponential type,*

$$h(\pm\pi/2) = 0, \qquad h(0) = a, \qquad h(\pi) = b, \qquad e^a - e^{-b} < \sqrt{5},$$

and $f(n)$ is an integer for $n = 0, 1, 2, \cdots$, then $f(z)$ has the form

$$P_0(z) + P_1(z)2^z + P_2(z)3^z + \cdots,$$

where $P_j(z)$ are polynomials.

There are corresponding results when $f(z)$ is required to take integral values at all the integers, positive and negative;[d] the role of 2 is then played by $\frac{1}{2}(3 + \sqrt{5})$.

Notes for Chapter 9

9.1a. See, e.g., Whittaker [2], p. 3; Germay [1].

9.1b. See the references in note 10.2b.

9.1c. For further results of this kind, with more refined hypotheses, see Ganapathy Iyer [13], Levin [1], Hervé [1], Shah [23]. In the rest of this chapter we shall consider mainly theorems in which the zeros are on or close to a single line.

9.2a. See Titchmarsh [3], p. 186.

9.2b. Pólya [2], p. 606, for entire functions.

9.3a. Although formulated differently, theorem and proof are effectively in F. and R. Nevanlinna [1]. More refined results of similar character are given by Siddiqi [1]. Cf. also Brunk [1].

9.3b. Cf. Levinson [4], pp. 3–5.

9.3c. Cf. Levinson [4], p. 119, where the function is used for a different purpose.

9.3d. This fact is generalized by Boas [7].

9.4a. The special case 9.4.2 is due to Pólya (see [2], p. 607) and Valiron [6]; Valiron's proof also covers the statement given here.

9.5a. Fuchs [1]. A generalization is given by Kahane [1]; Fuchs's proof, unlike the other proofs we have given, makes it possible to take into account the size of $h(\theta)$ in all directions instead of just in the direction of the imaginary axis.

9.5b. For generalizations of Carlson's theorem in different directions see Levin [1], Buck [1], Brunk [1].

9.6a. Valiron [6].

9.6b. The content of this theorem was extracted by Boas and Pollard [1] from reasoning of Levinson [4].

9.7a. For the statement cf. Levinson [4], p. 13; see also the notes on Chapter 8.

9.8a. Cartwright [8]. Cf. Levinson [4], p. 17. A short proof of a similar but weaker theorem is given by A. J. Macintyre [3].

9.8b. Levinson [4], p. 19.

9.9a. Valiron [6].

9.9b. Dufresnoy and Pisot [1].

9.9c. Buck [2].

9.10a. The approach used in this section was developed by Gelfond [2] and Buck [3], [4]. The introduction of summability methods is due to Buck; the same idea was used later by Lohin [1], [2]. For other general accounts of parts of the theory see Whittaker [2] and Gontcharoff [1].

9.10b. A. J. Macintyre [4] gives an alternative method in which $\phi(w)$ is expanded instead of the kernel e^{zw}.

9.10c. Buck [3], Lohin [1], [2].

9.10d. Buck [4].

9.10e. For Newton series, Stirling series, and generalizations see the references in note 9.10a; also Lammel [1], A. J. Macintyre [4], Ogura [1], Schmidli [1], Shteinberg [1], Uhl [1].

9.10f. For Abel series see the references in note 9.10a; also Schmidli [1], Eweida [1], [2], A. J. Macintyre [4], S. S. Macintyre [3], A. J. and S. S. Macintyre [1].

9.10g. Whittaker [4], Boas [12], Schmidli [1], S. S. Macintyre [4], [5].

9.10h. Schoenberg [1]; another proof by Boas [10].

9.10i. Schmidli [1], Dzhrbashyan [1], Gelfond and Ibragimov [1], Gurin [1], S. S. Macintyre [4], [5].

9.11a. Theorems and references on generalized Abel series are given by Whittaker [4], Gontcharoff [1], Shteinberg [1]; cf. also the references in note 9.10i and in the other notes on §9.11.

9.11b. This theorem is due to Kakeya; the proof of the text is that of Ibragimov [1]. For other proofs and discussion of the associated expansion theorems see Whittaker [4], Boas [9], Eweida [1], [2], Gelfond [3], S. S. Macintyre [6]. The hypotheses on the location of the a_n can be varied to give theorems for functions of other finite orders or even for functions which are regular only in a finite circle. A theorem in which $f^{(n)}(a_n)$ is not zero, but "small," is given by Boas [9]. The best results on the constant W are due to S. S. Macintyre [2], [3]. The constant $\pi/4$ for real a_n was obtained by Schoenberg, and in the more general case when $\{a_n\}$ has only real limit points, by S. S. Macintyre [3]. For a discussion of problems of this kind in a more general setting see Pólya [3].

9.11c. The theorem was proved by Boas [8] and another proof was given by Levinson [5].

9.11d. Generalization of a theorem of Ganapathy Iyer [8]; cf. Boas [11], p. 842.

9.11e. Davis [1] proves this and other similar results.

9.12a. Theorem 9.12.1 is due to Pólya and Hardy: see Whittaker [4] for references and the proof of the text.

9.12b. The discovery of theorems of the form 9.12.2 was made by A. Selberg [1]; the specific results stated are due to Pisot [1], [2], [3]. See also Buck [5] for a unified discussion, some further results, and references to related problems. The example showing that $\tau = \log 5$ is inadmissible in 9.12.2 is constructed by Buck [5]. For functions which are integral-valued together with their derivatives see A. Selberg [2], where earlier work of Gelfond is cited. Functions which are "almost" integral-valued are considered by Gelfond [1] and Buck [2].

9.12c. Buck [5].

9.12d. For discussion and references see Buck [5].

CHAPTER 10

GROWTH THEOREMS

10.1. Exponential growth on an arithmetic progression. In this chapter we consider the problem of determining the rate of growth of a function of exponential type along a line from its rate of growth along a sequence of points on (or near) the line. The problem may be thought of as a generalization of that considered in Chapter 9, where the function was zero at the points and the conclusion was that it was zero all along the line (and hence always zero). The method which we shall chiefly use is the comparison of $f(z)$ with an interpolation series based on the given sequence of points, and requires a fairly exact knowledge of the behavior of an entire function of exponential type which is defined by a simple infinite product having zeros at the given points.[a] We illustrate the method first by some special cases in which the given sequence of points consists of integers and the function used in forming the interpolation series is simply $\sin \pi z$, whose behavior is both simple and exactly known.

10.1.1. *Pólya's theorem.*[b] *Let $f(z)$ be regular in $|\arg z| \le \alpha \le \pi/2$ and let*

$$h(\theta) \le a \cos \theta + b|\sin \theta|, \qquad |\theta| \le \alpha, \tag{10.1.2}$$

where a and b are finite. If $b < \pi$, we have

$$\limsup_{n\to\infty} n^{-1} \log |f(n)| = \limsup_{r\to\infty} r^{-1} \log |f(r)| = h(0). \tag{10.1.3}$$

Here $h(\theta)$ is the Phragmén-Lindelöf indicator function of $f(z)$ (5.1.1); according to 5.1.2, if $\alpha = \pi/2$ (10.1.2) is equivalent to saying that $f(z)$ is of exponential type with $h(\pm\pi/2) \le b$.

This is another theorem which contains Carlson's theorem (9.2.1).

The example $f(z) = \sin \pi z$ shows that $b = \pi$ is inadmissible; nevertheless, it is true that the condition $b < \pi$ can be weakened to $b \le \pi$ if an additional restriction is imposed on $f(z)$ on the rays $\theta = \pm\alpha$. For entire functions there is a theorem which generalizes 10.1.1 in the same way that 9.2.2 generalizes Carlson's theorem.

10.1.4.* *Theorem.*[c] *If $f(z)$ is an entire function of exponential type and its indicator diagram is not bounded on the right by a vertical line segment of length 2π or more,* (10.1.3) *holds.*

Other theorems in which $f(z)$ is of type π and additional restrictions are imposed will be discussed in a more general setting in §10.7.

It is enough to prove 10.1.1 when $h(0) = 0$, since we may consider $f(z)e^{-h(0)z}$ instead of $f(z)$; we may then take $a = 0$ in (10.1.2). We start by constructing a simple entire function which has the values $f(n)$ at the points

$z = n$. Suppose (without loss of generality) that $\alpha < \pi/2$, and take

$$c > \limsup n^{-1} \log |f(n)|.$$

Then we may put

$$g(z) = \sum_{n=0}^{\infty} \frac{(-1)^n f(n) e^{-cn}}{\pi(z - n)} e^{cz} \sin \pi z. \tag{10.1.5}$$

Since the series converges uniformly in any bounded closed region which excludes the points $z = n$, and $\lim_{z\to n} g(z) = g(n)$, $g(z)$ is an entire function coinciding with $f(z)$ at $z = n$. Hence

$$\phi(z) = \{f(z) - g(z)\}/\sin \pi z \tag{10.1.6}$$

is regular for $|\arg z| \leq \alpha$. If we exclude circles of radius ¼ around $z = n$ from this region,

$$|\phi(z)| \leq \left| \frac{f(z)}{\sin \pi z} \right| + \sum_{n=0}^{\infty} \frac{|f(n)|\, e^{-cn}}{|z - n|} e^{cx},$$

and so

$$r^{-1} \log |\phi(z)| \leq \max \{(b - \pi) \sin \theta,\ c \cos \theta\} \tag{10.1.7}$$

for $|\arg z| \leq \alpha$ and z outside the excluded circles. Hence (10.1.7) holds, by the maximum principle, inside the circles also. In other words, if $h_\phi(\theta)$ is the indicator of $\phi(z)$,

$$h_\phi(\pm\alpha) \leq \max \{(b - \pi) \sin \alpha,\ c \cos \alpha\}.$$

By 5.1.2, it follows that, for $|\theta| \leq \alpha$, $h_\phi(\theta)$ does not exceed the (unique) function of the form $A \cos \theta + B \sin \theta$ which takes the value

$$\max \{(b - \pi) \sin \alpha,\ c \cos \alpha\}$$

at $\theta = \alpha$ and at $\theta = -\alpha$; by inspection this function is

$$\max \{(b - \pi) \tan \alpha,\ c\} \cdot \cos \theta.$$

Thus in particular

$$h_\phi(0) \leq \max \{(b - \pi) \tan \alpha,\ c\}. \tag{10.1.8}$$

By (10.1.6),

$$|f(x)| \leq |\phi(x) \sin \pi x| + |g(x)|$$

and so

$$0 = h(0) \leq \max \{h_\phi(0),\ c\} \tag{10.1.9}$$

because $|g(x)|e^{-cx}$ is bounded. We have, in fact,

$$\pi \mid g(x) \mid e^{-cx} \leq \sum_{\mid n-x \mid > 1} \frac{\mid f(n) \mid e^{-cn}}{\mid x - n \mid} + \sum_{\mid n-x \mid \leq 1} \mid f(n) \mid e^{-cn} \left| \frac{\sin \pi x}{x - n} \right|,$$

and the right-hand side is bounded because c was chosen so that

$$\sum \mid f(n) \mid e^{-cn}$$

converges. By (10.1.8), $h_\phi(0)$ is either 0 or less than 0, so by (10.1.9), $c \geq 0$. Since c can be arbitrarily close to the left-hand side of (10.1.3), $\limsup n^{-1} \log \mid f(n) \mid \geq 0$, and (10.1.3) follows.

10.2. Boundedness on an arithmetic progression. In Pólya's theorem 10.1.1 we compared the rate of growth of $f(x)$ with that of an exponential, i.e. with as rapidly an increasing (or decreasing) function as possible. Now we shall compare $f(x)$ with a function which changes as slowly as possible, namely a constant.

10.2.1. *Cartwright's theorem.*[a] *Let $f(z)$ be regular in $\mid \arg z \mid \leq \alpha \leq \pi/2$, and let*

$$h(\theta) \leq a \mid \cos \theta \mid + b \mid \sin \theta \mid, \qquad \mid \theta \mid \leq \alpha, \tag{10.2.2}$$

where a and b are finite. If $b < \pi$ and $\mid f(n) \mid \leq M$, $n = 0, 1, 2, \cdots$, then $\mid f(x) \mid$ is bounded for $x \geq 0$; if $f(n) \to 0$ as $n \to \infty$, $f(x) \to 0$ as $x \to \infty$.

It is convenient to begin by proving the following more special but more precise theorem.

10.2.3. *Theorem.*[a] *Let $f(z)$ be entire and satisfy (10.2.2) for all θ. If $b < \pi$ and $\mid f(n) \mid < M$, $n = 0, \pm 1, \pm 2, \cdots$, then $\mid f(x) \mid \leq KM$ for all real x, where K depends only on b.*

We may take

$$K = 4 + 2e \log (\pi/(\pi - b)); \tag{10.2.4}$$

the smallest possible K, as a function of b, is asymptotically

$$2\pi^{-1} \log (1/(\pi - b)).$$

We notice first that, by Pólya's theorem (10.1.1) $h(0) \leq 0$ and so we may take $a = 0$ in (10.2.2). Let $0 < \delta < (\pi - b)/\pi$, and consider

$$g(z) = \sum_{n=-\infty}^{\infty} \frac{(-1)^n f(n) \sin \delta(n - z_0)}{\pi(z - n)\delta(n - z_0)} \sin \pi z; \tag{10.2.5}$$

since $\{f(n)\}$ is bounded, the series converges and is an entire function, coinciding at $z = n$ with $f(z)\{\delta(z - z_0)\}^{-1} \sin \delta(z - z_0)$. Hence

$$\phi(z) = \frac{1}{\sin \pi z} \left\{ f(z) \frac{\sin \delta(z - z_0)}{\delta(z - z_0)} - g(z) \right\} \tag{10.2.6}$$

is entire, and, just as in the proof of 10.1.1, $\phi(z)$ is of exponential type.

For $\theta \neq 0, \pi$,

$$(10.2.7)\quad \phi(z) \leq \left|\frac{f(z)}{\sin \pi z}\frac{\sin \delta(z - z_0)}{\delta(z - z_0)}\right| + \sum_{n=-\infty}^{\infty} \frac{|f(n)|}{\pi|z - n|}\left|\frac{\sin \delta(n - z_0)}{\delta(n - z_0)}\right|;$$

since $b + \delta < \pi$, the right-hand side is bounded for each θ. Hence $\phi(z)$ is bounded in particular for $\theta = \pm\pi/4, \pm 3\pi/4$, and so by the Phragmén-Lindelöf theorem (1.4.2), $\phi(z)$ is bounded for all θ. That is, $\phi(z)$ is a bounded entire function, and so a constant. For $z = iy$, $y \to \infty$, the first term on the right of (10.2.7) approaches zero. So does the second, since it does not exceed

$$\frac{M}{\delta\pi}\sum_{n=-\infty}^{\infty}\frac{1}{|iy - n||z_0 - n|} \leq \frac{M}{\delta\pi y^{1/2}}\sum_{n\neq 0}\frac{1}{|n|^{1/2}|z_0 - n|} + \frac{|f(0)|}{\pi|y|}.$$

Therefore $\phi(z) \equiv 0$, and this means that

$$(10.2.8)\quad f(z)\frac{\sin \delta(z - z_0)}{\delta(z - z_0)} = g(z) = \sum_{n=-\infty}^{\infty}\frac{(-1)^n f(n)}{\pi(z - n)}\frac{\sin \delta(z_0 - n)}{\delta(z_0 - n)}\sin \pi z.$$

This is an identity in z; if we take $z = z_0$ and then drop the subscript, we have[b]

$$(10.2.9)\qquad f(z) = \sum_{n=-\infty}^{\infty}\frac{(-1)^n f(n)}{\pi\delta(z - n)^2}\sin \delta(z - n)\sin \pi z.$$

From (10.2.9) Theorem 10.2.3 follows easily, since

$$|f(x)| \leq 2M + M\left\{\sum_{-\infty}^{[x]-1} + \sum_{[x]+2}^{\infty}\right\}\left|\frac{\sin \pi(x - n)\sin \delta(x - n)}{\pi\delta(x - n)^2}\right|;$$

and since $|\theta^{-1}\sin\theta| \leq 1$, if $0 < \epsilon < 1$ we have

$$\left|\frac{\sin \delta(x - n)}{\delta(x - n)}\right| = \left|\frac{\sin \delta(x - n)}{\delta(x - n)}\right|^{1-\epsilon}\left|\frac{\sin \delta(x - n)}{\delta(x - n)}\right|^{\epsilon} \leq \delta^{-\epsilon}|x - n|^{-\epsilon},$$

$$|f(x)| \leq 2M\left\{1 + \pi^{-1}\delta^{-\epsilon}\sum_{1}^{\infty} n^{-1-\epsilon}\right\} \leq 2M\{1 + \pi^{-1}\delta^{-\epsilon} + \epsilon^{-1}\delta^{-\epsilon}\};$$

taking $1/\epsilon = \log(1/\delta)$, we obtain

$$|f(x)| \leq 2M\{2 + e\log(1/\delta)\},$$

and since $\delta = (\pi - b)/\pi$ this is (10.2.4).

There are now several ways of deriving 10.2.1. The shortest is to deduce it from 10.2.3 with the help of Macintyre's lemma (5.6.8). This states that we can write $f(z) = f_1(z) + f_2(z)$, where $f_1(z)$ is entire and of exponential type less than π, and $f_1(x) = O(1/|x|)$ as $x \to -\infty$, while $f_2(z)$ is regular in $|\arg z| < \alpha$ and is $O(1/x)$ as $x \to +\infty$. Then 10.2.3 applies to $f_1(z)$

and shows that $f_1(x)$ is bounded for positive x, so $f(x)$ is bounded (since $f_2(x)$ is bounded).

To complete the proof of 10.2.1 we have to consider the case when $f(n) \to 0$. It is sufficient, by what we have just said, to consider an entire $f(z)$. Returning to (10.2.9), we have, with $N = [x]$, $x > 0$, $M = \sup | f(n) |$,

$$| f(x) | \leq | f(N) | + | f(N+1) | + (\pi\delta)^{-1} \sum_{-\infty}^{N/2} \frac{M}{| x - n |^2}$$

$$+ (\pi\delta)^{-1} \left\{ \sum_{N/2}^{N-1} \frac{| f(n) |}{| x - n |^2} + \sum_{N+2}^{\infty} \frac{| f(n) |}{| x - n |^2} \right\}$$

$$\leq | f(N) | + | f(N+1) | + M(\pi\delta)^{-1} \sum_{N/2}^{\infty} n^{-2}$$

$$+ \left\{ \sum_{N/2}^{N} \frac{1}{(x-n)^2} + \sum_{N+2}^{\infty} \frac{1}{(x-n)^2} \right\} \sup_{n > N/2} | f(n) |,$$

and the right-hand side approaches zero as $x \to \infty$ if $f(n) \to 0$.

Since the proof of Macintyre's lemma is somewhat complicated, there is some interest in having a proof of Cartwright's theorem which does not depend on it. The proof of 10.2.3 can in fact be modified to work, but special devices have to be used because $f(z)$ is not necessarily entire and so we cannot show that the function $\phi(z)$ of the proof of 10.2.3 is constant.[c] (Theorem 10.2.1 can also be proved by different methods depending on the theory of normal families.[d])

We may assume that $\alpha < \pi/2$ and that $| f(z) | \leq Ae^{c| \sin \theta | r}$, $b < c < \pi$, for $\alpha/2 \leq | \arg z | \leq \alpha$, since we may take $a = 0$ in (10.1.2) because of (10.1.1); here A is independent of θ in the specified range. We may also assume that $f(0) = 0$. Let $0 < \delta < (\pi - c)/2$, and let x_0 be an arbitrary positive number. Put $t(z) = z^{-1}(z+1) \sin \pi z$ and

$$s(z) = \delta^{-2}(z - x_0)^{-2} \sin^2 \delta(z - x_0),$$

and consider the functions

$$g(z) = \sum_{n=1}^{\infty} \frac{f(n)s(n)}{(z-n)t'(n)}\, t(z),$$

$$\phi(z) = \{f(z)s(z) - g(z)\}/t(z).$$

Then as in the proof of (10.2.3), $g(z)$ is an entire function coinciding at $z = n$ with $f(z)s(z)$. Hence $\phi(z)$ is regular and of exponential type in $| \arg z | \leq \alpha$. We now show that $\phi(z)$ is bounded on $\arg z = \pm\alpha$, with a bound which is independent of x_0. We have, for $| \arg z | = \alpha$,

$$| \phi(z) | \leq \left| \frac{f(z)s(z)}{t(z)} \right| + \left| \sum_{n=1}^{\infty} \frac{f(n)s(n)}{t'(n)(z - n)} \right|$$
$$= S_1 + S_2.$$

For $| \arg z | = \alpha$, $| s(z) | \leq e^{2\delta r \sin \alpha}$, and $| f(z) | \leq Ae^{cr \sin \alpha}$, while $| t(z) | \geq | \sin \pi z | \geq Be^{\pi r \sin \alpha}$ for $| z | \geq 1$, and $| t(z) |$ has a positive minimum for $| z | < 1$. Hence S_1 has an upper bound independent of x_0 for $| \arg z | = \alpha$, since $c + 2\delta - \pi < 0$. In S_2, $| z - n | \geq \sin \alpha$, and so

$$S_2 \leq \frac{M}{\pi \sin \alpha} \sum_{n=1}^{\infty} s(n),$$

(10.2.10) $$\sum_{n=1}^{\infty} s(n) \leq \delta^{-2} \left\{ \sum_{n=1}^{[x_0]-1} + \sum_{[x_0]+2}^{\infty} \right\} (n - x_0)^{-2} + 2$$
$$\leq 2\delta^{-2} \sum_{n=1}^{\infty} n^{-2} + 2 < 12\delta^{-2}.$$

Hence $| \phi(z) | \leq K$ on $| \arg z | = \alpha$ with K independent of x_0. Therefore, by the Phragmén-Lindelöf theorem (1.4.2), $| \phi(x) | \leq K$. Hence $| f(x)\, s(x) | \leq K + | g(x) |$. Since $| t(x)/(x-n) | \leq 4\pi$,

$$| g(x) | \leq \sum_{n=1}^{\infty} \frac{| f(n) | \, s(n)}{\pi \, | x - n |} \, | t(x) | \leq 4M \sum_{n=1}^{\infty} s(n),$$

whence by (10.2.10), $| g(x) | \leq 48M\delta^{-2}$. So $| f(x)s(x) | \leq K + 48M\delta^{-2}$, $x \geq 0$. Since $s(x_0) = 1$, $| f(x_0) | \leq K + 48M\delta^{-2}$, and since x_0 is arbitrary, 10.2.1 is proved.

An interesting application of 10.2.3 is to entire functions of zero exponential type. According to 6.2.13, such a function reduces to a constant if it is bounded on the real axis. Now we shall show that merely being bounded on the integers (and even somewhat less) suffices.

10.2.11. *Theorem.*[e] *If $f(z)$ is an entire function of zero exponential type and $f(n) = o(| n |)$ |, as $| n | \to \infty$, $f(z)$ is a constant. More generally, if $f(n) = O(| n |^p)$, $f(z)$ is a polynomial of degree not exceeding p.*

Since $f(az)$ is of zero exponential type when $f(z)$ is, any arithmetic progression may be substituted for $\{n\}$. The question of how general a sequence $\{\lambda_n\}$ can be used instead will be discussed in §10.7.

To prove 10.2.11, suppose first that $f(n) = O(1)$; then $| f(x) | = O(1)$ by 10.2.3, and so by 6.2.13, $f(z)$ is a constant. In the general case, let $P(z)$ denote the sum of the Maclaurin expansion of $f(z)$ up to terms of order $[p]$; then $f(z) - P(z)$ has a zero of order $[p] + 1$ at 0. Hence

$$\{f(z) - P(z)\}/z^{[p]+1}$$

is entire, of zero exponential type, and bounded at $z = n$, and therefore a constant. Thus $f(z) = cz^{[p]+1} + P(z)$, and since $P(n) = O(|n|^{[p]}) = o(|n|^{[p]+1})$ and $f(n) = O(|n|^{p}) = o(|n|^{[p]+1})$, c must be 0.

Finally, if $f(n) = o(|n|)$, the case $p = 1$ shows that $f(z)$ is a polynomial $P(z)$ of degree 1 at most, but $P(n) = o(|n|)$ is possible only if $P(z)$ is a constant.

10.2.12.* *Corollary.*[f] *If $f(z)$ is an entire function of zero exponential type and the sequences $\{f(\pm n)\}$ are summable (C, k) to 0, $f(z) \equiv 0$.*

There is a simple direct proof of 10.2.11 in the case when $\{f(n)\}$ is bounded.[g] We may assume without loss of generality that $\sum |f(n)| < \infty$, since $z^{-1}\{f(z) - f(0)\} = g(z)$ is still of zero type and has $g(n) = O(1/n)$, and $G(z) = \{g(z)\}^2$ is still of zero type and has $\sum |G(n)| < \infty$. Let

$$f(z) = \int_C \phi(w)e^{zw}dw,$$

where C is any contour surrounding the origin. We have

$$\Delta^n f(0) = \int_C \phi(w)(e^w - 1)^n \, dw,$$

and if $|z|$ is sufficiently small,

$$\begin{aligned} H(z) = \sum_{n=0}^{\infty} \Delta^n f(0) z^n &= \int_C \phi(w) \sum_{n=0}^{\infty} z^n (e^w - 1)^n \, dw \\ &= \int_C \phi(w)\{1 - (e^w - 1)z\}^{-1} \, dw \\ &= \int_C \phi(w)(1 + z)^{-1} \sum_{n=0}^{\infty} e^{nw}\{z/(1 + z)\}^n \, dw \\ &= (1 + z)^{-1} \sum_{n=0}^{\infty} f(n)\{z/(1 + z)\}^n. \end{aligned}$$

The last series converges if $|z| < |1 + z|$, so that $H(z)$ can be continued analytically into the half plane $x > -\frac{1}{2}$, in which this inequality holds. Moreover, since $\sum |f(n)| < \infty$, $H(z)$ is bounded for $x > -\frac{1}{2}$. On the other hand, we also have

$$\begin{aligned} H(z) &= \int_C \phi(w)(-1/z) \sum_{n=0}^{\infty} e^{-(n+1)w}\{(z + 1)/z\}^n \, dw \\ &= (-1/z) \sum_{n=0}^{\infty} f(-n - 1)\{(1 + z)/z\}^n, \end{aligned}$$

so that $H(z)$ can also be continued analytically into $x < \frac{1}{2}$ and is bounded there. Hence $H(z)$ is entire and bounded and therefore a constant.

Hence $f(z)$ is constant, for example by an application of Carlson's theorem.

We cannot of course allow $b = \pi$ in Cartwright's theorem 10.2.1 (or 10.2.3) any more than in Carlson's theorem or in Pólya's theorem 10.1.1 unless we impose some additional restrictions. The problem has not received much attention; we quote one result.

10.2.13.* *Theorem.*[h] *Let* $f(z)$ *be an entire function satisfying* $|f(z)| \leq r\epsilon(z)e^{\pi r}$, *where* $\epsilon(z)$ *is uniformly* $O(1)$ *and* $\epsilon(\pm iy) = o(1)$. *Let* $|f(n)| \leq M$ $(n = 0, \pm 1, \pm 2, \cdots)$, *and let there be numbers* L, N *such that every interval of length* L *contains an integer* m *for which* $|f'(m)| < N$. *Then* $f(z)$ *is bounded on the real axis.*

10.3. Exponential growth on a sequence having a density. The theorems of §§10.1 and 10.2 can be generalized by replacing the special sequence $\{n\}$ by much less severely restricted sequences. The restrictions which have to be imposed vary according to the kind of growth which is being considered; in this section we generalize 10.1.1.

10.3.1. *V. Bernstein's theorem.*[a] *Let* $f(z)$ *be regular in* $|\arg z| \leq \alpha \leq \pi/2$, *and let*

$$h(\theta) \leq a \cos \theta + b\,|\sin \theta|, \qquad |\theta| \leq \alpha, \tag{10.3.2}$$

where a *and* b *are finite. If* $b < \pi$, *we have*

$$\limsup_{n\to\infty} |\lambda_n|^{-1} \log |f(\lambda_n)| = \limsup_{r\to\infty} r^{-1} \log |f(r)| = h(0) \tag{10.3.3}$$

provided that $\{\lambda_n\}$ *is a (complex) sequence such that* $n/\lambda_n \to 1$ *and*

$$|\lambda_m - \lambda_n| \geq \delta|n - m|, \qquad n \neq m,\ \delta > 0.$$

No necessary and sufficient condition for (10.3.3), analogous to Fuchs's theorem (9.5.1), is known; the hypothesis of 9.5.1 is not sufficient.[b] In §10.7 we shall discuss generalizations of 10.3.1 in various directions; in particular, we shall see that the condition $b < \pi$ is somewhat more restrictive than necessary. The theorem can also be extended by allowing the λ_n to be farther from the real axis:

10.3.4.* *Theorem.*[c] *Let* $f(z)$ *be regular and of exponential type in* $|\arg z| \leq \alpha$; $h(\pm\alpha) < \pi B \sin \alpha$, $B > 0$; and $\limsup |\lambda_n|^{-1} \log |f(\lambda_n)| \leq 0$, *where* $n/|\lambda_n| \to D \geq B \sin \alpha \csc (\alpha - \beta)$, $|\lambda_{n+1}| - |\lambda_n| \geq \delta > 0$, $|\arg \lambda_n| \leq \beta$. *Then* $h(0) \leq \pi B \sin \alpha \sin \beta \csc (\alpha - \beta)$.

The proof of (10.3.1) is almost the same as that of 10.1.1, except that trivial estimates of $|\sin \pi z|$ have to be replaced by corresponding estimates for

$$F(z) = \prod_{n=1}^{\infty} (1 - z^2/\lambda_n^2) \tag{10.3.5}$$

which are by no means trivial. When the λ_n are real the estimates are somewhat simpler.

10.3.6. *Lemma.*[d] *Let* $F(z)$ *be defined by* (10.3.5), *where* $n/\lambda_n \to D > 0$ *and* $|\lambda_m - \lambda_n| \geq \delta|n - m|$, $\delta > 0$. *Then for every positive* ϵ,

$$(10.3.7) \qquad F(re^{i\theta}) = O\{\exp[\pi(D|\sin\theta| + \epsilon)r]\},$$

$$(10.3.8) \qquad 1/F(re^{i\theta}) = O\{\exp[\pi(-D|\sin\theta| + \epsilon)r]\}, \quad |z - \lambda_n| \geq \delta/8,$$

$$(10.3.9) \qquad 1/F'(\lambda_n) = O\{\exp(\epsilon\lambda_n)\}.$$

Conclusion (10.3.7) follows from 7.2.9 if $\sum |\lambda_n|^{-1} |\arg \lambda_n|$ converges. The lemma is also true when $D = 0$, but requires a somewhat different (and simpler) proof. Conclusion (10.3.8) implies (10.3.9), since

$$(10.3.10) \qquad F'(\lambda_m) = \lim_{z\to\lambda_m} \frac{F(z)}{z - \lambda_m} = -\lim_{z\to\lambda_m} \prod_{n\neq m} \left(1 - \frac{z^2}{\lambda_n^2}\right) \frac{z + \lambda_m}{\lambda_m^2},$$

and the product on the right satisfies the same conditions as $F(z)$ itself.

Before proving (10.3.8) we use 10.3.6 to prove V. Bernstein's theorem. We follow the proof of Pólya's theorem 10.1.1, using $F(z)$ instead of $\sin \pi z$. As before, we may assume $a = h(0) = 0$, $\alpha < \pi/2$; let

$$c > \limsup |\lambda_n|^{-1} \log |f(\lambda_n)|,$$

and put

$$g(z) = \sum_{n=0}^{\infty} \frac{f(\lambda_n)\exp(-c\lambda_n)}{(z - \lambda_n)F'(\lambda_n)} e^{cz} F(z),$$

an entire function coinciding with $f(z)$ at $z = n$. Then

$$\phi(z) = \{f(z) - g(z)\}/F(z)$$

is regular for $|\arg z| \leq \alpha$. It satisfies

$$h_\phi(\theta) \leq \max\{(b - \pi)|\sin\theta|, c\cos\theta\}$$

(here (10.3.8) is required). Therefore, as in the proof of 10.1.1,

$$h_\phi(0) \leq \max\{(b - \pi)\tan\alpha, c\};$$

but $|f(x)| \leq |\phi(x)F(x)| + |g(x)|$, so that $0 = h(0) \leq \max\{h_\phi(0), c\}$, and hence $c \geq 0$. Since we may take c arbitrarily close to

$$\limsup |\lambda_n|^{-1} \log |f(\lambda_n)|,$$

(10.3.3) follows.

We now prove 10.3.6. Since $F(z)$ is even, we may restrict z to the right-hand half plane. If $D > 0$, the "typical" $F(z)$ is $\sin \pi Dz$ or $\cos \pi Dz$, with zeros n/D or $(n + \tfrac{1}{2})/D$, and we shall proceed by showing that replacing

λ_n by n/D or $(n + \frac{1}{2})/D$ changes the asymptotic behavior of $F(z)$ at most to the extent asserted by (10.3.7) and (10.3.8). Since we are considering only the case when $D > 0$, we may suppose by a change of variable that $D = 1$.

Let ϵ be a (small) positive number, independent of z. Break the product (10.3.5) for $F(z)$ into three products Π_1, Π_2, Π_3 extending respectively over

$$|\lambda_n| \leq (1 - \epsilon)|z|,$$

$$(1 - \epsilon)|z| < |\lambda_n| < (1 + \epsilon)|z|,$$

$$|\lambda_n| \geq (1 + \epsilon)|z|.$$

Our aim is to show that Π_2 is nearly 1 and that $\Pi_1\Pi_3$ is nearly $|\sin \pi z|$.

We begin with Π_2. Since $n/\lambda_n \to 1$, if $|z|$ is large enough there are at most $2\epsilon|z|(1 + \epsilon)$ terms in Π_2, and so

$$\prod_2 |1 - z^2/\lambda_n^2| \leq \prod_2 3 \leq e^{4\epsilon|z|(1+\epsilon)}. \tag{10.3.11}$$

This gives us an upper bound for Π_2.

To obtain a lower bound for Π_2 we start from the fact that for large $|z|$ (and hence large $|\lambda_n|$) we have in Π_2

$$|\lambda_n^2/(z^2 - \lambda_n^2)| \leq |\lambda_n/(z - \lambda_n)|.$$

With z fixed, let $|z - \lambda_n|$ be a minimum when $n = N$; then

$$|\lambda_n - \lambda_N| \leq |z - \lambda_N| + |\lambda_n - z| \leq 2|z - \lambda_n|,$$

and so

$$|\lambda_n^2/(z^2 - \lambda_n^2)| \leq 2|\lambda_n|/|\lambda_n - \lambda_N|.$$

Consequently, if a prime denotes omission of the term for which $n = N$ (if it occurs in Π_2), we have, since $|\lambda_n - \lambda_N| \geq \delta|n - N|$,

$$\begin{aligned}\prod_2{}' \left|\frac{\lambda_n^2}{z^2 - \lambda_n^2}\right| &\leq \prod_2{}' \frac{2|\lambda_n|}{|\lambda_n - \lambda_N|} \\ &\leq 2^M(1 + \epsilon)^M |z|^M \delta^{-M} \prod_2{}' \frac{1}{|n - N|},\end{aligned} \tag{10.3.12}$$

where M is the number of terms in Π_2. Since $n/\lambda_n \to 1$, $M \leq 2\epsilon|z|(1 + \epsilon)$ if $|z|$ is large enough. We increase the last product in (10.3.12) by increasing some of its terms, and since there are at least $[(M - 1)/2]$ consecutive integers $|n - N|$ occurring in the product,

$$\prod_2{}' \left|\frac{\lambda_n^2}{z^2 - \lambda_n^2}\right| \leq (4/\delta)^M |z|^M/[(M - 1)/2)]!^2.$$

We estimate the factorial by using Stirling's formula in the form $k! \geq e^{k \log k - k}$, so that

$$\prod_2{}' \left| \frac{\lambda_n^2}{z^2 - \lambda_n^2} \right| \leq (4e^2/\delta)^M e^{M \log(|z|/M)}$$
$$\leq (4e^2/\delta)^{2\epsilon|z|(1+\epsilon)} e^{2\epsilon|z|(1+\epsilon) \log\{2\epsilon(1+\epsilon)\}}.$$

Since $\epsilon \log \epsilon \to 0$ as $\epsilon \to 0$, we have

$$\prod_2{}' \left| \frac{\lambda_n^2}{z^2 - \lambda_n^2} \right| \leq e^{\eta |z|},$$

where η is a function of ϵ which approaches 0 as $\epsilon \to 0$. If $n = N$ occurs in Π_2 we still have to consider the term

$$| \lambda_N^2/(z^2 - \lambda_N^2) | \leq 4 | z | / | z - \lambda_N |$$

if $| z |$ is large enough; since this upper bound exceeds 1, we have in all cases

$$\prod_2 \left| \frac{\lambda_n^2}{z^2 - \lambda_n^2} \right| \leq \frac{4 | z |}{| z - \lambda_N |} e^{\eta |z|},$$

or

$$\prod_2 | 1 - z^2/\lambda_n^2 | \geq \frac{| z - \lambda_N |}{4 | z |} e^{-\eta |z|}. \tag{10.3.13}$$

This is our lower bound for Π_2.

In estimating Π_1 and Π_3 we use the inequality

$$| (\lambda_n^2 - n^2)/\lambda_n^2 | \leq \epsilon^2/8, \qquad n \geq N_0, \tag{10.3.14}$$

where N_0 depends only on ϵ. Consider first λ_n which occur in Π_1 or Π_3 and for which $| \lambda_n | \leq 2 | z |$. Then $|| \lambda_n | - | z || \geq \epsilon | z |$, and

$$| n^2 - z^2 | \geq | \lambda_n^2 - z^2 | - | n^2 - \lambda_n^2 | \geq \epsilon | z || \lambda_n + z | - \epsilon^2 \lambda_n^2/8$$
$$\geq \epsilon | z || z/2 | - \epsilon^2 | z |^2/2,$$

so that if ϵ is small enough

$$| (n^2 - z^2)/z^2 | \geq \epsilon/4. \tag{10.3.15}$$

Now

$$\frac{1 - z^2/\lambda_n^2}{1 - z^2/n^2} = 1 + \frac{z^2(\lambda_n^2 - n^2)}{\lambda_n^2(n^2 - z^2)}, \tag{10.3.16}$$

and by using (10.3.14) and (10.3.15) we have

$$1 - \epsilon \leq \left| \frac{1 - z^2/\lambda_n^2}{1 - z^2/n^2} \right| \leq 1 + \epsilon \tag{10.3.17}$$

for $|\lambda_n| \leq 2|z|$, λ_n in Π_1 or Π_3, and $n > N_0$. We may then estimate Π_1 by using (10.3.17) for the at most $2|z|$ terms for which it applies, and introducing a number C (independent of z) such that

$$1/C \leq \prod_{n=1}^{N_0} \left| \frac{1 - z^2/\lambda_n^2}{1 - z^2/n^2} \right| \leq C.$$

We then have

$$C^{-1}(1 - \epsilon)^{2|z|} \prod_1 |1 - z^2/n^2| \leq \prod_1 |1 - z^2/\lambda_n^2| \leq C(1 + \epsilon)^{2|z|} \prod_1 |1 - z^2/n^2|. \tag{10.3.18}$$

For Π_3 the terms with $|\lambda_n| \leq 2|z|$ are covered by (10.3.17). For those with $|\lambda_n| > 2|z|$, we have by using (10.3.14) and (10.3.16) again,

$$1 - \frac{\epsilon^2 |z|^2}{8(n^2 - |z|^2)} \leq \left| \frac{1 - z^2/\lambda_n^2}{1 - z^2/n^2} \right| \leq 1 + \frac{\epsilon^2 |z|^2}{8(n^2 - |z|^2)}.$$

Since $\lambda_n/n \to 1$, for large $|z|$ and $|\lambda_n| > 2|z|$ we have

$$n^2 - |z|^2 \geq n^2 - |\lambda_n|^2/4 \geq \epsilon n^2/8.$$

Hence

$$(1 - \epsilon)^{2|z|} \prod_{|\lambda_n| > 2|z|} \left(1 - \frac{\epsilon |z|^2}{n^2}\right) \leq \prod_3 \left| \frac{1 - z^2/\lambda_n^2}{1 - z^2/n^2} \right| \leq (1 + \epsilon)^{2|z|} \prod_{|\lambda_n| > 2|z|} \left(1 + \frac{\epsilon |z|^2}{n^2}\right).$$

Now

$$\log \prod_{|\lambda_n| > 2|z|} (1 - \epsilon |z|^2/n^2) \geq - \sum_{|\lambda_n| > 2|z|} 2\epsilon |z|^2/n^2 \geq -4\epsilon |z|,$$

$$\log \prod_{|\lambda_n| > 2|z|} (1 + \epsilon |z|^2/n^2) \leq \sum_{|\lambda_n| > 2|z|} \epsilon^2 |z|^2/n^2 \leq 2\epsilon |z|,$$

$$2|z| \log (1 - \epsilon) \geq -4\epsilon |z|, \qquad 2|z| \log (1 + \epsilon) \leq 2\epsilon |z|,$$

and so

$$e^{-8\epsilon|z|} \prod_3 |1 - z^2/n^2| \leq \prod_3 |1 - z^2/\lambda_n^2| \leq e^{4\epsilon|z|} \prod_3 |1 - z^2/n^2|.$$

Combining this with (10.3.11), (10.3.13) and (10.3.18), we have

$$C^{-1} \frac{|z - \lambda_N|}{4|z|} e^{-(12\epsilon+\eta)|z|} \prod_{1,3} |1 - z^2/n^2| \leq |F(z)| \leq e^{(6\epsilon+2\epsilon(1+\epsilon))|z|} \prod_{1,3} |1 - z^2/n^2|. \tag{10.3.19}$$

The product $\Pi_{1,3}$ which occurs in (10.3.19) is

$$\left| \frac{\sin \pi z}{\pi z} \right| \div \prod_2 | 1 - z^2/n^2 |,$$

and this Π_2 is subject to the same estimates as the Π_2 considered earlier (with a different N, of course we could give more precise estimates here but there is no advantage in doing so). Combining these estimates with (10.3.19), and taking a new η, we have (10.3.7), and also

$$(10.3.20) \qquad | F(z) | / | z - \lambda_N | \geq e^{-\eta |z|} | \sin \pi z |,$$

where $\eta \to 0$ when $\epsilon \to 0$.

Since $\lambda_n/n \to 1$ means that $\lambda_n/(n + \frac{1}{2}) \to 1$ also, by modifying our calculations to compare $F(z)$ with $\cos \pi z$ we also have

$$| F(z) | / | z - \lambda_N | \geq e^{-\eta |z|} | \cos \pi z |.$$

This, with (10.3.20), establishes (10.3.8).

10.4. An application. As an application of V. Bernstein's theorem we prove a theorem whose content is roughly that along a ray from the origin a function of exponential type must frequently be almost as large as its indicator allows. (Compare the theorems of §§7.2, 7.3, which give stronger conclusions under more restrictive hypotheses; also the theorems of §3.7.)

10.4.1. *Theorem.*[a] *Let $f(z)$ be of exponential type in the angle $| \arg z | \leq \alpha \leq \pi$. Then for each θ, $| \theta | < \alpha$, and each positive δ and ϵ, there are a sequence $r_n \uparrow \infty$ and a positive η such that the subset of $(r_n, r_n + \delta r_n)$ in which*

$$r^{-1} \log | f(re^{i\theta}) | \geq h(\theta) - \epsilon$$

has measure at least ηr_n.

In other words, $r^{-1} \log | f(re^{i\theta}) | \geq h(\theta) - \epsilon$ for a set of positive upper linear density on each ray.

It is enough to prove 10.4.1 for $\theta = 0$, and for $\alpha < \pi/2$ and

$$h(\pm\alpha) < \pi \sin \alpha.$$

Suppose that with given δ and ϵ we have $r^{-1} \log | f(r) | \leq h(0) - \epsilon$ in all of the intervals $(n, n + \delta n)$ in sets of measure

$$\delta n - o(n).$$

Then we can find a real increasing sequence $\{\lambda_n\}$ such that

$$\lambda_n^{-1} \log | f(\lambda_n) | \leq h(0) - \epsilon, \qquad n - \lambda_n = o(n),$$

and $| \lambda_n - \lambda_m | > d > 0$ for $n \neq m$. Applying 10.3.1 to this sequence we get $h(0) < h(0) - \epsilon$, a contradiction.

By using 10.4.1 we can prove the theorem on the indicator diagram of a product which we stated as 5.4.12.

10.4.2. *Theorem.*[b] *If $f_1(z)$ is an entire function of exponential type whose indicator diagram is a point, and $f_2(z)$ is any entire function of exponential type, the indicator diagram of $f_1(z)f_2(z)$ is the sum of the indicator diagrams of $f_1(z)$ and $f_2(z)$.*

What we have to prove is that $h(\theta) \geq h_1(\theta) + h_2(\theta)$, where h, h_1, h_2 are the indicators of f_1f_2, f_1, f_2. By multiplying $f_1(z)$ by some exponential $e^{\alpha z}$ we can bring its indicator diagram to the origin, so that it is sufficient to consider the case when $h_1(\theta) = 0$. In this case $f_1(z)$ is of exponential type 0, and by (3.7.1), for every positive ϵ there are circles, the sum of whose radii in $|z| \leq r$ is $o(r)$, such that $|f_1(z)| \geq e^{-\epsilon r}$ outside them. On the other hand, we have $\log|f_2(z)| \geq r\{h_2(\theta) - \epsilon\}$ on a set of positive upper density on $\arg z = \theta$. Hence there are points with arbitrarily large r on $\arg z = \theta$ such that $\log|f_1(z)f_2(z)| \geq h_2(\theta) - 2\epsilon$, so that $h(\theta) \geq h_2(\theta)$.

Another application of 10.3.1 is to the proof of 9.8.1.

10.5. Boundedness on a sequence which is close to an arithmetic progression. We now extend Cartwright's theorems of §10.2 by allowing a more general sequence $\{\lambda_n\}$, but keeping the same rate of growth for $f(x)$. It can then be proved that if $f(z)$ is of exponential type for $x \geq 0$, $h(\pm\pi/2) < \pi$, and $f(\lambda_n)$ is bounded for a complex sequence $\{\lambda_n\}$ with $|\lambda_n - \lambda_m| \geq \delta > 0$ $(n \neq m)$, and $|\lambda_n - n| < L$, then $f(x)$ is bounded. The sequence $\{\lambda_n\}$ is thus much more restricted than in §10.3, where the chief requirement was that $|\lambda_n - n| = o(n)$. The theorem is not true without some condition more restrictive than this, but it is an open question whether the condition $|\lambda_n - n| < L$ can be weakened to $|\lambda_n - n| < \epsilon(n)$ with some $\epsilon(n)$ which becomes infinite. We shall prove the theorem only for real $\{\lambda_n\}$, in order to keep the proof from becoming too complicated; the proof given here follows the pattern of §10.2 as closely as possible, but there are proofs depending on entirely different ideas.

10.5.1. *Theorem of Duffin and Schaeffer.*[a] *Let $f(z)$ be regular in $|\arg z| \leq \alpha \leq \pi/2$ and let*

$$(10.5.2) \qquad h(\theta) \leq a|\cos\theta| + b|\sin\theta|, \qquad |\theta| \leq \alpha,$$

where $b < \pi$. If $\{\lambda_n\}$ is an increasing sequence of real numbers such that $\lambda_{n+1} - \lambda_n \geq 2\delta > 0$, $|n - \lambda_n| \leq L$, $n = 1, 2, \cdots$, and $\{f(\lambda_n)\}$ is bounded, then $f(x)$ is bounded for $x > 0$; if $f(\lambda_n) \to 0$, $f(x) \to 0$ as $x \to \infty$.

We may deduce 10.5.1 by Macintyre's lemma, just as in §10.2, from the following theorem about entire functions.

10.5.3. *Theorem.*[a] *If $f(z)$ is entire, and satisfies* (10.5.2) *for all θ, with $b < \pi$; if $\{\lambda_n\}_{-\infty}^{\infty}$ is an increasing sequence of real numbers satisfying the* *conditions of* 10.5.1 *for* $-\infty < n < \infty$, *and* $|f(\lambda_n)| \le M$, *then* $|f(x)| \le KM$ *for* $-\infty < x < \infty$, *where K depends only on b, L and* δ.

We have, just as in §10.2, the following corollary.

10.5.4. *Corollary. If* $f(z)$ *is an entire function of zero exponential type and* $|f(\lambda_n)| \le M$, *where the sequence* $\{\lambda_n\}$ *satisfies the conditions of* 10.5.1 *for* $-\infty < n < \infty$, $f(z)$ *is a constant.*

See also §10.7.

The proof of 10.5.3 depends on obtaining sufficiently sharp estimates for the product

$$(10.5.5) \qquad \psi(z) = \lim_{N\to\infty} (z - \lambda_0) \prod_{-N}^{N}{}' (1 - z/\lambda_n);$$

the term corresponding to $n = 0$ is omitted from $\prod'$. The following lemma provides estimates which are sufficient for our purposes; the constants are by no means the best possible.[b]

10.5.6. *Lemma. There are numbers* $C_1, \cdots, C_5, Q$, *depending only on L and* δ, *such that* $\psi(z)$, *defined by* (10.5.5), *satisfies*

$$(10.5.7) \qquad 1/|\psi(x + iy)| \le C_1 y^{2Q}/\sinh \pi(|y| - |x| - Q), \qquad |y| > |x| + Q;$$

$$(10.5.8) \qquad |\psi(x)| \le C_2, \qquad 0 \le x \le 1;$$

$$(10.5.9) \qquad 1/|\psi'(\lambda_n)| \le C_3 + C_4 |\lambda_n|^{C_5}.$$

We first use 10.5.6 to establish 10.5.3. Choose $\epsilon > 0$ and a positive integer k so that $k\epsilon < \pi - b$, $k > C_5 + 1$, $k > 2Q$. Let

$$(10.5.10) \qquad \omega(z) = (\epsilon z)^{-k} \sin^k(\epsilon z),$$

and consider the function

$$(10.5.11) \qquad H(z) = f(z)\omega(z - z_0) - \sum_{n=-\infty}^{\infty} \frac{f(\lambda_n)}{z - \lambda_n} \frac{\omega(\lambda_n - z_0)}{\psi'(\lambda_n)} \psi(z).$$

Our choice of k makes the series converge to an entire function coinciding with $f(z)\omega(z - z_0)$ at $z = \lambda_n$, and so $H(z)/\psi(z)$ is an entire function, which is of exponential type as in §10.2. For $\arg z = \pm\theta$, $\pi \pm \theta$, $\pi/4 < \theta < \pi/2$, we have by using (10.5.2) and (10.5.7)

$$\log |H(z)/\psi(z)| = O\{r(a \cos\theta + [(b + \eta) + k\epsilon - \pi] \sin\theta)\}$$

with η arbitrarily small; so if θ is close enough to $\pi/2$, $H(z)/\psi(z)$ is bounded on these four rays. Hence it is bounded for all z, and so is a constant. Since $b < \pi$, by (10.5.7) $H(z)/\psi(z) \to 0$ as $y \to \infty$, and therefore the constant is zero. Thus

$$f(z)\omega(z - z_0) = \psi(z) \sum_{-\infty}^{\infty} \frac{f(\lambda_n)}{z - \lambda_n} \frac{\omega(\lambda_n - z_0)}{\psi'(\lambda_n)}.$$

This is true for all z, and we may therefore take $z = z_0$ and drop the subscript; since $\omega(0) = 1$,

$$f(z) = \sum_{-\infty}^{\infty} \frac{f(\lambda_n)}{z - \lambda_n} \frac{\omega(\lambda_n - z)}{\psi'(\lambda_n)} \psi(z). \tag{10.5.12}$$

Now for $0 \leq x \leq 1$,

$$| f(x) | \leq M \left(\sum_1 + \sum_2\right) \frac{| \omega(\lambda_n - x) | \, | \psi(x) |}{| x - \lambda_n | \, | \psi'(\lambda_n) |},$$

where $\sum_1$ refers to λ_n in (say) $-\frac{1}{2} \leq x \leq \frac{3}{2}$, and $\sum_2$ to the remaining λ_n. In $\sum_1$, $| \omega(\lambda_n - x) |/| x - \lambda_n |$ has a bound depending only on L and δ, and so has the number of terms in $\sum_1$. In $\sum_2$,

$$| \omega(\lambda_n - x) |/| x - \lambda_n | \leq \epsilon^{-k} | 1 - \lambda_n |^{-k-1},$$

and so $\sum_2$ also has a bound depending only on L and δ. That is, there is a function $K(\delta, L)$ such that $| f(x) | \leq MK(\delta, L)$, $0 \leq x \leq 1$.

Now consider the function $f(z + m)$, which is bounded by M at

$$\lambda_n^{(m)} = \lambda_{n+m} - m;$$

we have $\lambda_n^{(m)} - \lambda_{n-1}^{(m)} = \lambda_{n+m} - \lambda_{n+m-1} \geq 2\delta$ and

$$| \lambda_n^{(m)} - n | = | \lambda_{n+m} - (n + m) | \leq L;$$

so for $0 \leq x \leq 1$, $| f(x + m) | \leq MK(\delta, L)$ for $m \leq x \leq m + 1$. Since m is any integer, the conclusion of 10.5.3 follows.

The proof shows why the condition $| \lambda_n - n | < L$ is needed: this condition is preserved when the λ_n are translated, while a condition of the form $| \lambda_n - n | < \epsilon(n)$ is not preserved. We can prove 10.5.4 with weaker estimates because an inequality $| f(z) | \leq K(x)M$, with a function $K(x)$ depending only on b, L and δ, can be applied to $f(x)^m$, when $f(z)$ is of zero exponential type, to yield $| f(x) | \leq \{K(x)\}^{1/m}M$, and so (if we let $m \to \infty$) $| f(x) | \leq M$ for all real x; but a function of zero exponential type, bounded on the real axis, is a constant. This seems to account for the fact that 10.5.4 can be extended to more general $\{\lambda_n\}$, whereas 10.5.3 cannot (or at least has not been; no counterexample is known).

We now establish 10.5.6. Since we do not need the most precise estimates possible, we shall make some simplifications in the data to make the discussion easier. It is convenient to have $\lambda_n > 0$ for $n > 0$ and $\lambda_n < 0$ for $n < 0$; if this does not happen to be the case, we re-index the λ's so that λ_0 is the one closest to 0. Since $| \lambda_0 | < L$ and there are at most $L/(2\delta)$ λ's between 0 and λ_0, we change the indices by at most $L/(2\delta)$ and so our original conditions are still satisfied with $Q = L + L/(2\delta)$. By increasing Q we can make it an integer. Since λ_0 is the λ closest to 0, $|\lambda_n| > \delta$ for $n \neq 0$. We then have, for $n > 0$,

$$\left|\left(1 - \frac{z}{\lambda_n}\right)\left(1 - \frac{z}{\lambda_{-n}}\right)\right|$$

$$\geq \left|\Re\left(1 - \frac{z}{\lambda_n}\right)\left(1 - \frac{z}{\lambda_{-n}}\right)\right| = \left|1 + \frac{y^2 - x^2 + x(\lambda_n + \lambda_{-n})}{-\lambda_n\lambda_{-n}}\right|$$

$$\geq 1 + \frac{y^2 - x^2 - 2Q\,|x|}{(n+Q)^2}$$

$$> 1 + \frac{y^2 - (x+Q)^2}{(n+Q)^2}$$

for $|y| > |x| + 2Q$. Hence

$$|\psi(x+iy)| \geq (|y| - |x| - Q) \prod_{n=1}^{\infty} \left\{1 + \frac{y^2 - (|x|+Q)^2}{(n+Q)^2}\right\}$$

$$= (|y| - |x| - Q) \sinh\{\pi[y^2 - (|x|+Q)^2]^{\frac{1}{2}}\}$$

$$\div \prod_{n=1}^{Q} \left(1 + \frac{y^2 - (|x|+Q)^2}{n^2}\right)$$

Since $\sinh t$ increases and $y^2 - (|x|+Q)^2 \geq (|y| - |x| - Q)^2$,

$$|\psi(x+iy)|$$

$$\geq (|y| - |x| - Q) \sinh \pi(|y| - |x| - Q) \div \prod_{n=1}^{Q} (1 + y^2/n^2)$$

and (10.5.7) follows.

To prove (10.5.8), we observe that $|\lambda_n| > 1$ if $|n| > Q$, and we have $|\lambda_n| > \delta$ if $|n| > 0$. Then for $0 \leq x \leq 1$,

$$|\psi(x)| = |x - \lambda_0| \prod_{n=1}^{\infty} \left|1 - \frac{x}{\lambda_n}\right|\left|1 + \frac{x}{\lambda_{-n}}\right|$$

$$\leq (1+Q)\left(\frac{1+\delta}{\delta}\right)^{2Q} \prod_{n=Q+1}^{\infty} \left(1 - \frac{x}{n+Q}\right)\left(1 + \frac{x}{n-Q}\right)$$

$$= (1+Q)\left(\frac{1+\delta}{\delta}\right)^{2Q} \left|\frac{\sin \pi x}{\pi x}\right| \div \prod_{n=1}^{2Q} \left(1 - \frac{x}{n}\right)$$

and (10.5.8) follows.

For (10.5.9), we observe that

$$\psi'(\lambda_m) = \frac{\lambda_m - \lambda_0}{\lambda_m} \prod_{0<n\neq m} \left(1 - \frac{\lambda_m}{\lambda_n}\right)\left(1 + \frac{\lambda_m}{-\lambda_{-n}}\right), \qquad m > 0.$$

Suppose for definiteness that $\lambda_m > 0$. Since the λ_n increase,

$$| \psi'(\lambda_m) |$$
$$\geq \frac{\lambda_m - \lambda_0}{\lambda_m} \prod_{0<n<m} \left(\frac{\lambda_m}{\lambda_n} - 1\right)\left(1 + \frac{\lambda_m}{-\lambda_{-n}}\right) \prod_{n>m} \left(1 - \frac{\lambda_m}{\lambda_n}\right)\left(1 + \frac{\lambda_m}{-\lambda_{-n}}\right).$$

We write the product as $\Pi_1\Pi_2\Pi_3$, where Π_1 contains the terms (if any) with $n < [\lambda_m] - Q - 1$; Π_3, the terms with $n > [\lambda_m] + Q + 1$; Π_2, the rest. Then

$$\Pi_1 \geq \prod_1 \left(\frac{\lambda_m}{n + Q} - 1\right)\left(1 + \frac{\lambda_m}{n + Q}\right) = \prod_1 \left(\frac{\lambda_m^2}{(n + Q)^2} - 1\right)$$

and $n + Q < [\lambda_m] - 1$ in the term of highest index, $n + Q < [\lambda_m] - 2$ in the next lower term, etc., so that

$$\Pi_1 \geq \prod_{k=Q+1}^{[\lambda_m]-1} \left(\frac{\lambda_m^2}{k^2} - 1\right) = \prod_{k=1}^{[\lambda_m]-1} \left(\frac{\lambda_m^2}{k^2} - 1\right) \div \prod_{k=1}^{Q} \left(\frac{\lambda_m^2}{k^2} - 1\right).$$

For Π_3, we have

$$\Pi_3 \geq \prod_3 \left(1 - \frac{\lambda_m}{n - Q}\right)\left(1 + \frac{\lambda_m}{n + Q}\right)$$
$$\geq \prod_{[\lambda_m]+2}^{\infty} \left(1 - \frac{\lambda_m^2}{k^2}\right) \div \prod_{[\lambda_m]+2}^{[\lambda_m]+2Q+1} \left(1 + \frac{\lambda_m}{k}\right).$$

Thus

$$\Pi_1\Pi_3 \geq \prod_{1\leq k\neq[\lambda_m],[\lambda_m]+1} \left| 1 - \frac{\lambda_m^2}{k^2} \right|$$
$$\div \left\{\prod_{k=1}^{Q} \left(\frac{\lambda_m^2}{k^2} - 1\right) \prod_{[\lambda_m]+2}^{[\lambda_m]+2Q+1} \left(1 + \frac{\lambda_m}{k}\right)\right\}.$$

The denominator is increased if we replace $(\lambda_m^2/k^2) - 1$ by λ_m^2/k^2 and $1 + \lambda_m/k$ by 2; so, with numbers A_1, A_2 depending only on L and δ,

$$\Pi_1\Pi_3 \geq A_1\lambda_m^{-2Q-1} | \sin \pi\lambda_m | \div \left| \left(1 - \frac{\lambda_m^2}{[\lambda_m]^2}\right)\left(1 - \frac{\lambda_m^2}{([\lambda_m] + 1)^2}\right) \right|$$
$$\geq A_2\lambda_m^{-2Q-1}.$$

In Π_2 we have $2Q + 3$ terms of the form

$$\left| 1 - \frac{\lambda_m}{\lambda_n} \right| \cdot \left| 1 + \frac{\lambda_m}{-\lambda_{-n}} \right| \geq \frac{| n - m | \delta}{n - Q};$$

$[\lambda_m] - Q - 1 \leq n \leq [\lambda_m] + Q + 1$, so $n - Q \leq \lambda_m + 1$, and

$$\Pi_1 \geq A_5(\lambda_m + 1)^{-2Q-3}.$$

This establishes (10.5.9) when $[\lambda_m] > Q + 2$. For smaller positive values of $[\lambda_m]$ the product Π_1 does not occur and $\Pi_3 \geq A_4$, so (10.5.9) is true generally.

10.6. Applications to L^p and l^p problems. In 6.7.1 we proved incidentally that an entire function of exponential type, belonging to L^p on the real axis, approaches zero as $| x | \to \infty$. A simple deduction from 10.5.1 is the following generalization of this fact.

10.6.1. *Theorem.*[a] *If $\phi(t)$ is a nonnegative unbounded nondecreasing function, $f(z)$ is a function regular and of exponential type in an angle $| \arg z | \leq \alpha$, $0 < \mu < \nu$, and*

$$\int_\mu^\nu \phi(| f(x + t | \cdot) \, dt$$

is bounded, $0 < x < \infty$, then $f(x)$ is bounded; if also

$$\int_\mu^\nu \phi(| f(x + t) |) \, dt \to 0, \qquad x \to \infty,$$

and $\phi(t) > 0$ for $t > 0$, then $f(x) \to 0$.

We may suppose that $\alpha < \pi/2$, and then (by considering $f(cz)$, if necessary) that $h(\theta) \leq a \cos \theta + b | \sin \theta |$, $b < \pi$. We have

$$\int_\mu^{\mu+\gamma} \phi(| f(n + t) |) \, dt$$

bounded, $\gamma = \min (\nu - \mu, \frac{1}{4})$, and hence $f(\lambda_n)$ is bounded for some sequence $\{\lambda_n\}$ such that $| \lambda_n - n | < \frac{1}{4}$ or $f(\lambda_n) \to 0$ for a similar sequence. By 10.5.1, $f(x)$ is bounded or approaches zero, respectively.

It follows, in particular, that 6.7.9 and 6.7.10 can be generalized as follows.

10.6.2. *Theorem. If $f(z)$ is an entire function of exponential type, and if, for some fixed numbers μ and ν,*

$$\int_\mu^\nu | f\{(r + s)e^{i\theta}\} | \, ds \text{ is bounded} \tag{10.6.3}$$

for three values of θ, no two of which differ by as much as π; or if (10.6.3) holds for some two values θ, $\theta + \pi$, and $f(z)$ is of zero type; then $f(z)$ is a constant.

In §6.7 we also proved that the convergence of $\int_{-\infty}^{\infty} | f(x) |^p \, dx$, for an entire function $f(z)$ of exponential type, implies the convergence of

$$\sum_{-\infty}^{\infty} |f(\lambda_n)|^p,$$

for a fairly general sequence $\{\lambda_n\}$. The results of §10.5 suggest a converse result in which, if $f(z)$ is of sufficiently small type, the convergence of a series implies the convergence of the corresponding integral. There are a number of theorems of this kind; we shall establish one which corresponds to 10.2.3 and illustrates one of the kinds of proof which can be used.

10.6.4. *Theorem.*[b] *If $f(z)$ is an entire function of exponential type, $h(\pi/2) + h(-\pi/2) = 2c < 2\pi$, $\phi(x)$ is an increasing convex nonnegative function, and $\sum_{-\infty}^{\infty} \phi(|f(n)|) < \infty$, then there is a positive number H (depending only on c and ϕ) such that*

$$\int_{-\infty}^{\infty} \phi(H|f(x)|)\,dx \le \sum_{n=-\infty}^{\infty} \phi(|f(n)|). \tag{10.6.5}$$

In particular, for $\phi(t) = t^p$, $p \ge 1$, we have

10.6.6. *Corollary.*[c] *If $f(z)$ is an entire function of exponential type,*

$$h(\pi/2) + h(-\pi/2) = 2c < 2\pi,$$

and $\sum_{-\infty}^{\infty} |f(n)|^p = M < \infty$, then there is a function $K(p, c)$ (depending only on c and p) such that

$$\int_{-\infty}^{\infty} |f(x)|^p\,dx \le K(p, c)M. \tag{10.6.7}$$

The function $K(p, c)$ has the form $K(c)^p$ when $p > 1$. Cartwright's theorem (10.2.3) may be regarded as the limiting case $p = \infty$. It can be shown[d] that, when $p > 1$ and the hypothesis on $h(\theta)$ is replaced by

$$\lim_{r\to\infty} M(r)e^{-\pi r} = 0,$$

(10.6.7) still holds with some $K(p)$.

It can also be shown[e] that 10.6.4 remains true if $\phi(t)$ is an increasing nonnegative continuous function such that $\phi(0) = 0$, $\phi(\infty) = \infty$, $\phi(x)/x$ decreases and $\phi(xy) \le \phi(x)\phi(y)$ (in particular, $\phi(t) = t^p$, $0 < p < 1$), provided that we replace (10.6.5) by

$$\int_{-\infty}^{\infty} \phi(|f(x)|)\,dx \le K \sum_{-\infty}^{\infty} \phi(|f(n)|),$$

so that 10.6.6 remains true when $0 < p < 1$.

In addition, 10.6.4 and the modification just stated remain true when $\{n\}$ is replaced by an increasing sequence $\{\lambda_n\}$ satisfying the hypotheses of (10.5.3) ($\lambda_{n+1} - \lambda_n \ge 2\delta > 0$ and $|\lambda_n - n| \le L$).[f]

There are similar theorems involving mean values,[g]

$$\limsup_{T\to\infty} (2T)^{-1} \int_{-T}^{T} \phi(|\, f(x) \,|)\, dx, \qquad \limsup_{N\to\infty} (2N)^{-1} \sum_{-N}^{N} \phi(|\, f(n) \,|).$$

In proving 10.6.4 we may suppose that $h(\pm\,\pi/2) = c < \pi$, since $e^{i\alpha z}f(z)$, with a suitable real α, satisfies this and has the same modulus for real z. Since $\sum \phi\{|\, f(n) \,|\}$ converges, $\{f(n)\}$ is bounded, and so we have the formula (10.2.9),

$$f(x) = \sum_{n=-\infty}^{\infty} \frac{(-1)^n f(n)}{\pi\delta(z-n)^2} \sin \delta(z-n) \sin \pi z, \qquad 0 < \delta < \pi - c.$$

Applying this to $f(z+m)$, we have

$$f(z+m) = \sum_{n=-\infty}^{\infty} \frac{(-1)^n f(n+m)}{\pi\delta\,(z-n)^2} \sin \delta(z-n) \sin \pi z.$$

If Γ denotes the square of side 1 with center at 0, and μ_m is the maximum of $|\, f(z+m) \,|$ on this square, we have

$$\mu_m \le |\, f(m) \,| \max_{\Gamma} \left| \frac{\sin \delta z}{\delta z} \frac{\sin \pi z}{\pi z} \right| + \sum_{n\neq 0} \frac{|\, f(n+m) \,|}{\pi\delta(\,|\, n \,| - \frac{1}{2})^2}$$

$$= \sum_{n=-\infty}^{\infty} |\, f(n+m) \,|\, b_n = \sum_{n=-\infty}^{\infty} |\, f(n) \,|\, b_{n-m}\,,$$

where b_n are numbers independent of n such that $\sum_{-\infty}^{\infty} b_n$ converges. Let $1/H = \sum_{-\infty}^{\infty} b_n$. Then because $\phi(t)$ is convex,

$$\phi(H\mu_m) = \phi\left\{H \sum_{n=-\infty}^{\infty} |\, f(n) \,|\, b_{n-m}\right\} \le H \sum_{n=-\infty}^{\infty} b_{n-m}\phi(\,|\, f(n) \,|\,),$$

and hence

$$\sum_{m=-\infty}^{\infty} \phi(H\mu_m) \le H \sum_{m=-\infty}^{\infty} b_{n-m} \sum_{n=-\infty}^{\infty} \phi(\,|\, f(n) \,|\,) = \sum_{n=-\infty}^{\infty} \phi(\,|\, f(n) \,|\,),$$

$$\int_{-\infty}^{\infty} \phi(H\,|\, f(x) \,|\,)\, dx \le \sum_{m=-\infty}^{\infty} \int_{-\frac{1}{2}}^{\frac{1}{2}} \phi\{H\,|\, f(x+m) \,|\,\}\, dx$$

$$\le \sum_{m=-\infty}^{\infty} \phi\left\{\max_{-\frac{1}{2}\le x \le \frac{1}{2}} H\,|\, f(x+m) \,|\right\}$$

$$\le \sum_{m=-\infty}^{\infty} \phi(H\mu_m) \le \sum_{n=-\infty}^{\infty} \phi(\,|\, f(n) \,|\,).$$

The case of a more general $\{\lambda_n\}$ can be handled in a similar way by formula 10.5.12, where in the definition (10.5.10) of $\omega(z)$ we take k so large that

$k > C_5 + 2$, and so that $\sum_n \phi(n^{C_5-k})$ converges if the alternative hypotheses are imposed on $\phi(t)$. The resulting theorem reads as follows.

10.6.8.* *Theorem. Let $f(z)$ be an entire function of exponential type such that $h(\pi/2) + h(-\pi/2) \leq 2c \leq 2\pi$. Let $\{\lambda_n\}$ be a sequence of complex numbers, with $\lambda_0 = 0$, $|\lambda_n - n| \leq L$, $|\lambda_{n+m} - \lambda_n| > 2\delta > 0$ $(m \neq 0)$. Let $\beta(x)$ be a nondecreasing function such that $|\beta(x + L) - \beta(x - L)| \leq b$ and (if $L < 1$) $|\beta(x + \frac{1}{2}) - \beta(x - \frac{1}{2})| \leq b$. Let $\phi(x)$ be either* (i) *an increasing convex nonnegative function or* (ii) *an increasing nonnegative continuous function such that $\phi(0) = 0$, $\phi(\infty) = \infty$, $\phi(x)/x$ decreases, and $\phi(xy) \leq \phi(x)\phi(y)$. Then if $\sum_{-\infty}^{\infty} \phi(|f(\lambda_n)|)$ converges, there are numbers H and K, depending only on c, L, δ, b and ϕ, such that*

$$\int_{-\infty}^{\infty} \phi(H|f(x)|)\, d\beta(x) \leq K \sum_{n=-\infty}^{\infty} \phi(|f(\lambda_n)|);$$

in case (i), $K \leq b$; *in case* (ii), $H \geq 1$.

Another interesting special case results from taking $\phi(x) = e^{-1/x}$ for x near 0: if $\sum \exp\{-1/|f(\lambda_n)|\}$ converges, then

$$\int_{-\infty}^{\infty} \exp\{-H/|f(x)|\}\, dx$$

converges for some H.

10.7. Generalizations. The principal theorems of this chapter have had the following general form (taking for simplicity the case when the function is regular in the right-hand half plane): if $f(z)$ is of exponential type,

$$h(\pm \pi/2) < \pi,$$

and the growth of $f(z)$ is known on a sequence $\{\lambda_n\}$ which is nearly the sequence of positive integers, then the growth of $f(z)$ is known along the whole positive real axis. It is natural, especially after the refinements of Carlson's theorem given in Chapter 9, to ask how much the condition $h(\pm \pi/2) < \pi$ can be weakened (if some supplementary condition is added to exclude functions like $z^m \sin \pi z$). The general situation is that even if $h(\pm \pi/2) = \pi$, the conclusion

$$\limsup_{n\to\infty} |\lambda_n|^{-1} \log |f(\lambda_n)| = h(0) \tag{10.7.1}$$

holds provided that an excess of $\log|f(iy)|$ over $\pi|y|$, or a deficiency of $n_\lambda(t)$ under t, is compensated for by a corresponding excess or deficiency in the other quantity. Thus Theorem 10.3.1 allows $n_\lambda(t) - t$ to be as small as $-o(t)$, but compensates for this by demanding that $\log|f(iy)| < \pi|y| - \epsilon|y|$, $\epsilon > 0$. More delicate possibilities are illustrated by the following statements: (10.7.1) holds if[a] either

$$(10.7.2)\quad \begin{cases} \log | f(iy) | \le \pi | y | + \epsilon(| y |), & \int_1^\infty y^{-2}\epsilon(y)\, dy < \infty, \\ n_\lambda(t) - t \ge \delta(t), \quad \int_1^\infty t^{-2}\delta^+(t) = \infty, & \int_1^\infty t^{-2}\delta^-(t)\, dt > -\infty; \end{cases}$$

or

$$(10.7.3)\quad \begin{cases} \log | f(iy) | \le \pi | y | - \epsilon(| y |), & \int_1^\infty y^{-2}\epsilon(y)\, dy = \infty, \\ n_\lambda(t) - t \ge \delta(t), & \int_1^\infty t^{-2}\delta^-(t)\, dt > -\infty. \end{cases}$$

Other combinations of conditions producing the same general effect can also be used. A similar result, which is interesting because it contains a necessary and sufficient condition, is

10.7.4.* *Theorem.*[b] *Let* $\{\lambda_n\}$ *be a sequence of complex numbers such that* $n/\lambda_n \to 0$, $\arg \lambda_n \to 0$, *and* $| \lambda_n - \lambda_m | \ge | n - m | \delta$, $\delta > 0$, $n \ne m$. *Then*

$$\limsup_{n\to\infty} | \lambda_n |^{-1} \log | f(\lambda_n) | = h(0),$$

for all functions $f(z)$ *which are regular and of exponential type in* $x \ge 0$ *and satisfy*

$$(10.7.5)\qquad \int_{-\infty}^{\infty} \frac{\log^+ | f(iy) |}{1 + y^2}\, dy < \infty,$$

if and only if $\sum 1/| \lambda_n |$ *diverges.*

If $\{\lambda_n\}$ satisfies the stronger separation condition

$$|| \lambda_n | - | \lambda_m || \ge | n - m | \delta,$$

the sufficiency part of this result follows from 7.2.6, which shows that $\lim_{x\to\infty} x^{-1} \log | f(x) |$ exists if x is excluded from a set of finite logarithmic length; the conclusion can then be deduced from Lemma 1.5.3. By 7.2.9, (10.7.5) can be replaced by weaker conditions. The necessity part is established by the second part of 9.3.8.

It would be interesting to find a similar necessary and sufficient condition for $f(\lambda_n) = O(1)$ to imply $f(x) = O(1)$.

In §§10.3 and 10.5 we compared the growth of $f(z)$ with that of an exponential and a constant, respectively. Another question which arises naturally is whether other rates of growth can be used. There are theorems in this direction, but the existing results are rather fragmentary. When $|\lambda_n - n | \le L$ (or even when a somewhat more general condition is satisfied), more general comparison functions can be used;[c] but it is naturally to be expected that a more rapidly increasing comparison function can be used also when $|\lambda_n - n| \leq \epsilon(n)$, where $\epsilon(n) \to \infty$ but $\epsilon(n) = o(n)$. One might expect (always assuming the spacing condition

$$|\lambda_n - \lambda_m| \geq \delta |n - m|)$$

that $|\lambda_n - n| \leq \epsilon(n)$ would imply

$$(10.7.6) \quad \limsup_{n\to\infty} \epsilon(\lambda_n)^{-1} \log |f(\lambda_n)| = \limsup_{x\to\infty} \epsilon(x)^{-1} \log |f(x)|,$$

or at least that

$$(10.7.7) \quad \limsup_{x\to\infty} \epsilon(x)^{-1} \log |f(x)| < \infty \text{ if } \limsup_{n\to\infty} \epsilon(\lambda_n)^{-1} \log |f(\lambda_n)| < \infty.$$

The most that is known, however, is that[d] (10.7.7) holds if

$$|\lambda_n - n| \leq \epsilon(n)/\log \{n/\epsilon(n)\},$$

and $\epsilon(x)$ satisfies a number of auxiliary conditions which keep its behavior from being too irregular. The greater simplicity of the condition

$$|\lambda_n - n| < L$$

in the Duffin-Schaeffer theorem is possible because this condition is invariant under translation; on the other hand the greater simplicity of V. Bernstein's theorem comes from the fact that it uses $|\lambda_n - n| = o(n)$ instead of using a specific $\epsilon(n)$. In the same general direction it is also known that[e] for the class of entire functions $\omega(z)$ of genus zero, with zeros z_n, for which $|\omega(x)| \geq 1$ and $\sum 1/|\Im(z_n)| < \infty$, there exist a sequence $\{\lambda_n\}$ and an entire function $\psi(z)$ of exponential type c such that, for every entire function $f(z)$ of exponential type τ less than c, $|f(\lambda_n)| \leq |\omega(\lambda_n)|$ implies

$$|f(x)| \leq C |\psi(x)|,$$

where C depends only on ω, c and τ.

Results of this kind are evidently related to the theorems of §9.8, where the hypothesis on an entire function that $f(\lambda_n)$ is small led to the conclusion that $f(z) \equiv 0$, i.e. to the strongest possible restriction on the growth of the function.

Rather more can be said in the special case when the function in question is entire and of zero exponential type. Here we know from the Duffin-Schaeffer theorem (see 10.5.4) that boundedness on a set $\{\lambda_n\}$ with

$$|\lambda_n - n| < L$$

(and the usual spacing condition) makes the function a constant. We can say definitely that a more irregular sequence, but not every sequence of unit density, will suffice: if $|\lambda_n - n| < \epsilon(n)$ and $\int^{\infty} t^{-2}\epsilon(t)\,dt$ diverges, there is an entire function of zero exponential type, not a constant, which is bounded on $\{\lambda_n\}$; but if $\int^{\infty} t^{-2}\,\epsilon(t) \log \{t/\epsilon(t)\}\,dt$ converges, an entire function of zero exponential type, bounded on $\{\lambda_n\}$, is necessarily constant.[f]

Another way of generalizing 10.5.4 is to assume less about the rate of growth of $f(z)$ at the points $\{\lambda_n\}$ and to obtain a weaker conclusion. For example,[g] if $f(z)$ is an entire function of exponential type zero, and if

$$\limsup |\lambda_n|^{-\rho} \log |f(\lambda_n)|$$

is finite, where $|\lambda_n - n| < n^{\rho}$, $0 < \rho < 1$ (with the usual spacing condition), then $f(z)$ is actually at most of order ρ.

Throughout this chapter we have supposed that the points λ_n are subject to a spacing condition which keeps them from being too close together. Our results naturally do not hold unless some sort of spacing condition is assumed,[h] but it is possible to let the points coalesce in groups, provided that we assume something in the nature of "multiple boundedness." For example, we have

10.7.8.* *Theorem.*[i] *If $f(z)$ is regular and of exponential type in $x \geq 0$, $h(\pm\,\pi/2) < k\pi$, and*

$$|f(n)| < C,\ |f'(n)| < C, \cdots, |f^{(k-1)}(n)| < C, \qquad n = 1,2,\cdots,$$

then $f(x)$ is bounded for $x \geq 0$.

Another line of generalization consists in letting the sequence $\{\lambda_n\}$ spread out; in the simplest case after that in which all the λ_n are close to one line, $\{\lambda_n\}$ breaks into two sequences each of which is close to a different line. We quote two theorems in this direction, the first of which is related to the second somewhat as V. Bernstein's theorem is related to the theorem of Duffin and Schaeffer.

10.7.9.* *Theorem.*[j] *Let $f(z)$ be regular and of exponential type in*

$$|\arg z| \leq \omega, \qquad \omega < \pi/2.$$

Let $\limsup |z|^{-1} \log |f(z)| \leq 0$ *as $z \to \infty$ on two sequences $\{\lambda_n\}$, $\{\mu_n\}$ satisfying* $\lim n/\lambda_n = \lambda \neq 0$, $\lim n/\mu_n = \mu \neq 0$,

$$-\,\omega < \phi = -\arg\lambda < \psi = -\arg\mu < \omega,$$

and $|\lambda_n - \lambda_m| \geq c\,|n - m|$, $\quad |\mu_n - \mu_m| \geq c\,|n - m|$, $n \neq m$. *Let* $g(z) = \sin \pi\lambda z \sin \pi\mu z\, e^{\pi i(\lambda-\mu)z}$. *Then if $h_f(\pm\,\omega) \leq h_g(\pm\,\omega)$, it follows that $h_f(\theta) \leq h_g(\theta)$ for $|\theta| < \omega$.*

10.7.10.* *Corollary.*[k] *If $f(z)$ is an entire function of exponential type less than that of* $\sin \pi\lambda z \sin \pi\mu z$, *and is bounded on sequences* $\{\pm \lambda_n\}$, $\{\pm \mu_n\}$ *satisfying the conditions of* 10.7.9, *$f(z)$ is a constant.*

10.7.11.* *Theorem.*[l] *If the sequences* $\{\lambda_n\}$, $\{\mu_n\}$ *in* 10.7.9 *are of the form* n/λ, n/μ, *then $f(z)$ is bounded for* $\phi \le |\arg z| \le \psi$ *if it is bounded on* $\{\lambda_n\}$ *and* $\{\mu_n\}$ *and satisfies the other hypotheses of* 10.7.9.

There are also many theorems in which the behavior of $f(z)$ is prescribed on the lattice points in an angle (or in the whole plane), or on some other set of more or less evenly distributed points.[m]

Finally, analogy with §9.10 suggests that we should consider theorems in which the growth of a sequence $T[f(\lambda_n)]$ or even $T_n[f(\lambda_n)]$, where T and T_n are appropriate operators, determines the growth of $f(z)$. The introduction of operators $T[f]$ also introduces a new question, that of determining the growth of $T[f(x)]$ when the growth of $f(x)$ is known. Results of this kind will be discussed in the next chapter.

Notes for Chapter 10

10.1a. For other methods see V. Bernstein [1], Duffin and Schaeffer [3], Agmon [1], Kjellberg [1], [2].

10.1b. Pólya [2].

10.1c. Pólya [2].

10.2a. Cartwright [6]. Other proofs have been given by Pfluger [1], A. J. Macintyre [1], Boas [6], Korevaar [3]; still others are obtainable as special cases of the theorems of §10.5 (references in note 10.5a). The proof of the text is that of Boas and Schaeffer [1], who obtain the asymptotic order of K; S. Bernstein [12] obtains the constant $2/\pi$. For trigonometric polynomials the result (with a change of variable) was obtained for special sequences of points by Gronwall [1] and Grandjot [1]: if a trigonometric polynomial of degree n is bounded by M at the points $2k\pi/(2n+1)$, it is bounded everywhere by $K(n)M$, where $K(n) \sim 2\pi^{-1} \log n$.

10.2b. For other interpolation formulas of a similar character see later sections of this chapter and also §11.4. For the general question of interpolation to entire functions see especially Pfluger [4] and Noble [2], [3], and references given by these authors. For functions of higher order there are results which have no counterpart for functions of order 1; for example, an entire function of order 2 and type less than $\pi/2$ is a constant if it is bounded at the lattice points $m + in$: Ganapathy Iyer [5], Pfluger [1].

10.2c. The proof given in the text is similar to one given by Korevaar [3]. The theorem is still true in a curvilinear angle bounded by $\arg z = \pm\theta(r)$ provided that $\theta(r)$ is continuous, $\pi/2 > \alpha > \theta(r) > 0$ and $\liminf r\theta(r) > 0$, if we suppose that

$$|f(z)| < A \exp\{br \sin \theta(r)\}$$

on the boundary.

10.2d. Duffin and Schaeffer [3], Agmon [1].

10.2e. The theorem is contained in work of Valiron [6]; it was later set as a problem by Pólya and attracted considerable attention.

10.2f. H. L. Selberg [1].

10.2g. Buck [7]; another simple proof is given by Korevaar [2].

10.2h. A. J. Macintyre [1], Ahiezer [5].

10.3a. V. Bernstein [1] proved this and similar theorems by using the theory of Dirichlet series. The proof of the text originated with Levinson [2] and Pfluger [2]. For generalizations in various directions see Cartwright [10], Ganapathy Iyer [2]–[7], [10]–[12], Junnila [1], Levin [2], Pfluger [4], Noble [illegible], [illegible], and references given by these authors.

10.3b. Fuchs [1].

10.3c. Pfluger [2].

10.3d. This is a sharpened form, due to Levinson [4], of a theorem of Carlson (for which see V. Bernstein [1]). For an alternative proof for real λ_n see Redheffer [1], [2]. Similar results are given by Mandelbrojt [2].

10.4a. V. Bernstein [2].

10.4b. Pólya [2], p. 593.

10.5a. The theorem was first proved by Duffin and Schaeffer [3]. The proof of the text is essentially that of Levin [3]. See also Ahiezer [5]. The interpolation formula used here and by Levin goes back to Valiron [6]. Another proof (of a more general result) is given by Agmon [1]. It should be noted that Levinson [1] gave estimates for (10.5.5) which would suffice to prove the theorem. See also Kjellberg [1], [2], Macphail [1], S. Bernstein [12].

10.5b. I quote the best estimates (Levin [3]) since they are useful, for example, in formulating the uniqueness theorem suggested in §9.6: if λ_0 is one of the λ_n which is closest to 0,

$$| \psi(z) | > O(1) \exp \{\pi | y |\} | y/z^2 |^{2L}, \qquad | y | > L;$$

$$| \psi(z) | < O(1) \exp \{\pi | y |\} | z^2/y |^{2L}, \qquad | y | > L;$$

$$| \psi'(\lambda_k) | > C(L, \delta)(1 + | \lambda_k |)^{-4L-1}.$$

10.6a. Boas [6].

10.6b. Boas [24].

10.6c. Plancherel and Pólya [1], Boas [6].

10.6d. Plancherel and Pólya [1].

10.6e. Boas [24]; for $\phi(t) = t^p$, Plancherel and Pólya [1].

10.6f. Duffin and Schaeffer [4] for $\phi(t) = t^2$; Boas [24].

10.6g. Harvey [1].

10.7a. In (10.7.2) and (10.7.3) I have omitted some minor conditions which keep the functions ϵ and δ from being too irregular. For details see Levinson [4] and Boas [16].

10.7b. Levinson [3], [4]; Boas [16]. In the second reference Levinson inadvertently omits the essential condition arg $\lambda_n \to 0$.

10.7c. Agmon [1].

10.7d. Boas [23] announced stronger results, but his proofs turned out to be incorrect. One of the estimates on which the proofs depended has been shown by Redheffer [2] to be wrong, and the results stated in the text are all that can be proved using the correct estimates. Details will appear in Boas [31].

10.7e. Ahiezer [7].

10.7f. Levinson [4]. An alternative approach to the sufficiency part is given by Boas [31]. Levinson's example for the necessity part occupies more than thirty pages;

this at least suggests why no examples are known to show that most of the theorems discussed here are best possible.

10.7g. Boas [31].

10.7h. Cf. Levinson [4], where the spacing is taken into account.

10.7i. Korevaar [3].

10.7j. Korevaar [3].

10.7k. Levinson [4], generalizing a result of Ganapathy Iyer.

10.7l. Korevaar [3].

10.7m. See the references in note 10.3a.

CHAPTER 11

OPERATORS AND THEIR EXTREMAL PROPERTIES

11.1. Introduction. In Chapter 9 we had theorems which said that an entire function of exponential type vanishes identically if it vanishes on a certain sequence of points; in Chapter 10, we had theorems which said more generally that the rate of growth of the function along a line is determined by its growth along a sequence of points on the line. Here we shall ask a more general question: if $g(z)$ is an entire function of exponential type, thought of as obtained from the function $f(z)$ by applying an operator L (e.g., $g(z) = f'(z)$), does the rate of growth of $g(z)$ along a line follow from that of $f(z)$ on a sequence of points or (if not) from the growth of $f(z)$ along the whole line? That some limitation on $L[f(z)]$ is deducible from a restriction on $f(z)$ is fairly clear for a wide class of operators, and the chief interest lies in obtaining theorems which are quantitatively rather than qualitatively exact. For example, it is clear from 6.2.4, combined with Cauchy's integral formula for $f'(z)$, that if $f(z)$ is an entire function of exponential type which is bounded on the real axis, $f'(z)$ is also bounded on the real axis; but it is not at all clear that (as is in fact true: see 11.1.2) $|f'(x)| \le \tau \sup |f(x)|$ when $f(z)$ is of type τ. It is also clear that a bound for $f'(x)$ cannot follow from a bound for $\{f(n\pi/\tau\}_{n=-\infty}^{\infty}$, as the example $f(z) = \sin \tau z$ shows, but that a bound for $f'(x)$ does follow from a bound for $\{f(n\pi/(\tau + \delta)\}$, $\delta > 0$, since a bound for the latter sequence implies a bound for $|f(x)|$.

The oldest theorem dealing with bounds for an operator is S. Bernstein's theorem, originally proved for trigonometric polynomials:

11.1.1. *Theorem. If* $f(x) = \sum_{k=-n}^{n} c_k e^{ikz}$ *is a trigonometric polynomial of degree n, and M is the maximum of* $|f(x)|$, *then* $|f'(x)| \le Mn$.

Equality in the conclusion of 11.1.1 is evidently possible, with $f(x) \not\equiv 0$, even in the class of trigonometric polynomials which are real for real z (example: $f(z) = \sin (nz + \alpha)$).

Theorem 11.1.1 is a special case of the following theorem, which we shall cite as "Bernstein's theorem."

11.1.2. *Theorem.*[a] *If $f(z)$ is an entire function of exponential type τ and* $|f(x)| \le M$ *for all real x, then* $|f'(x)| \le M\tau$.

This includes, for example, almost periodic exponential sums and more generally all functions of the form

$$f(z) = \int_{-\tau}^{\tau} e^{izt}\, d\alpha(t), \tag{11.1.3}$$

where $\alpha(t)$ is of bounded variation.

Since we can apply 11.1.2 to the successive derivatives of $f(z)$, we also have $|f^{(n)}(x)| \leq M\tau^n (n = 1, 2, \cdots)$. Consequently if D stands for d/dx and $L(D)$ is a polynomial in D, $L(D) = \sum_{k=0}^{n} a_k D^k$, we have

$$| L(D)f(x) | \leq M \sum_{k=0}^{n} | a_k | \tau^k;$$

this bound may, however, be improved in special cases. We may also consider differential operators of infinite order with constant coefficients, and the most general operators which we shall consider can be thought of, at least heuristically, in this way.

Besides generalizing 11.1.2 by generalizing the operator which acts on $f(z)$, we can, in some circumstances, replace the hypothesis that $|f(x)| \leq M$ for all x by the hypothesis that $|f(x)| \leq M$ on some sequence of points; the interesting cases are naturally those in which we get a better inequality than that resulting from applying §10.2 or 10.5 to obtain a bound for $f(x)$ as an intermediate step. In §11.7 we shall also generalize 11.1.2 by replacing the hypothesis by $|f(x)| \leq |\omega(x)|$ and the conclusion by $|L[f(x)]| \leq |L[\omega(x)]|$.

11.2. A general method. A large number of inequalities can be obtained by first considering functions (11.1.3)[a] and then proceeding to more general functions by approximation. This leads in particular to simple proofs of 11.1.2 and of a number of related theorems.

Consider an operator L which carries

$$f(z) = \int_{-\tau}^{\tau} e^{izt}\, d\alpha(t) \tag{11.2.1}$$

into

$$L[f(z)] = \int_{-\tau}^{\tau} \lambda(t) e^{izt}\, d\alpha(t); \tag{11.2.2}$$

if L is to be defined for all $f(z)$ of the form (11.2.1) with $\alpha(t)$ of bounded variation, we need to require that $\lambda(t)$ is continuous in $-\tau \leq t \leq \tau$; but sometimes we shall consider only a subclass of functions (11.2.1) for which $\alpha(t)$ is constant in intervals containing the discontinuities of $\lambda(t)$. We do not try to characterize this class of operators by intrinsic properties, but we note one important subclass, characterized by the following theorem.

11.2.3.* *Theorem.*[b] *The class of operators L which are distributive, carry* *entire functions of exponential type τ into entire functions of exponential type τ, are permutable with differentiation, and have the continuity property*

$$\lim_{n\to\infty} L[s_n(z)] = L[f(z)], \qquad f(z) = \sum_{k=0}^{\infty} a_k z^k, \qquad s_n(z) = \sum_{k=0}^{n} a_k z^k,$$

is identical with the class of operators of the form

$$\lambda(D) = \sum_{n=0}^{\infty} \lambda_n D^n, \qquad D = d/dz,$$

with $\lambda(z) = \sum_{n=0}^{\infty} \lambda_n z^n$ *regular in* $|z| \le \tau$.

Suppose that, for some real s, $\lambda(t)e^{ist}$ has an absolutely convergent Fourier series on $(-\tau, \tau)$,

$$\lambda(t)e^{ist} = \sum_{n=-\infty}^{\infty} c_n(s) e^{n\pi i t/\tau}, \tag{11.2.4}$$

$$C(s) \equiv \sum_{-\infty}^{\infty} |c_n(s)| < \infty. \tag{11.2.5}$$

(A necessary condition for this is that $|\lambda(\tau)| = |\lambda(-\tau)|$.) Then we may substitute (11.2.4) into (11.2.2) and integrate term by term, obtaining the interpolation formula

$$L[f(x)] = \sum_{n=-\infty}^{\infty} c_n(s) f(x - s + n\pi/\tau). \tag{11.2.6}$$

If

$$|f(x)| \le M, \qquad -\infty < x < \infty, \tag{11.2.7}$$

we then have

$$|L[f(x)]| \le MC(s). \tag{11.2.8}$$

If it happens that $c_n(s) \ge 0$ we also have $L[f(x)] \ge 0$ whenever $f(x) \ge 0$ for all real x. We also have

$$\sup_k |L[f(s + k\pi/\tau)]| \le C(s) \sup_n |f(n\pi/\tau)|, \tag{11.2.9}$$

which asserts more than (11.2.8) but asserts it for fewer values of x. If $C(s)$ is a bounded function of s, (11.2.9) yields the stronger inequality

$$|L[f(x)]| \le \sup_s |C(s)| \cdot \sup_n |f(n\pi/\tau)|, \tag{11.2.10}$$

giving a bound for $L[f(x)]$ in terms of the values of $f(x)$ on the sequence $\{n\pi/\tau\}$ instead in terms of all the values of $f(x)$. (A necessary condition for $C(s)$ to be bounded is that $\lambda(\tau) = \lambda(-\tau) = 0$.) In this case we may ake $s = x$ in (11.2.6) and obtain the interpolation formula

$$L[f(x)] = \sum_{n=-\infty}^{\infty} c_n(x) f(n\pi/\tau). \tag{11.2.11}$$

Furthermore, let $\phi(t)$ be a nondecreasing convex function. From (11.2.6) we obtain

$$\phi\left\{\frac{|L[f(x)]|}{C(s)}\right\} \leq \sum_{n=-\infty}^{\infty} \frac{|c_n(s)|\, \phi\{|f(x-s+n\pi/\tau)|\}}{C(s)},$$

and so

$$\int_{-\infty}^{\infty} \phi\left\{\frac{|L[f(x)]|}{C(s)}\right\} dx \leq \int_{-\infty}^{\infty} \phi\{|f(x)|\}\, dx \tag{11.2.12}$$

and

$$\limsup_{T\to\infty} \frac{1}{2T}\int_{-T}^{T} \phi\left\{\frac{|L[f(x)]|}{C(s)}\right\} dx \leq \limsup_{T\to\infty} \frac{1}{2T}\int_{-T}^{T} \phi\{|f(x)|\}\, dx. \tag{11.2.13}$$

In particular, we may take $\phi(t) = t^p$, $p \geq 1$, and obtain

$$\int_{-\infty}^{\infty} |L[f(x)]|^p\, dx \leq \{C(s)\}^p \int_{-\infty}^{\infty} |f(x)|^p\, dx.$$

It should be noted that (11.2.10) is not obtainable in general from §10.2 combined with (11.2.8), since the boundedness of $\{f(n\pi/\tau)\}$ does not in general imply that $f(x)$ is bounded unless $h_f(\pm\pi/2) < \tau$, which means that $\alpha(t)$ is constant in neighborhoods of $\pm\tau$ (6.9.5).

In any special case of (11.2.8) or (11.2.10) we should like to have an explicit value for the constant on the right. We also want to know whether the constant is the best possible, i.e. whether the inequality becomes an equality for some $f(z)$ which is not identically zero. The explicit calculation of $C(s)$ is easily done in special cases when the signs of the coefficients $c_n(s)$ have sufficient regularity. For example, if $(-1)^n c_n(s) \geq 0$, we have, by (11.2.6) with $x = s$,

$$L[f(s)] = \sum_{n=-\infty}^{\infty} c_n(s) f(n\pi/\tau). \tag{11.2.14}$$

If we take $f(z) = \cos \tau z$, we have $f(n\pi/\tau) = (-1)^n$ and so

$$C(s) = \sum_{n=-\infty}^{\infty} |c_n(s)| = L[\cos z\tau]|_{z=s} = \{\lambda(\tau)e^{is\tau} + \lambda(-\tau)e^{-is\tau}\}/2. \tag{11.2.15}$$

Similarly, when $c_n(s) \geq 0$ we have

$$C(s) = L[1]\,|_{z=s} = \lambda(0).$$

If $(-1)^{n+1}c_n(s) \geq 0$ for $n \neq 0$ and $c_0(s) > 0$, we have

$$C(s) = L[2(z\tau)^{-1} \sin z\tau - \cos z\tau] \,|_{z=s}$$

$$= \tau^{-1} \int_{-\tau}^{\tau} \lambda(t) e^{ist} \, dt - \{\lambda(t) e^{is\tau} + \lambda(-\tau) e^{-is\tau}\}/2.$$

However, we do not necessarily obtain the best possible inequality in this way. We do if $(-1)^n c_n(s) \geq 0$ or if $c_n(s) \geq 0$, since in both cases the particular $f(x)$ which enables us to evaluate $C(s)$ takes its maximum absolute value for real x at the points $n\pi/\tau$, so that (11.2.8) becomes an equality. On the other hand, in the third case mentioned the function

$$f(x) = 2(x\tau)^{-1} \sin x\tau - \cos x\tau$$

does not take its maximum absolute value for $x = n\pi/\tau$, and we do not obtain the best possible constant in (11.2.8). Nevertheless, in this case we do have

$$L[f(s)] = C(s) \sup_n | f(n\pi/\tau) |.$$

and so the constant in (11.2.10) is the smallest possible if $C(s)$ happens to be maximal for the value of s in question.

The apparently more general problem of deducing a bound for an operator L from a given bound for another operator M can be reduced to the one we are considering by regarding $L[f]$ as obtained from $M[f]$ by applying the operator LM^{-1}, which amounts to considering properties of the function $\lambda(t)/\mu(t)$; note that the method may still work even when $\mu(t)$ has zeros (so that M^{-1} is not always defined) if $\lambda(t)$ has zeros at the same points; cf. (11.4.15), (11.4.16), (11.4.17).

Further inequalities can be obtained from (11.2.6) or (11.2.11), under other assumptions about $f(x)$ and $\lambda(t)$, by the use of Hölder's inequality.

Before discussing the extension of the method to functions more general than (11.2.1), we apply it to the proof of Theorem 11.1.2 (Bernstein's theorem on $f'(z)$).

11.3. Bounds for derivatives. We prove 11.1.2, thus obtaining incidentally a simple proof of 11.1.1.[a]

For a function of the form (11.2.1) we have

$$f'(z) = \int_{-\tau}^{\tau} ite^{izt} \, d\alpha(t),$$

and

$$ite^{-i\pi t/(2\tau)} = 4\tau\pi^{-2} \sum_{k=-\infty}^{\infty} \frac{(-1)^k}{(2k+1)^2} e^{k\pi it/\tau}, \qquad -\tau \leq t \leq \tau.$$

Hence in this case (11.2.6) becomes

$$(11.3.1) \qquad f'(x) = 4\tau\pi^{-2} \sum_{k=-\infty}^{\infty} \frac{(-1)^k}{(2k+1)^2} f\left(x + \frac{2k+1}{2\tau}\pi\right);$$

and, either by taking the value of $\sum (2n+1)^{-2}$ as known or by taking $f(z) = \sin \tau z$ and using (11.3.1) for $x = 0$, we find

$$(11.3.2) \qquad | f'(x) | \leq \tau \sup_{-\infty < x < \infty} | f(x) |,$$

with equality possible.

We have therefore proved 11.1.1 (by taking $\tau = n$ and $\alpha(t)$ a step-function with jumps at the integers); we have actually proved more, since for any fixed x we have obtained

$$| f'(x) | \leq \tau \max_{0 \leq k \leq 2n} \left| f\left(x + \frac{2k+1}{2n}\pi\right)\right|.$$

To extend (11.3.2) to an entire function $f(z)$ of exponential type τ which is merely bounded on the real axis, consider the functions

$$g_\delta(z) = f(z)(\delta z)^{-1} \sin \delta z.$$

We have $g_\delta(z)$ of type $\tau + \delta$, $| g_\delta(x) | \leq M$ for real x, and $g_\delta(x)$ belonging to L^2 on the real axis. Hence (Paley-Wiener theorem) $g_\delta(z)$ has the form (11.2.1) with $\tau + \delta$ in place of τ, and so satisfies (11.3.2) with $\tau + \delta$ in place of τ, i.e. $| g_\delta'(x) | \leq (\tau + \delta)M$. Since $g_\delta'(z) \to f'(z)$ as $\delta \to 0$, 11.1.2 follows.

Similarly, by using (11.2.12) and (11.2.13) we obtain

11.3.3. *Theorem.*[b] *If $f(z)$ is an entire function of exponential type τ, and $\phi(t)$ is a nondecreasing convex function,*

$$\int_{-\infty}^{\infty} \phi\{| f'(x) |/\tau\}\, dx \leq \int_{-\infty}^{\infty} \phi\{| f(x) |\}\, dx$$

and

$$\limsup_{T\to\infty} (2T)^{-1} \int_{-T}^{T} \phi\{| f'(x) |/\tau\}\, dx \leq \limsup_{T\to\infty} (2T)^{-1} \int_{-T}^{T} \phi\{| f(x) |\}\, dx.$$

In particular, for $p \geq 1$

$$\int_{-\infty}^{\infty} | f'(x) |^p\, dx \leq \tau^p \int_{-\infty}^{\infty} | f(x) |^p\, dx,$$

and if $f(z)$ has period 2π (and so is a trigonometric polynomial)

$$\int_{-\pi}^{\pi} | f'(x) |^p\, dx \leq \tau^p \int_{-\pi}^{\pi} | f(x) |^p\, dx.$$

We also obtain a simple proof of a special case of 6.2.13 by using Bernstein's theorem: if $f(z)$ is an entire function of zero exponential type, and is bounded for real x, then $f'(x) \equiv 0$ by 11.1.2.

We note the following analogue of Bernstein's theorem which follows from 6.2.8.

11.3.4.* *Theorem. If $f(z)$ is regular and of exponential type for $x \geq 0$ then* $\lim_{x \to \infty} f'(x) = 0$ *if* $\lim_{x \to \infty} f(x)$ *exists.*

The method suggested at the end of §11.2 leads to inequalities of a different sort. For example,

11.3.5.* *Theorem. If $f(z)$ is an entire function of exponential type τ and belongs to L^2 on the real axis then*

$$| f'(x) |^2 \leq \frac{\tau^2}{3} \sum_{n=-\infty}^{\infty} | f(n\pi/\tau) |^2.$$

Some generalizations of another kind are given in the next two sections. A still different generalization which requires much deeper methods is

11.3.6.* *Theorem.*[c] *Let E be a closed set on the real axis such that, for some positive numbers A and α, every interval of length A intersects E in a set of measure at least α. Then at almost all points x of E there is a number $q(x)$ with the property that if $f(z)$ is an entire function of exponential type τ such that $| f(x) | \leq 1$ in E, $| f'(x) | \leq \tau q(x)$.*

It is also possible to sharpen Bernstein's theorem by imposing additional restrictions on the function.

11.3.7. *Theorem.*[d] *If $f(z)$ is an entire function of exponential type τ which is bounded and monotonic on the real axis, then*

$$| f'(x) | \leq \pi^{-1} \tau \sup | f(x) |, \tag{11.3.8}$$

and equality is attained for the function

$$f(z) = \int_0^z t^{-2} \sin^2 (\tau t/2)\, dt. \tag{11.3.9}$$

Suppose, for definiteness, that $f(z)$ is nondecreasing. We know from 11.1.2 that $f'(z)$ is bounded, and then from 7.5.1 that there is an entire function $g(z)$ of exponential type $\tau/2$ such that $f'(x) = | g(x) |^2$ for real x. Since $f'(x) \geq 0$ and $f(x)$ is bounded, $f'(x)$ is integrable and therefore $g(x)$ belongs to L^2 on $(-\infty, \infty)$. By the Paley-Wiener theorem (6.8.1), we have

$$g(x) = \int_{-\tau/2}^{\tau/2} G(t) e^{ixt}\, dt,$$

where $G(t)$ belongs to L^2 and

$$\int_{-\tau/2}^{\tau/2} | G(t) |^2 \, dt = (2\pi)^{-1} \int_{-\infty}^{\infty} | g(x) |^2 \, dx = (2\pi)^{-1} \int_{-\infty}^{\infty} f'(x) \, dx$$
$$= (2\pi)^{-1}[f(\infty) - f(-\infty)].$$

Thus Schwarz's inequality yields

$$(2\pi)^{-1} \cdot 2 \sup | f(x) | \geq \int_{-\tau/2}^{\tau/2} | G(t) |^2 \, dt \geq \tau^{-1} \left\{ \int_{-\tau/2}^{\tau/2} G(t) \, dt \right\}^2$$
$$= \tau^{-1} \{g(0)\}^2 = \tau^{-1} f'(0),$$

with equality if $f(\infty) = -f(-\infty)$ and $G(t)$ is constant. Since $f(z + x_0)$ (x_0 real) satisfies the same conditions as $f(z)$, we may replace $f'(0)$ by $f'(x_0)$. We have therefore established (11.3.8). Equality occurs when $G(t)$ is constant, $g(x) = A \sin \tau x/2$, and $f'(x) = A^2 \sin^2 (\tau x/2)$; the choice (11.3.9) for $f(z)$ makes $f(\infty) = -f(-\infty)$.

There are also inequalities for $f'(x)$ in which the hypothesis that $f(x)$ is an entire function is replaced by something weaker.[e]

11.4. Inequalities involving the maximum of $| f(x) |$. We suppose in this section that $f(z)$ has the form $\int_{-\tau}^{\tau} e^{izt} \, d\alpha(t)$ with $\alpha(t)$ of bounded variation. In most of the examples the extension to more general functions presents no difficulty, and can be made as in §11.3; see §11.6 for further discussion of the extension problem. We shall consider only inequalities for

$$\sup | L[f(x)] |;$$

inequalities involving integrals or mean values can be obtained as indicated in §11.2 or illustrated in §11.3.

As a first illustration consider the problem of obtaining a bound for $f(x + \delta) - f(x - \delta)$. From Bernstein's theorem we have

$$| f(x + \delta) - f(x - \delta) | \leq 2\delta \sup | f'(x) | \leq 2\tau\delta \sup | f(x) |,$$

and we also have the trivial inequality $| f(x + \delta) - f(x - \delta) | \leq 2 \sup | f(x) |$. To apply the method of §11.2, take $\lambda(t) = -2 \sin \delta t$, and calculate the Fourier coefficients of $\lambda(t) e^{ist}$. When $2s = \pi/\tau$, these are

$$c_n = \frac{2(-1)^n \delta \cos \delta\tau}{\tau\{(n - \tfrac{1}{2})^2\pi^2/\tau^2 - \delta^2\}}.$$

If $0 < 2\delta < \pi/\tau$, we have $(-1)^n c_n \geq 0$; hence if we take $f(z) = \cos z\tau$ and $2x = \pi/\tau$ we have equality in (11.2.8), so that we have at the same time the explicit value $2 \sin \delta\tau$ for the constant $C(s)$ in (11.2.8) and a function for which equality is attained. Since our inequality is easily extended to the general case by the method of §11.3, we have proved the following theorem.

11.4.1. *Theorem.*[a] *If* $f(z)$ *is an entire function of exponential type* τ, *bounded on the real axis, then*

$$(11.4.2) \qquad | f(x+\delta) - f(x-\delta) | \le 2 \sin \delta\tau \cdot \sup | f(x) |, \quad 0 < 2\delta < \pi/\tau.$$

The constant in (11.4.2) cannot be improved. Letting $\delta \to 0$ we have Bernstein's theorem 11.1.2 again.

Next we consider some more complicated generalizations of Bernstein's theorem which involve derivatives explicitly.

11.4.3. *Theorem. For any real numbers* ω *and* μ,

$$(11.4.4) \quad | \tau \sin \mu \cdot f(x) + (\sin \omega - \sin \mu)\tilde{f}'(x) + \cos \omega \cdot f'(x) | \le \tau \sup | f(x) |.$$

Here

$$\tilde{f}(z) = -i \int_{-\tau}^{\tau} \operatorname{sgn} t \, e^{izt} \, d\alpha(t)$$

is the integral conjugate to $f(z)$, defined by analogy with the conjugate of a trigonometric polynomial; $\tilde{f}(z)$ is not defined in general if $\alpha(t)$ is discontinuous at $t = 0$, but $\tilde{f}'(z)$ is, since

$$\tilde{f}'(z) = \int_{-\tau}^{\tau} | t | \, e^{izt} \, d\alpha(t).$$

The inequality (11.4.4) is (11.2.8) with

$$\lambda(t) = \tau \sin \mu + (\sin \omega - \sin \mu) \, | t | + it \cos \omega;$$

when $s\tau = \omega - \pi/2$, the Fourier coefficients of $\lambda(t)e^{ist}$ are

$$c_n(s) = \frac{(-1)^n \tau}{(s\tau - n\pi)^2} \{(-1)^n + \sin \mu\}\{(-1)^n - \sin \omega\},$$

and so $(-1)^n c_n(s) \ge 0$. We get interesting special cases by choosing $\mu = \omega$, when (11.4.4) becomes

$$(11.4.5) \qquad | f'(x) \cos \omega + \tau f(x) \sin \omega | \le \tau \sup | f(x) |;$$

or $\mu = \pi/2$, when it becomes

$$(11.4.6) \qquad | \tau\sigma(f) + f'(x) \cos \omega + \tilde{f}'(x) \sin \omega | \le \tau \sup | f(x) |,$$

where

$$(11.4.7) \qquad \sigma(f) = \int_{-\tau}^{\tau} (1 - | t |/\tau) e^{ixt} \, d\alpha(t),$$

the first-order arithmetic mean of the "partial integrals" of $f(z)$. If $f(z)$ is real for real z, we get, by choosing ω appropriately,

(11.4.8)[b] $$\{ |f'(x)|^2 + \tau^2 |f(x)|^2 \}^{1/2} \le \tau \sup |f(x)|$$

and

(11.4.9)[c] $$\tau |\sigma(f)| + \{ |f'(x)|^2 + |\tilde{f}'(x)|^2 \}^{1/2} \le \tau \sup |f(x)|.$$

The fact that (11.4.4) is a significant generalization of Bernstein's theorem is most clearly seen from (11.4.8) and (11.4.9). If $f(z)$ is real for real z, the first of these shows that not only is $|f'(x)| \le \tau \sup |f(x)|$, but $|f(x)|$ is large only when $|f'(x)|$ is correspondingly small (as is qualitatively clear). Inequality (11.4.9) shows in particular that $|\tilde{f}'(x)|$ has the same upper bound as $|f'(x)|$, which is especially interesting because $|\tilde{f}(x)|$ does not have the same upper bound as $|f(x)|$ (for trigonometric polynomials of degree n the best bound for $|\tilde{f}(x)|$ is asymptotically $2\pi^{-1} \log n \cdot \sup |f(x)|$). It is well known from the theory of Fourier series that, for trigonometric polynomials, $|\sigma(f)| \le \sup |f(x)|$ ($\sigma(f)$ being just one of the $(C, 1)$ means of the partial sums of the Fourier series for $f(x)$).

If we choose $\mu = 0$, $\omega = \tau x + \pi/2$, we have $\sum |c_n(x)| < \infty$ for all x, and so

(11.4.10) $$|\tilde{f}'(x) \cos \tau x - f'(x) \sin \tau x| \le \tau \sup_{-\infty < n < \infty} |f(n\pi/\tau)|, \qquad -\infty < x < \infty.$$

In particular, then, although we naturally cannot assert that

$$|f'(x)| \le \tau \sup |f(n\pi/\tau)|,$$

(cf. $f(z) = \sin \tau z$), we can assert that[d]

(11.4.11) $$|f'((n + \tfrac{1}{2})\pi/\tau)| \le \tau \sup |f(n\pi/\tau)|;$$

this also follows directly from (11.3.1), or from (11.5.12).

If $\mu = \pi/2$, $c_n(s)$ is zero for odd n and so $c_n(s) \ge 0$. Thus we have, if $f(x) \ge 0$ for real x,

(11.4.12) $$\tau\sigma(f(x)) + \sin \omega \cdot \tilde{f}'(x) + \cos \omega \cdot f'(x) \ge 0.$$

Choosing ω appropriately, we obtain

11.4.13. *Theorem.*[e] *If $f(x) \ge 0$ for all real x, then*

$$\{ |f'(x)|^2 + |\tilde{f}'(x)|^2 \}^{1/2} \le \tau\sigma(f(x)).$$

This is again a sharpening of Bernstein's theorem (for the special case of a nonnegative function), not only because $\sigma(f) \le \sup f(x)$ but because it is an inequality between functional values instead of between maxima. It can be still further sharpened, for example as follows. If we replace $\lambda(t)$ by $\lambda(t) - c_0(s)e^{-ist}$, and $c_n(s) \ge 0$, the coefficients $c_n(s)$ for the new $\lambda(t)$ are still nonnegative, and (11.4.12) is replaced by

$$\tau\sigma(f(x)) - c_0(s)f(x - s) + \sin \omega\cdot\tilde{f}'(x) + \cos \omega\cdot f'(x) \geq 0.$$

We have $\mu = \pi/2$, $s\tau = \omega - \pi/2$, and $c_0(s) = 2\tau/(\omega - \pi/2)^2$. Taking $\mu \to 0$ and $\omega \to \pi$, we obtain

$$\tau\sigma(f(x)) - 8\tau\pi^{-2}f(x + \pi/(2\tau)) \pm f'(x) \geq 0,$$

so that for a nonnegative $f(x)$,

$$| f'(x) | \leq \tau\sigma(f(x)) - 8\tau\pi^{-2} \min \{f(x + \pi/(2\tau)), f(x - \pi/(2\tau))\}.$$

Another group of results with exact constants can be found by applying the following lemma.

11.4.14. *Lemma. If $\lambda(t)$ is even, and convex or concave on $(-\tau, \tau)$, $(-1)^n c_n(0) = (-1)^n c_n \geq 0$ or ≤ 0, respectively, for $n \neq 0$; and if $\lambda'(t)$ is bounded, $\sum | c_n |$ converges.*

We have, for $n \neq 0$,

$$\tau c_n = \frac{1}{2}\int_{-\tau}^{\tau} e^{-in\pi t/\tau}\lambda(t)\, dt = \int_0^{\tau} \lambda(t) \cos n\pi t/\tau \, dt$$

$$= -\tau(n\pi)^{-1}\int_0^{\tau} \lambda'(t) \sin n\pi t/\tau \, dt.$$

Since $\lambda'(0) = 0$ and $\lambda'(t)$ increases when $\lambda(t)$ is convex, in this case $-c_n$ has the sign that $\sin n\pi t/\tau$ has in a left-hand neighborhood of τ, namely the sign of $(-1)^{n+1}$; when $\lambda(t)$ is concave, $-\lambda(t)$ is convex. If $\lambda'(t)$ is bounded, a further integration by parts shows that $\sum | c_n | < \infty$.

Hence[f] if $\lambda(t)$ is convex and satisfies the other hypotheses of (11.4.14) and

$$\tau c_0 = \int_0^{\tau} \lambda(t)\, dt \geq 0,$$

we have

$$| L[f(x)] | \leq \lambda(\tau) \sup | f(x) |;$$

the constant is best possible. In fact, if $\lambda(t)$ is convex we have $(-1)^n c_n \geq 0$ for all n and hence equality in (11.2.6) for $f(z) = \cos \tau z$, $s = 0$, $x = 0$, so that $C(0) = \lambda(\tau)$.

As an example, consider $\lambda(t) = \sin at/\sin bt$, $b > a$. A function is convex if its logarithm is convex, and the second derivative of $\log \lambda(t)$ is $b^2 \csc^2 bt - a^2 \csc^2 at$, which is positive if $bt < \pi$. Hence if $0 < a < b \leq \pi/\tau$, $\lambda(t)$ is convex. Now $f(z + b) - f(z - b)$ is the operator corresponding to $\lambda(t) = 2 \sin bt$, so if we start from this function instead of from $f(z)$ itself, we obtain[g]

$$|f(x+a) - f(x-a)| \le \frac{\sin a\tau}{\sin b\tau} \sup_x |f(x+b) - f(x-b)|, \tag{11.4.15}$$
$$0 < a < b \le \pi/\tau,$$

first for finite Fourier transforms and then by approximation (as in §11.3) for all entire functions of exponential type for which the right-hand side is finite. Letting $a \to 0$ (which corresponds to considering $\lambda(t) = (t/2) \csc bt$), we obtain[h]

$$2\,|f'(x)| \le \tau \csc b\tau \cdot \sup |f(x+b) - f(x-b)|, \quad 0 < b < \pi/\tau; \tag{11.4.16}$$

and if then $b = \pi/(2\tau)$, we have Bernstein's theorem 11.1.2 again. Taking the same value of b in (11.4.15) leads to (11.4.2) again.

Similarly, $\sec \delta t$ is convex if $0 < \delta t < \pi/2$, so

$$2\,|f(x)| \le \sec \delta\tau \cdot \sup |f(x+\delta) + f(x-\delta)|, \quad 0 < \delta < \pi/(2\tau). \tag{11.4.17}$$

These inequalities can be further sharpened by replacing $\lambda(t)$ by $\lambda(t) - c_0$ since this leaves $\lambda(t)$ convex and makes $c_0 = 0$, so that we still have $(-1)^n c_n \ge 0$.

As an example of the concave case of 11.4.14, consider $\lambda(t) = t^{-1} \sin at$. This is concave in $(0, \tau)$ provided $0 < a < c/\tau$, where c (about $2.08 < 2\pi/3$) is the smallest positive root of $2 - t^2 = 2t \cot t$. We obtain the inequality (which should be compared with (11.4.16))

$$|f(x+a) - f(x-a)| \le \tau^{-1}\{2\,\mathrm{Si}\,(a\tau) - \sin a\tau\} \sup |f'(x)|, \quad 0 < a < c/\tau.$$

The constant is presumably not the best possible, although it is less than the trivial constant $2a$.

As another example, we can seek a bound for the conjugate of a trigonometric polynomial[i], or more generally for the conjugate of an integral of the form

$$\int_{1 \le |t| \le \tau} e^{izt}\, d\alpha(t),$$

by taking $\lambda(t) = -i \operatorname{sgn} t$ for $|t| \ge 1$, $\lambda(t) = -it$ for $|t| < 1$. The resulting inequality will be asymptotically correct as the degree of the trigonometric polynomial becomes infinite.

Still further examples can be obtained as follows.

11.4.18.* *Theorem.*[j] *If $\lambda(t)$ is even, and increasing and concave on $(0, \tau)$, then*

$$|L[f(x)]| \le \left\{2\tau^{-1} \int_0^\tau \lambda(t)\, dt - 2\lambda(0)\right\} \sup |f(x)|.$$

There is equality for $f(z) = 2(\tau z)^{-1} \sin \tau z - 1$.

Here we have $c_n < 0$ for $n \neq 0$, $c_0 > 0$.

There are other operators which cannot be handled by the method used here; an example is the operator which corresponds to $\lambda(t) = (it)^{-1}$, and so is defined in general only if $\alpha(t)$ in (11.2.1) is constant in a neighborhood of 0. Then, in fact, the "integral" theorem is not really one about entire functions at all, since for either periodic or almost periodic functions, the integral (with vanishing mean value) has a bound which depends on the smallest, rather than on the largest, Fourier exponent. Thus[k] for a trigonometric series without constant term, and a fortiori for such a trigonometric polynomial, the maximum of the integral does not exceed

$$(\pi/2) \sup | f(x) |.$$

However, the extremal function is not an entire function, and there remains the question of whether an improved constant can be found for entire functions. For trigonometric polynomials, the best bound has been determined.[1]

11.5. Inequalities involving $\sup | f(n\pi/\tau) |$. We give some illustrations of the case in which (11.2.9) can be applied to give inequalities of the form (11.2.10), in which $L[f]$ is estimated in terms of the maximum of $| f(x) |$ over the sequence $\{n\pi/\tau\}$. Since this maximum may be finite even when $f(x)$ is unbounded, our method is not well adapted to obtaining results which require as little as possible about the growth of $f(x)$, but it is useful for suggesting the right form for the inequalities. We require the function $\lambda(t)e^{ist}$ to have an absolutely convergent Fourier series for all real s, and so in particular $\lambda(\tau) = \lambda(-\tau) = 0$. In the first place, if $\lambda(t)$ satisfies suitable conditions for the absolute convergence of its Fourier series (for instance if it is piecewise convex or concave, possibly with corners but without cusps), except that the condition $\lambda(\tau) = \lambda(-\tau) = 0$ is not satisfied, we can extend $\lambda(t)$ (for instance, linearly) to $(-\tau - \delta, \tau + \delta)$, $\delta > 0$, so that it vanishes at $\pm(\tau + \delta)$. We can then consider $f(z)$ as being of the form

$$\int_{-\tau-\delta}^{\tau+\delta} e^{izt}\, d\alpha(t), \tag{11.5.1}$$

with $\alpha(t)$ constant outside $(-\tau, \tau)$; the operator defined by using the extended $\lambda(t)$ coincides with the original operator for $f(z)$ (though naturally not for all functions (11.5.1) with unrestricted $\alpha(t)$). We thus have an interpolation formula

$$\begin{aligned} L[f(x)] &= \sum_{n=-\infty}^{\infty} c_n^{(\delta)}(x) f(n\pi/(\tau + \delta)), \\ c_n^{(\delta)}(x) &= \frac{1}{2(\tau + \delta)} \int_{-\tau-\delta}^{\tau+\delta} e^{it\{x - n\pi/(\tau+\delta)\}} \lambda(t)\, dt, \end{aligned} \tag{11.5.2}$$

and an inequality

$$(11.5.3)\qquad | L[f(x)] | \leq \sum_{-\infty}^{\infty} | c_n^{(\delta)}(x) | \sup_n | f(n\pi/(\tau + \delta)) |.$$

As an example, take $\lambda(t) = 1$ on $(-\tau, \tau)$; extending this function linearly leads to an absolutely convergent interpolation formula[a] which is the same as the formula obtained otherwise in (10.2.9). Extending $\lambda(t) = it$ similarly leads to no better an estimate for $f'(x)$ in terms of $| f(n\pi/(\tau + \delta) |$ than would follow from estimating $| f(x) |$ by (10.2.3) and then applying 11.1.2.

As an illustration of the original case where $\lambda(t)$ has an absolutely convergent Fourier series on $(-\tau, \tau)$ and $\lambda(\tau) = \lambda(-\tau) = 0$, consider the operator corresponding to $\lambda(t) = 2 \cos \pi t/(2\tau)$. We have

$$c_n(s) = \frac{(-1)^{n+1}\pi \cos s\tau}{(s\tau - n\pi)^2 - \pi^2/4},$$

and $C(s) = \sum | c_n(s) | \leq C(0)$. Since $(-1)^{n+1}c_n(0) \geq 0 (n \neq 0)$, $c_0(0) > 0$, we have equality in (11.2.10) for $f(z) = 2(\tau z)^{-1} \sin \tau z - \cos \tau z$ and $x = 0$. Hence we have the "interference theorem:"[b]

11.5.4. *Theorem. If $f(z)$ is an entire function of exponential type τ which is bounded on the real axis,*

$$(11.5.5)\qquad | f(x + \pi/(2\tau)) + f(x - \pi/(2\tau)) | \leq (8/\pi) \sup_n | f(n\pi/\tau) |,$$

and the constant cannot be improved.

Compare (11.4.17), which goes in the opposite direction.

The interest of 11.5.4 comes from the fact that in general nothing can be inferred about $| f(x) |$ for a single x from the boundedness of $\{f(n\pi/\tau)\}$. We shall show below that 11.5.4 remains true even when $f(x)$ is assumed to be bounded only on the sequence $\{n\pi/\tau\}$, being required otherwise only to be $o(| x |)$ as $| x | \to \infty$.

Next consider the function $\lambda(t) = \tau^2 - t^2$, corresponding to the differential operator $I + D^2$. We obtain[c]

$$(11.5.6)\qquad | \tau^2 f(x) + f''(x) | \leq A(\tau) \sup_n | f(n\pi/\tau) |,$$

with an explicit but complicated (and not best possible) $A(\tau)$; and there is a similar result for the operator $P(D)$ whenever the polynomial $P(x)$ vanishes at $x = \pm\tau$.

Inequalities (11.5.5), (11.5.6), and some similar ones involving only the values of $f(x)$ at the points of an arithmetic progression, can be generalized to entire functions of exponential type which need not even be bounded on the real axis, but are bounded at $\{n\pi/\tau\}$ and satisfy a weaker growth condition on the whole real axis. The method of §11.3 cannot be used here, since it involves approximation by functions of type greater than τ, about which the sequence $\{n\pi/\tau\}$ gives insufficient information. From our present point of view, a natural way to proceed is as follows. The Fourier series of e^{izt} over $(-\tau, \tau)$ has coefficients

$$c_n(z) = (-1)^n(\tau z - n\pi)^{-1} \sin \tau z.$$

If

$$f(z) = \int_{-\tau}^{\tau} e^{izt}\phi(t)\, dt, \tag{11.5.7}$$

and $\phi(t)$ is integrable, we can replace e^{izt} by its Fourier series, integrate term by term, and obtain

$$f(z) = \sin \tau z \sum_{-\infty}^{\infty} \frac{(-1)^n f(n\pi/\tau)}{\tau z - n\pi} \tag{11.5.8}$$

(the "cardinal series"[d]), the series converging uniformly in any bounded set in the z-plane. If $f(x) = O(x^\alpha)$, $\alpha < \frac{1}{2}$, $z^{-1}\{f(z) - f(0)\}$ belongs to L^2 on the real axis and so has the form (11.5.7); hence

$$f(z) - f(0) = \tau z \sin \tau z \sum_{n \neq 0} \frac{(-1)^n\{f(n\pi/\tau) - f(0)\}}{(\tau z - n\pi)n\pi} + \tau^{-1} f'(0) \sin \tau z.$$

If $\{f(n\pi/\tau)\}$ is bounded, the series on the right may be broken into two series, and

$$\sum_{n \neq 0} \frac{(-1)^n}{n\pi(\tau z - n\pi)} = \frac{1}{\tau z \sin \tau z} - \frac{1}{\tau^2 z^2},$$

so that

$$f(z) - f'(0) \sin \tau z = f(0) \frac{\sin \tau z}{\tau z} + \tau z \sin \tau z \sum_{n \neq 0} \frac{(-1)^n f(n\pi/\tau)}{n\pi(\tau z - n\pi)}. \tag{11.5.9}$$

By subtracting more and more terms of the Maclaurin series of $f(z)$ we could obtain similar series when $|f(x)|$ grows at most as rapidly as a power of $|x|$ on the real axis.

Actually (11.5.9) holds for functions of still more rapid growth on the real axis, provided that $\{f(n\pi/\tau)\}$ is bounded, and we may see this independently of our derivation of (11.5.9), once its form has been discovered. Let $f(x) = o(|x|)$ for real x. The series on the right of (11.5.9) converges, if $\{f(n\pi/\tau)\}$ is bounded, to an entire function $g(z)$ of exponential type τ, which is $O(e^{\tau|y|})$ for $z = iy$, takes the values $f(n\pi/\tau)$ at the points $\{n\pi/\tau\}$, and is $O(|x|)$ for $z = x$. The left-hand side of (11.5.9) is an entire function of exponential type which is $o(x)$ for real x, and consequently by 6.2.4 and 1.4.4 is $o(|y|\, e^{\tau|y|})$ for $z = iy$. By a Phragmén-Lindelöf argument, both $g(z)$ and $f(z) - f'(0) \sin \tau z$ are $O(re^{\tau r})$ as $r \to \infty$. By 9.4.2, applied to the difference of these two functions,

$$f(z) - f'(0) \sin \tau z = cz \sin \tau z + g(z)$$

for some constant c. Since everything except the term $cz \sin \tau z$ is $o(|y| e^{\tau|y|})$ for $z = iy$, c must be zero.

We have therefore proved

11.5.10. *Theorem.*[e] *If $f(z)$ is an entire function of exponential type τ which is $o(|x|)$ on the real axis and has $\{f(n\pi/\tau)\}$ bounded, $f(z)$ has the representation*

$$(11.5.11)\quad f(z) = \tau^{-1} f'(0) \sin \tau z + f(0) \frac{\sin \tau z}{\tau z} + \tau z \sin \tau z \sum_{n \neq 0} \frac{(-1)^n f(n\pi/\tau)}{n\pi(\tau z - n\pi)}.$$

There is a corresponding (more complicated) result when $f(x) = O(|x|^k)$, $k > 1$.

We can now prove 11.5.4 when $f(x) = o(|x|)$ and $\{f(n\pi/\tau)\}$ is bounded by expressing the left-hand side of (11.5.5) by (11.5.11) and estimating the resulting series; and a similar proof can be given when the left-hand side is replaced by other operators of similar character. To show that the left-hand side is still bounded when $f(x) = o(|x|^2)$, we observe that the right-hand side of (11.5.11) is an entire function $g(z)$ of exponential type τ which is $o(|x|)$. By an application of 9.4.2, $g(z)$ differs from $f(z)$ at most by a function of the form $(az + b) \sin \tau z$, for which the expression in question is indeed bounded. To establish (11.5.6) under the same hypotheses on $f(z)$, we observe that (11.5.10) can be differentiated (after division by $\sin \tau z$) to give formulas expressing the derivatives of $f(z)$ in terms of $f(z)$ and $\{f(n\pi/\tau)\}$. Thus

$$(11.5.12)\qquad f'(z) = -\tau \sin \tau z \sum_{n=-\infty}^{\infty} \frac{(-1)^n f(n\pi/\tau)}{(\tau z - n\pi)^2} + \tau f(z) \cot \tau z,$$

and for $\tau^2 f(x) + f''(x)$ we obtain the formula leading to (11.5.6). From (11.5.12) we can obtain (11.4.11) (under less restrictive hypotheses on $f(z)$).

The following theorem generalizes these results.

11.5.13.* *Theorem. If $f(z)$ is an entire function of exponential type τ which is $o(|x|^2)$ on the real axis, if* $\sup |f(n\pi/\tau)|$ *is finite, and if*

$$\lambda(z) = \sum_{n=0}^{\infty} \lambda_n z^n$$

is regular for $|z| \leq \tau$, then $\lambda(D)f(z) = \sum_{n=0}^{\infty} \lambda_n f^{(n)}(z)$ is bounded for real z if $\lambda(i\tau) = \lambda(-i\tau) = 0$, but not necessarily bounded if this condition is not satisfied.

11.5.14.* *Corollary.*[f] *Under the hypotheses of* 11.5.13 *on* $f(z)$, *the transform* $\int_{-\infty}^{\infty} f(z+t)\,d\rho(t)$, *with* $\int_{-\infty}^{\infty} e^{\sigma|t|}\,|d\rho(t)|$ *finite for some* $\sigma > \tau$, *is bounded for real* z *if* $\int_{-\infty}^{\infty} e^{\pm i\tau t}\,d\rho(t) = 0$, *but not necessarily if this hypothesis is not satisfied.*

11.6. Extension of operators. The inequalities of §11.1 which involve only $f(x)$ and its derivatives are readily extended to functions more general than those of the form

$$\int_{-\tau}^{\tau} e^{izt}\,d\alpha(t) \tag{11.6.1}$$

by the methods used in §§11.3 and 11.5. In order to extend the others it is necessary first to have transforms like $\tilde{f}'(z)$ and $\sigma(f(z))$ defined for entire functions of exponential type which are, for example, bounded on the real axis without being of the form (11.6.1). We could, if $L[f]$ is defined for $f(z)$ of the form (11.6.1) by

$$L[f(z)] = \int_{-\tau}^{\tau} \lambda(t) e^{izt}\,d\alpha(t), \tag{11.6.2}$$

and $e^{ist}\lambda(t)$ has a Fourier series which is absolutely convergent either for one s or for all s, simply define $L[f(z)]$ for a more general function $f(z)$ by (11.2.6) or (11.2.9):

$$L[f(z)] = \sum_{n=-\infty}^{\infty} c_n(s) f(z - s + n\pi/\tau) \tag{11.6.3}$$

or

$$L[f(z)] = \sum_{n=-\infty}^{\infty} c_n(z) f(n\pi/\tau). \tag{11.6.4}$$

This would make the extension of the inequalities trivial, but would make it difficult to verify, for some operators which have a natural intrinsic definition, that the extended operator coincides with the original one outside the class of functions (11.6.1); examples are the operators considered in 11.2.3. We shall show, for a class of operators which includes all the special ones which have been used as illustrations, that an extension can be made by continuity from functions (11.6.1) to all entire functions of exponential type which are bounded on the real axis, in such a way that (11.6.3) is preserved. Thus all the inequalities of §11.4 remain valid for this wider class of functions. The problem of extending an operator so that (11.6.4) will still hold, in cases when $f(x)$ is not necessarily bounded for real x, is more difficult and we shall not go beyond what was done in §11.5.

We shall require that the operator L is defined for functions (11.6.1) by

(11.6.2) with $\lambda(t)$ absolutely continuous and $\lambda'(t)$ in L^2; this covers most of the cases of interest.[a] If $f(z)$ has the form (11.6.1), we may write it in the form

$$f(z) = f(0) + \int_{-\tau}^{\tau} e^{izt}\, d\beta(t),$$

where $\beta(t) = \alpha(t) - f(0)\operatorname{sgn} t - \alpha(-\tau)$, and $\beta(-\tau) = \beta(\tau) = 0$; we have

$$L[f(z)] = f(0)\lambda(0) + \int_{-\tau}^{\tau} \lambda(t) e^{izt}\, d\beta(t).$$

We may write

$$f(z) - f(0) = -iz \int_{-\tau}^{\tau} e^{izt}\beta(t)\, dt,$$

$$L[f(z)] = f(0)\lambda(0) - \int_{-\tau}^{\tau} \beta(t)e^{izt}\lambda'(t)\, dt - iz\int_{-\tau}^{\tau} \beta(t)e^{izt}\lambda(t)\, dt.$$

On the other hand, if $f(z)$ is an entire function of exponential type which is bounded on the real axis, we have

(11.6.5) $$f(z) = f(0) + z\int_{-\tau}^{\tau} e^{izt}\phi(t)\, dt, \qquad \phi(t)\epsilon L^2.$$

This leads us to define the operator L when $\lambda'(t) \epsilon L^2$ by writing $f(z)$ in the form (11.6.5) and putting

(11.6.6) $$L[f(z)] = \lambda(0)f(0) - i\int_{-\tau}^{\tau} \phi(t)\lambda'(t)e^{izt}\, dt + z\int_{-\tau}^{\tau} \phi(t)\lambda(t)e^{izt}\, dt.$$

The operator L is then defined for all $f(z)$ of the class under consideration, and it agrees with the original definition for functions (11.6.1). Moreover, L has the following continuity property. If $f_n(z)$ are entire functions of exponential type, uniformly bounded on the real axis, such that $f_n(z) \to f(z)$ uniformly in each bounded region, then $L[f_n(z)] \to L[f(z)]$ uniformly in each bounded region. To prove this, let $f_n(z)$ have the representation (11.6.5) with corresponding functions $\phi_n(t)$. Our conclusion follows from (11.6.6) provided we show that

(11.6.7) $$\int_{-\tau}^{\tau} |\, \phi(t) - \phi_n(t)\,|^2\, dt \to 0.$$

Now $\phi(t) - \phi_n(t)$ is the inverse Fourier transform of $z^{-1}\{f(z) - f_n(z)\}$, and so (11.6.7) is a consequence of

$$\int_{-\infty}^{\infty} x^{-2}\,|\, f(x) - f_n(x)\,|^2\, dx \to 0,$$

which follows from our hypothesis on $\{f_n\}$.

It remains to show that L also has the representation (11.6.3). To do this, construct a sequence $\{\phi_n(t)\}$ such that $\phi_n'(t)$ is continuous, $\phi_n(\pm\tau) = 0$, and (11.6.7) holds. Then $f(z) = \lim_{n\to\infty} f_n(z)$, where

$$\begin{aligned} f_n(z) &= f(0) + z \int_{-\tau}^{\tau} e^{izt}\phi_n(t)\, dt \\ &= f(0) + \int_{-\tau}^{\tau} \phi_n(t)\, d\,(-\, ie^{izt}) \\ &= f(0) - i \int_{-\tau}^{\tau} e^{izt}\phi_n'(t)\, dt. \end{aligned}$$

Thus $f_n(z)$ has the form (11.6.1), and consequently $L[f_n(z)]$ has both the representations (11.6.6) and (11.6.3). Now (11.6.7) shows that $L[f_n(z)] \to L[f(z)]$, where $L[f(z)]$ is defined by (11.6.6). Let $g(x)$ denote the sum of the series in (11.6.3), so that

$$g(x) = \sum_{k=-\infty}^{\infty} c_k(s) f(x - s + k\pi/\tau),$$

where we are assuming that $\sum | c_k(s) | < \infty$. (This follows from our hypothesis on $\lambda(t)$ provided that $\lambda(\tau)e^{i\tau s} = \lambda(-\tau)e^{-i\tau s}$, since then $\lambda(t)e^{ist}$, extended with period 2τ, will be absolutely continuous and have its derivative in L^2; in this case $\sum | nc_n(s) |^2$ converges and by Cauchy's inequality $\sum | c_n(s) | < \infty$.) Then we can write

$$L[f_n] - g(x) = \sum_{k=-\infty}^{\infty} c_k(s)\{f_n(x - s + k\pi/\tau) - f(x - s + k\pi/\tau)\},$$

and if we show that this approaches zero, it follows that $L[f(x)]$ has the representation (11.6.3). Now

$$\begin{aligned} | L[f_n] - g(x) | &\le \left| \sum_{k=-N}^{N} c_k(s)\{f_n(x - s + k\pi/\tau) - f(x - s + k\pi/\tau)\}\right| \\ &\quad + \sum_{|k|>N} | c_k(s) |\, | f_n(x - s + k\pi/\tau) | \\ &\quad + \sum_{|k|>N} | c_k(s) |\, | f(x - s + k\pi/\tau) | \\ &= S_1 + S_2 + S_3 . \end{aligned}$$

We have $S_1 \to 0$ with a fixed N; and S_3 is arbitrarily small, independently of n, if N is large (since $f(x)$ is bounded). Finally,

$$S_2^2 \le \sum_{|k|>N} k^2 | c_k(s) |^2 \sum_{|k|>N} k^{-2} | f_n(x - s + k\pi/\tau)|^2.$$

The first sum on the right is small when N is large, so that we have only to show that the second sum is bounded, uniformly in n, with a fixed x and s. We have

$$\frac{f_n(x - s + k\pi/\tau) - f(0)}{x - s + k\pi/\tau} = \int_{-\tau}^{\tau} e^{ik\pi t/\tau} e^{i(x-s)t} \phi_n(t)\, dt,$$

$$\sum_{k=-\infty}^{\infty} \left| \frac{f_n(x - s + k\pi/\tau) - f(0)}{x - s + k\pi/\tau} \right|^2 = 2\tau \int_{-\tau}^{\tau} | \phi_n(t) |^2\, dt;$$

by (11.6.7) this is bounded, and the conclusion follows.

We therefore have the following theorem.

11.6.8. *Theorem. If $\lambda(t)$ is absolutely continuous, $\lambda'(t) \in L^2$, and $\lambda(\tau)e^{is\tau} = \lambda(-\tau)e^{-is\tau}$, the operator L defined on functions of the form*

$$f(z) = \int_{-\tau}^{\tau} e^{izt}\, d\alpha(t)$$

by

$$L[f(z)] = \int_{-\tau}^{\tau} \lambda(t)e^{izt}\, d\alpha(t)$$

can be extended to all entire functions $f(z)$ of exponential type which are bounded on the real axis so that

$$L[f(z)] = \sum_{k=-\infty}^{\infty} c_k(s) f(x - s + k\pi/\tau),$$

$$c_k(s) = (2\tau)^{-1} \int_{-\tau}^{\tau} e^{i(s-k\pi/\tau)t} \lambda(t)\, dt.$$

Moreover, if $g_n(z) \to g(z)$ uniformly in every bounded region, and $g_n(z)$ are uniformly bounded on the real axis, $L[g_n] \to L[g]$ uniformly in every bounded region.

This theorem allows us to extend $\tilde{f}'(z)$ (though not $\tilde{f}(z)$ itself), $\sigma(f)$, and the operators described in 11.2.3.

11.7. A class of operators.[a] The method described in the preceding sections does not work particularly well for proving inequalities of the form "$| f(x) | \le | \omega(x) |$ implies $| L[f(x)] | \le | L[\omega(x)] |$." In order to obtain satisfactory results we introduce a new class of operators, whose definition involves the class P (7.8.2): an entire function $\omega(z)$ belongs to P if it is of exponential type, has no zeros for $y < 0$, and satisfies one of the (equivalent) conditions $h(-\alpha) \ge h(\alpha)$ for some α, $0 < \alpha < \pi$, or $| \omega(z) | \ge | \omega(\bar{z}) |$ for $y < 0$.

11.7.1. *Definition. An additive homogeneous operator $B[f(z)]$ which carries entire functions of exponential type into entire functions of exponential type and leaves the class P invariant is called a B-operator.*

From the results on the class P established in §7.8 we can easily derive a general inequality for B-operators.

11.7.2. *Theorem. If $f(z)$ is an entire function of exponential type τ, B is a B-operator, and $\omega(z)$ is an entire function of class P and of order 1, type $\sigma \geq \tau$, then*

(11.7.3) $$| f(x) | \leq | \omega(x) |, \qquad -\infty < x < \infty,$$

implies

(11.7.4) $$| B[f(x)] | \leq | B[\omega(x)] |, \qquad -\infty < x < \infty.$$

By 7.8.6, (11.7.3) implies that $f(z) - u\omega(z)$ belongs to P for $| u | \geq 1$. Since B leaves the class P invariant, $B[f] - u\, B[\omega]$ belongs to P for $| u | \geq 1$, and by the other half of 7.8.6, (11.7.4) follows.

The interest of 11.7.2 comes from the fact that B-operators include differentiation and other operators of simple form. The relevance of class P is shown by the fact that (11.7.3) does not imply (11.7.4) for all $f(z)$ of exponential type unless $\omega(z)$ or $\bar{\omega}(z)$ belongs to P (see 11.7.8).

11.7.5. *Theorem. Differentiation is a B-operator. More generally, the operator L defined by*

$$L[f(z)] = e^{\beta z} D[e^{-\beta z} f(z)] = f'(z) - \beta f(z) = (D - \beta I) f(z)$$

is a B-operator if $\Im(\beta) \leq 0$.

We want to show that $L[\omega(z)]$ is in P if $\omega(z)$ is in P. In the first place, the representation 7.8.3 for $\omega(z)$ shows that

(11.7.6) $$\Im\{\omega'(z)/\omega(z)\} = \nu + \sum_{k=1}^{\infty} \frac{-(y - b_k)}{(x - a_k)^2 + (y - b_k)^2},$$

where $z_k = a_k + ib_k$, $b_k \geq 0$, are the zeros of $\omega(z)$, and $\nu \geq 0$. The right-hand side is positive when $y < 0$ unless there are no z_k and $\nu = 0$ (a trivial case), and since $\Im(\beta) \leq 0$, the imaginary part of $\{\omega'(z) - \beta\omega(z)\}/\omega(z)$ is also not zero. Hence $\omega'(z) - \beta\omega(z)$ has no zeros in the lower half plane.

Let $h(\theta)$, $h_1(\theta)$ be the indicators of the functions $\omega(z)$, $\omega'(z) - \beta\omega(z)$, respectively. We have to show that $h_1(-\alpha) \geq h_1(\alpha)$ for some α, $0 < \alpha < \pi$. Suppose first that $h(-\theta) > h(\theta)$ for $0 < \theta < \pi$. The proof depends on the fact that, while the indicator of the derivative does not exceed the indicator of the function (5.4.13), the maximum value of the indicator is the type, which is not changed by differentiation (2.4.1). Hence the indicators of a function and its derivative must have the same value at at least one point where their common maximum is assumed. By assumption, the maximum of $h(\theta)$ is attained in $-\pi < \theta < 0$. The indicator diagram of $e^{-\beta z}\omega(z)$ is that of $\omega(z)$ translated by β, and since $\Im(\beta) \leq 0$, the indicator $h_\beta(\theta)$ of $e^{-\beta z}\omega(z)$ also attains its maximum in $-\pi < \theta < 0$, at some point θ_0 where the indicator of $\{e^{-\beta z}\omega(z)\}'$ has the same value. Let $k(\theta)$ be the indicator of $e^{-\beta z}$; then

$$h_\beta(\theta_0) = h(\theta_0) + k(\theta_0) > h_\beta(-\theta_0) = h(-\theta_0) + k(-\theta_0).$$

Now $h_\beta(\theta_0)$ is equal to the value at θ_0 of the indicator of $\{e^{-\beta z}\omega(z)\}'$; multiplication of this function by $e^{\beta z}$ subtracts $k(\theta)$ from its indicator; hence $h_1(\theta_0) = h(\theta_0) > h_1(-\theta_0)$.

Suppose on the other hand that $h(-\theta) = h(\theta)$ for some θ. Then if $p > 0$, $e^{ipz}\omega(z)$ belongs to P, and if its indicator is $h_p(\theta)$ we have

$$h_p(-\theta) > h_p(\theta), \qquad 0 < \theta < \pi.$$

Consequently $L[e^{ipz}\,\omega(z)]$ belongs to P. Letting $p \to 0$, we have $L[\omega(z)]$ in P because the limit of a sequence of functions of P is in P, as we remarked in §7.8.

Since the product of two B-operators is a B-operator, we have the following corollary.

11.7.7. *Corollary. If $Q(z)$ is a polynomial all of whose roots are in the lower half plane, $Q(D)$ is a B-operator.*

For the proof, factor $Q(z)$ into linear factors and apply 11.7.5.

We now prove that, at least for the operator of differentiation, the presence of the class P in 11.7.2 is essential.

11.7.8. *Theorem. Let $\omega(z)$ be an entire function of order 1 and type σ such that neither $\omega(z)$ nor $\bar{\omega}(z)$ belongs to P. Then there is an entire function $f(z)$ of exponential type $\tau \le \sigma$ such that $|f(x)| \le |\omega(x)|$ for $-\infty < x < \infty$ while, for some real x_0, $|f'(x_0)| > |\omega'(x_0)|$.*

If neither $\omega(z)$ nor $\bar{\omega}(z)$ belongs to P, then either (i) $\omega(z)$ has zeros in both the upper and lower half planes, or (ii) $\omega(z)$ has a zero in $y > 0$, none in $y < 0$, and $h(-\alpha) < h(\alpha)$ for $0 < \alpha < \pi$ (or correspondingly with upper and lower half planes interchanged).

In case (i) take x_0 so that $\omega(x_0) \ne 0$ and $\omega'(x_0) \ne 0$, supposing for definiteness that $\Im\{\omega(x_0)/\omega'(x_0)\} \ge 0$. Let α be a zero of $\omega(z)$ in $y > 0$, and put

$$f(z) = \frac{z - \bar{\alpha}}{z - \alpha}\,\omega(z).$$

Then $f(z)$ is an entire function of the same type as $\omega(z)$, and $|f(x)| = |\omega(x)|$ for all real x. On the other hand,

$$|f'(x)| = |\omega'(x)| \left| -2i\,\frac{\omega(x)}{\omega'(x)}\,\Im\left(\frac{1}{x - \alpha}\right) + 1 \right|.$$

Since $\omega(x_0)/\omega'(x_0)$ has a positive (or zero) imaginary part, the real part of

$$-2i\,\frac{\omega(x_0)}{\omega'(x_0)}\,\Im\left(\frac{1}{x_0 - \alpha}\right)$$

is not negative and the factor multiplying $\omega'(x)$ exceeds 1 when $x = x_0$. Hence 11.7.8 is proved in case (i).

In case (ii) put $f(z) = e^{2i\delta z}\omega(z)$, where δ is to be chosen later. Then $|f(x)| = |\omega(x)|$ and $\Re\{f'(x)/f(x)\} = \Re\{\omega'(x)/\omega(x)\}$. Moreover, since $\omega(z)$ has no zeros in $y < 0$, but does not belong to P, it has the representation 7.8.3 except that $\nu = \Im(c) < 0$; hence, as in (11.7.6),

$$\Im\{\omega'(x)/\omega(x)\} = -\nu + \sum_{k=1}^{\infty} \frac{b_k}{(x - a_k)^2 + b_k^2}, \tag{11.7.9}$$

where $a_k + ib_k$, $b_k \geq 0$, are the zeros of $\omega(z)$. Take $\delta = \nu$; then

$$\Im\{f'(x)/f(x)\} = \nu + \sum_{k=1}^{\infty} \frac{b_k}{(x - a_k)^2 + b_k^2}. \tag{11.7.10}$$

Since the sum in (11.7.9) and (11.7.10) is positive,

$$|\Im\{f'(x)/f(x)\}| > |\Im\{\omega'(x)/\omega(x)\}|,$$

and so $|f'(x)| > |\omega'(x)|$, since $\Re\{f'(x)/f(x)\} = \Re\{\omega'(x)/\omega(x)\}$.

By 7.7.2 we have (cf. 7.7.6)

$$h_\omega(\theta) = 2\nu \sin\theta + h_\omega(-\theta),$$

and we have chosen δ so that

$$h_f(\theta) = h_\omega(\theta) - 2\nu \sin\theta = h_\omega(-\theta).$$

Consequently $\tau = \max h_f(\theta) = \sigma = \max h_\omega(\theta)$.

We now identify a substantial class of B operators.

11.7.11. *Theorem. If* $F(z) = \sum_{k=0}^{\infty} c_k z^k$ *and* $\bar{F}(z)$ *is of class P, then*

$$F(D) = \sum_{k=0}^{\infty} c_k D^k, \qquad D = d/dz, \tag{11.7.12}$$

is a B-operator.

Theorem 11.7.5 is included because $F(z) = z - \bar{\beta}$ is in P.

The applicability of operators (11.7.12) to entire functions of exponential type is settled by the following theorem.

11.7.13. *Theorem.*[b] *If* $f(z)$ *is an entire function,* $F(D)$ *carries every entire function of exponential type into an entire function of the same exponential type.*

The converse is also true:

11.7.14.* *Theorem.*[b] *If* $F(z)$ *is a function, regular at* 0, *such that* $F(D)$ *carries every entire function of exponential type into an entire function of the same exponential type,* $F(z)$ *is entire.*

To prove 11.7.13, we represent the given entire function of exponential type τ in the form

$$f(z) = \int_C e^{zw}\phi(w)\,dw,$$

where C is the circle $|z| = \tau + \epsilon$, $\epsilon > 0$. Then

$$f^{(k)}(z) = \int_C w^k e^{zw}\phi(w)\,dw,$$

$$|f^{(k)}(z)| \le 2\pi(\tau+\epsilon)^{k+1}e^{(\tau+\epsilon)|z|}M, \qquad M = \max_{|w|=\tau+\epsilon} |\phi(w)|,$$

and the series $\sum_{k=0}^{\infty} c_k f^{(k)}(z)$ is dominated by

$$\sum_{k=0}^{\infty} 2\pi M\,|c_k|\,(\tau+\epsilon)^{k+1}e^{(\tau+\epsilon)|z|}, \tag{11.7.15}$$

and so converges uniformly in every bounded set if $F(z)$ is entire. Hence it represents an entire function. Moreover,

$$\begin{aligned} F(D)f(z) &= \sum_{k=0}^{\infty} c_k f^{(k)}(z) = \sum_{k=0}^{\infty} c_k \int_C w^k e^{zw}\phi(w)\,dw \\ &= \int_C e^{zw}\left\{\sum_{k=0}^{\infty} c_k w^k\right\}\phi(w)\,dw \\ &= \int_C e^{zw}F(w)\phi(w)\,dw, \end{aligned}$$

the rearrangement being justified by the convergence of the dominant (11.7.15). Since ϵ can be arbitrarily small, the last expression shows that $F(D)f(z)$ is of exponential type τ.

To complete the proof of 11.7.11 we have to show that the operator $F(D)$ leaves the class P invariant. For this it is enough to construct a sequence $P_n(z)$ of polynomials such that, for every $\omega(z)$ in P, $P_n(D)\omega(z)$ belongs to P and converges to $F(D)\omega(z)$, uniformly in every bounded set. The property of convergence is possessed by

$$P_n(z) = c_0 + \sum_{k=1}^{n}\left(1-\frac{1}{n}\right)\left(1-\frac{2}{n}\right)\cdots\left(1-\frac{k-1}{n}\right)c_k z^k.$$

For, if $\omega(z) = \int_C e^{zw}\phi(w)\,dw$, where C surrounds the conjugate indicator diagram of $\omega(z)$,

$$P_n(D)\omega(z) = \int_C e^{zw}P_n(w)\phi(w)\,dw,$$

and the conclusion follows if $P_n(w) \to F(w)$, uniformly in every bounded set. Now

$$\begin{aligned}
| P_n(w) - F(w) | &\leq \left| \sum_{k=2}^{p} \left\{1 - \left(1 - \frac{1}{n}\right) \cdots \left(1 - \frac{k-1}{n}\right)\right\} c_k z^k \right| \\
&+ \left| \sum_{k=p+1}^{n} \left(1 - \frac{1}{n}\right) \cdots \left(1 - \frac{k-1}{n}\right) c_k z^k \right| \\
&+ \left| \sum_{k=p+1}^{\infty} c_k z^k \right| \\
&\leq \left| \sum_{k=2}^{p} \left\{1 - \left(1 - \frac{1}{n}\right) \cdots \left(1 - \frac{k-1}{n}\right)\right\} c_k z^k \right| + 2 \sum_{p+1}^{\infty} \left| c_k z^k \right| \\
&= S_1 + S_2 .
\end{aligned}$$

If z is confined to a given bounded set, we can make S_2 small by taking p large, because $\sum_{k=0}^{\infty} | c_k | z^k$ is entire, and then we can make S_1 small by taking n large with a fixed p.

It is therefore enough to show that the operator $P_n(D)$ transforms functions $\omega(z)$ of P into functions of P. Now by 11.7.7, $Q(D)$ is a B-operator if $Q(z)$ is a polynomial with all its roots in the lower half plane. Consequently it is enough to show that all the roots of $P_n(z)$ are in the lower half plane. The fact that $P_n(z) \to F(z)$ and $F(z)$ has all its zeros in the lower half plane is of course not conclusive. However, there do exist polynomials

$$Q_m(z) = \sum_{k=0}^{m} a_{k,m} z^k$$

such that $Q_m(z) \to F(z)$ uniformly in every bounded set, as $m \to \infty$, so that $a_{k,m} \to c_k$ as $m \to \infty$, and such that $Q_m(z)$ has all its zeros in the lower half plane. To construct them, write $F(z)$ in the form

$$F(z) = cz^m e^{az} \prod_{k=1}^{\infty} (1 - z/\lambda_k) \exp \{z\Re(1/\lambda_k)\},$$

where $\Im(\lambda_k) \leq 0$ and $\Im(a) \geq 0$. Given a bounded region G and $\epsilon > 0$ choose an integer p such that

$$\left| F(z) - cz^m e^{bz} \prod_{k=1}^{p} (1 - z/\lambda_k) \right| < \epsilon/2$$

in G, where

$$b = a + \sum_{k=1}^{p} \Re(1/\lambda_k).$$

Now replace e^{bz} by the polynomial $(1 + bz/n)^n$, and consider

$$Q_m(z) = cz^m(1 + bz/n)^n \prod_{k=1}^{p} (1 - z/\lambda_k);$$

we can make $| Q_m(z) - F(z) | < \epsilon$ by choosing n sufficiently large. Since $\mathfrak{I}(b) \leq 0$ and $\mathfrak{I}(\lambda_k) \leq 0$, all the roots of $Q_m(z)$ are in the lower half plane. Now $Q_m(D)z^n = Q_{n,m}(z)$ is a polynomial; and all its zeros are in the upper half plane because z^n belongs to P and $Q_m(D)$ is, as we have just seen, a B-operator, so that $Q_{n,m}(z)$ belongs to P. Then

$$z^n Q_{n,m}(1/z) = a_{0,m} + a_{1,m} z + \cdots + n(n-1) \cdots (n-k+1)a_{k,m}z^k + \cdots + n!\, a_{n,m}z^n$$

has all its roots in the lower half plane. If we write z/n for z and let $m \to \infty$, with n fixed, this approaches $P_n(z)$, so $P_n(z)$ has all its roots in the lower half plane also.

Since $e^{-cz^2} = \lim (1 - cz^2/n)^n$, a limit of polynomials with real zeros when $c > 0$, e^{-cD^2} is also a B-operator. Hence we have

11.7.16. *Corollary. If $F(z)$ is of class P and $c > 0$, $e^{-cD^2} \bar{F}(D)$ is a B-operator.*

We cannot go any farther by the same method, since[c] a limit of polynomials all of whose zeros are in the lower half plane is necessarily of the form $e^{-cz^2} F(z)$ with $\bar{F}$ in P.

Notes for Chapter 11

11.1a. See, e.g., S. Bernstein [2], Ahiezer [6]; many of the papers cited later in this chapter contain further references.

11.2a. This part of the theory was worked out by Civin [1], who gives a number of examples not included here. The same idea was apparently used by Krein [1] in a paper which is completely inaccessible outside the USSR, although its contents can be reconstructed with some confidence from references in the Russian literature; see in particular Ahiezer [2], [3].

11.2b. Boas [14].

11.3a. Boas [2].

11.3b. Special cases are due to Zygmund [1] and Bellman [1].

11.3c. Schaeffer [2].

11.3d. S. Bernstein [4].

11.3e. See Schaeffer [1], Meiman [1].

11.4a. S. Bernstein [12].

11.4b. van der Corput and Schaake [1] for trigonometric polynomials; Duffin and Schaeffer [1], Ahiezer [3], Redheffer [3]; (11.4.8) is actually a special case of (11.4.9).

11.4c. Szegö [2] for trigonometric polynomials; Boas [1] for trigonometric integrals; Ahiezer [3] without the term $\tau| \sigma(f) |$.

11.4d. Pólya and Szegö [1], vol. 2, p. 219; Schaeffer and Szegö [1]; A. J. Macintyre [1]; cf. the remarks below in connection with 11.5.11.

11.4e. Schaeffer and Szegö [1], for trigonometric polynomials. Schaeffer and Szegö prove many other theorems.

11.4f. Boas [17].

11.4g. Boas [17].

11.4h. Stechkin [1], Nikolskii [1], S. Bernstein [11], in varying degrees of generality.

11.4i. If $f(x)$ is a trigonometric polynomial of degree n, $|\tilde{f}(x)| \leq K(n) \sup |f(x)|$, where $K(n) \sim (2/\pi) \log n$ as $n \to \infty$. The order $\log n$ was determined by Fekete; the constant $2/\pi$ by Boas [1] for trigonometric integrals (11.2.1) with $\alpha(t)$ constant in $(-1,1)$; the exact value of $K(n)$ was determined by Szegö [3].

11.4j. Civin [1].

11.4k. Bohr; for references and generalizations see Sz.-Nagy [1], [2], Sz.-Nagy and Strausz [1].

11.4l. Turetskii [1]. In a later paper [2] Turetskii calculates the bound for $|T_n(x)|$, given a bound for $|aT_n''(x) + bT_n'(x) + cT_n(x)|$ at $x = 2k\pi/(2n+1)$, where $T_n(x)$ is a trigonometric polynomial of degree n.

11.5a. Boas [6].

11.5b. The name and the theorem are due to S. Bernstein [12]; he proves more generally that (as will be shown later in this section) $f(x)$ needs only to be $o(|x|)$ on the real axis, instead of $O(1)$. If $f(x)$ is merely $o(|x|^2)$, the left-hand side of (11.5.5) is still bounded (Timan [1]); cf. 11.5.13, 11.5.14.

11.5c. A. J. Macintyre [1].

11.5d. Whittaker [2].

11.5e. The representation (11.5.11) and its generalizations were introduced by Valiron [6]; A. J. Macintyre [1] gives an alternative approach and other references.

11.5f. Timan [1].

11.6a. Ahiezer [2] gives a different method of extending operators which allows him to define $\tilde{f}(z)$ for $f(z)$ such that $f(x)/(1+|x|)$ belongs to L^2.

11.7a. The material of this section is due to Levin [6]. Still more general results are given by Meiman [1]. Levin [8] gives an intrinsic characterization of B-operators.

11.7b. See Muggli [1].

11.7c. Pólya; the first published proof seems to have been given by Obrechkoff. For more accessible proofs, and references and related results, see Korevaar [4]; Chebotarëv and Meiman [1], p. 106; Levin [7].

Chapter 12

APPLICATIONS

12.1. Introduction. In this chapter we shall give some applications of entire functions of exponential type. In most cases we shall prove one or two typical theorems and refer to the literature for more detailed expositions.

12.2. Asymptotic behavior of a mean value. Our first example is an ostensibly purely "real variable" theorem which has apparently not been treated by real variable methods.

Let $0 \leq a < b$; if $f(t)$ is integrable and nonnegative, and does not vanish almost everywhere in a neighborhood of b, we have

$$\lim_{n\to\infty} \left\{ \int_a^b t^n f(t)\, dt \right\}^{1/n} = b;$$

indeed, if $0 < \epsilon < b - a$,

$$(b - \epsilon)^n \int_{b-\epsilon}^b f(t)\, dt \leq \int_{b-\epsilon}^b t^n f(t)\, dt \leq \int_a^b t^n f(t)\, dt \leq b^n \int_a^b f(t)\, dt,$$

and taking nth roots we obtain

$$b - \epsilon \leq \liminf \left\{ \int_a^b t^n f(t)\, dt \right\}^{1/n} \leq \limsup \left\{ \int_a^b t^n f(t)\, dt \right\}^{1/n} \leq b.$$

If $f(t)$ is no longer required to be nonnegative, one is led to conjecture that

$$\limsup_{n\to\infty} \left| \int_a^b t^n f(t)\, dt \right|^{1/n} = b. \tag{12.2.1}$$

To prove this, set

$$F(z) = \int_a^b e^{izt} f(t)\, dt;$$

then by 6.9.1, $F(z)$ is of order 1 and type b (and not of smaller type). On the other hand,

$$| F^{(n)}(0) | = \left| \int_a^b t^n f(t)\, dt \right|$$

and if the lim sup in (12.2.1) is $c < b$, we have by (2.2.12) that $F(z)$ is of exponential type c. Hence $c < b$ is impossible.

As a corollary we have the following theorem on moments:[a] *If $f(t)$ is* *integrable over* (a, b), $0 \leq a < b$, *and if for some* $\delta > 0$ *and every* $\epsilon > 0$,

$$\int_a^b t^{n\delta} f(t)\, dt = O\{a + \epsilon)^{n\delta}\},$$

then $f(t) = 0$ *almost everywhere in* (a, b).

12.3. A theorem on convolutions. Our next example is another "real variable" theorem.[a]

12.3.1. *Theorem. Let* $f(x)$ *and* $g(x)$ *belong to* $L^2(0, c)$ *and let*

$$h(x) = \int_0^x f(t) g(x - t)\, dt = 0$$

for almost all x *in* $(0, c)$. *Then* $f(x) = 0$ *almost everywhere in* $(0, a)$ *and* $g(x) = 0$ *almost everywhere in* $(0, b)$, *where* $a + b \geq c$.

Define $f(x)$ and $g(x)$ to be 0 outside $(0, c)$, and suppose that $f(x) = 0$ almost everywhere in $(0, a)$ and $g(x) = 0$ almost everywhere in $(0, b)$, but not in longer intervals starting at 0 (where a or b or both may be 0). Then $h(x) = 0$ for $x < c$ and for $x > 2c$. Let F, G, H be the Fourier transforms of f, g, h; they are entire functions of exponential type. By 6.9.4 the indicator diagrams of F, G, H are segments of the imaginary axis, namely (ia, ic_1), (ib, ic_2), and a subset of $(ic, 2ic)$, where $c_j \leq c$ $(j = 1, 2)$. On the other hand, by 7.6.6 the indicator diagram of H is the sum of those of F and G, i.e. the segment $(i(a + b), i(c_1 + c_2))$. In particular, then, $a + b \geq c$.

12.4. Completeness of sets of functions.[a] One of the most important applications of functions of exponential type is to the problem of the completeness of a set of functions of the form $\{f(\lambda_n x)\}$, where $f(x)$ is a regular function of exponential type in an angle or in the whole plane. The usual terminology is as follows.[b] The set $\{f_n\}_1^\infty$ is complete with respect to the class $F(a, b)$, or simply complete $F(a, b)$, if

$$\int_a^b f_n(x) g(x)\, dx = 0, \qquad n = 1, 2, \cdots, \tag{12.4.1}$$

and $g(x) \epsilon F(a, b)$ imply that $g(x) = 0$ almost everywhere on (a, b). If $F = L^p$, $1 < p < \infty$, the appropriate class for the functions $f_n(x)$ is $L^{p'}$, $p' = p/(p - 1)$, and then completeness is equivalent to closure in $L^{p'}$, i.e. the property that every element of $L^{p'}$ can be approximated arbitrarily closely in the $L^{p'}$ metric by finite linear combinations of the $f_n(x)$.

When $f_n(x) = f(\lambda_n x)$, $f(z)$ is of exponential type, and (12.4.1) holds, we may introduce the function

$$\phi(z) = \int_a^b f(zt) g(t)\, dt, \tag{12.4.2}$$

which will also be of exponential type and will have zeros at the points

λ_n. If there are enough λ_n to make $\phi(z)$ vanish identically when $g(t) \epsilon F(a, b)$, and if

$$\int_a^b f(zt)g(t)\,dt \equiv 0 \tag{12.4.3}$$

implies $g(t) = 0$ almost everywhere, the set $\{f(\lambda_n x)\}$ will be complete $F(a, b)$. We illustrate the method with a few of the many examples to which it has been applied.

12.4.4. *Theorem.*[c] *The set* $\{t^{\lambda_n}\}$, $0 < \lambda_1 < \lambda_2 < \cdots$, *is complete L on any finite interval* (a, b), $a \geq 0$, *if* $\sum 1/\lambda_n$ *diverges.*

Completeness L implies completeness L^p for $1 \leq p \leq \infty$.

Here

$$\phi(z) = \int_a^b e^{z \log t} g(t)\,dt,$$

and we have $\phi(z)$ of exponential type in $x > 0$, continuous on the boundary, and bounded on the imaginary axis. It follows that $\phi(z) \equiv 0$ if $\sum 1/\lambda_n$ diverges, by 9.3.8. Hence in particular

$$\phi(n) = \int_a^b t^n g(t)\,dt = 0, \qquad n = 0, 1, 2, \cdots.$$

By §12.2, this implies that $g(t) = 0$ almost everywhere.

12.4.5. *Theorem.*[d] *The set* $\{e^{i\lambda_n x}\}$, $0 < \lambda_1 < \lambda_2 < \cdots$, *is complete L on the interval* $(-b, b)$ *(or on any interval of length* $2b$*) if* $b < \pi$ *and*

$$\liminf n/\lambda_n \geq 1.$$

Here

$$\phi(z) = \int_{-b}^b e^{izt} g(t)\,dt,$$

$$|\phi(iy)| \leq e^{b|y|} \int_{-b}^b |g(t)|\,dt,$$

and we can apply Theorem 9.3.4 to $\phi(\pi z/b)$ with $\sigma(y) = O(1)$, $\delta(x) = \epsilon > 0$, and conclude that $\phi(z) \equiv 0$, and hence $g(t) = 0$ almost everywhere by the uniqueness theorem for Fourier transforms. Theorem 9.3.4 shows more generally that $\liminf n/\lambda_n \geq 1$ can be generalized to $\Lambda(t) \geq t + \delta(t)$, where $\Lambda(t)$ is the number of λ_n less than t and $\int_1^\infty t^{-2}\,\delta(t)\,dt = +\infty$; the extension of Theorem 9.3.4 suggested on page 155 shows that it is even sufficient to have $\int_1^\infty t^{-3/2}\delta(t)\,dt = +\infty$.

The existence of a density for $\{\lambda_n\}$ is by no means necessary for the completeness of $\{e^{i\lambda_n x}\}$, as the following result shows.

12.4.6. *Theorem. The set* $\{e^{i\lambda_n x}\}$, $0 < \lambda_1 < \lambda_2 < \cdots$, *is complete L on the interval* $(-b, b)$ *if* $b < \pi$ *and* $\{\lambda_n\}$ *has maximum density at least* 1.

As in proving 12.4.5, consider

$$\phi(z) = \int_{-b}^{b} e^{izt} g(t)\, dt.$$

Then $\phi(z)$ vanishes identically by an application of Theorem 9.7.1 with $k = b$.

Our final completeness theorem deals with an interval of length 2π and includes the completeness of $\{e^{inx}\}_{n=-\infty}^{\infty}$.

12.4.7. *Theorem. If* $\Lambda(x)$ *denotes the number of* λ_n *with* $|\lambda_n| < x$, *the set* $\{e^{i\lambda_n x}\}_{-\infty}^{\infty}$ *is complete* L^p, $1 \leq p < \infty$, *on* $(-\pi, \pi)$, *if*

$$\int_1^r t^{-1}\Lambda(t)\, dt > 2r - \frac{p-1}{p}\log r - B, \tag{12.4.8}$$

with a constant B.

Here we have

$$\phi(z) = \int_{-\pi}^{\pi} e^{izt} g(t)\, dt, \qquad g(t) \in L^p.$$

We shall show that

$$|y|^{(p-1)/p} |\phi(x+iy)| e^{-\pi|y|} \to 0, \qquad |y| \to \infty, \tag{12.4.9}$$

and then it follows from Theorem 9.6.11 that $\phi(z) \equiv 0$ if $\phi(z)$ has zeros at points λ_n satisfying (12.4.8). We have

$$|\phi(x+iy)| \leq \int_{-\pi}^{\pi} e^{-yt} |g(t)|\, dt$$

$$\leq \int_0^{\pi} e^{|y|t} g_1(t)\, dt, \qquad g_1 \in L^p.$$

We give the detailed argument only for $p > 1$; the case $p = 1$ is simpler. Given a positive ϵ, choose δ so that

$$\int_{\pi-\delta}^{\pi} g_1(t)\, dt < \epsilon^p.$$

Then

$$|\phi(x+iy)| \leq$$

$$\left\{\int_0^{\pi-\delta} \{g_1(t)\}^p\, dt\right\}^{1/p} \left\{\int_0^{\pi-\delta} e^{p|y|t/(p-1)}\, dt\right\}^{(p-1)/p} + \epsilon\left\{\int_{\pi-\delta}^{\pi} e^{p|y|t/(p-1)}\, dt\right\}^{(p-1)/p}$$
$$\leq e^{\pi|y|}\,|\,y\,|^{-(p-1)/p}\{Ae^{-\delta|y|} + \epsilon\},$$

with some constant A. Since ϵ can be arbitrarily small, (12.4.9) follows

12.5. Fourier series. The completeness theorems of §12.4 for sequences $\{e^{i\lambda_n x}\}$ can be reformulated as theorems about Fourier series with gaps. For example,

12.5.1. *Theorem.*[a] *Let $f(x)$ be integrable on $(-\pi, \pi)$ and let its Fourier coefficients of negative index vanish except for a sequence of minimum density d. If $f(x)$ vanishes almost everywhere on an interval of length greater than $2\pi d$, it vanishes almost everywhere on $(-\pi, \pi)$.*

The hypothesis implies that

$$\int_{-\pi}^{\pi} f(x)e^{i\lambda_n x}\, dx = 0$$

for a sequence of integers λ_n of maximum density $1 - d$. By 12.4.6 (with a change of variable) $f(x)$ vanishes almost everywhere.

The source of theorems like 12.5.1 is the fact that

$$c_n = (2\pi)^{-1}\int_{-\pi}^{\pi} f(t)e^{-int}\, dt \tag{12.5.2}$$

are the values for $z = n$ of an entire function of exponential type, and this fact can be exploited in various ways. As another example, we have

12.5.3. *Theorem. If $f(x)$ is real and integrable on $(0, \pi)$, if its Fourier cosine coefficients a_n and a_{n+1} have the same sign for all n in a sequence of maximum density greater than $1 - \delta/\pi$, and if $f(x) = 0$ almost everywhere in $0 \leq x < \delta$, then $f(x) = 0$ almost everywhere in $(0, \pi)$.*

Write

$$\pi a_n/2 = \int_0^{\pi} f(t)\cos nt\, dt = (-1)^n\int_0^{\pi} f(\pi - t)\cos nt\, dt$$
$$= (-1)^n\int_0^{\pi-\delta} f(\pi - t)\cos nt\, dt.$$

Now put

$$g(z) = \int_0^{\pi-\delta} f(\pi - t)\cos zt\, dt.$$

Then $g(z)$ is of exponential type $\pi - \delta$, is bounded on the real axis, and has a zero between n and $n + 1$ if a_n and a_{n+1} have the same sign. Hence by 9.7.1, $g(z) \equiv 0$, so all the a_n vanish and $f(z) = 0$ almost everywhere.

Similarly one can prove

12.5.4*. *Theorem.*[b] *If $f(x)$ is real and integrable on $(0, \pi)$, if its Fourier cosine coefficients a_n and a_{n+1} have the same sign for all n in a sequence of density greater than $1 - \delta/\pi$, and if $f(x)$ coincides almost everywhere with a regular function in $0 \le x < \delta$, then $f(x)$ coincides almost everywhere with a regular function in $0 \le x < \pi$.*

Here we use 10.3.1 to show that $a_n = O(e^{-\epsilon n})$ for some positive ϵ.

In a similar way one can use the formula for the partial sums of a Fourier series to interpret them as the values for $z = n$ of an entire function, and so connect properties of the function and of its sequence of partial sums. As an illustration we prove the following theorem.

12.5.5. *Theorem.*[c] *Let $f(t)$ be a real even function of period 2π such that $f'(t)$ is absolutely continuous in a neighborhood of 0, $f''(t)/t$ is integrable and $f''(0) = 0$. If the partial sums $s_n(t)$ of the Fourier series of $f(t)$ satisfy $s_n(0) \ge f(0)$ for all n, $f(t)$ is a constant.*

We have

$$s_n(0) - f(0) = \pi^{-1} \int_0^\pi \{f(t) - f(0)\} \frac{\sin (n + \frac{1}{2})t}{\sin t/2} \, dt,$$

so

$$(-1)^n \{s_n(0) - f(0)\} = \pi^{-1} \int_0^\pi \frac{f(\pi - u) - f(0)}{\cos u/2} \cos (n + \tfrac{1}{2})u \, du$$

$$= \int_0^\pi \phi(u) \cos (n + \tfrac{1}{2})u \, du,$$

say. Then let

$$F(z) = \int_0^\pi \phi(u) \cos (z + \tfrac{1}{2})u \, du,$$

so that $F(z)$ is an entire function of exponential type π, real for real z, and even around $z = -\frac{1}{2}$. For $n = 0, 1, 2, \cdots$, $(-1)^n F(n) \ge 0$ and so $F(z)$ vanishes at least once between n and $n + 1$ (inclusive). Hence $F(z)$ vanishes at least once between $-n-2$ and $-n-1$. The function $zF(z)$ thus vanishes at least once in every interval $(n, n + 1)$, and hence will vanish identically, by 9.6.1, if it is uniformly $o(r^{-1}e^{\pi r})$. If we establish this, it follows that $\phi(u) = 0$ almost everywhere and so $f(t)$ is constant. Now it is easy to show that $\phi(\pi -) = \phi'(\pi -) = 0$, and that $\phi''(u)$ is integrable in (λ, π) for some λ, $0 < \lambda < \pi$. Then we have

$$F(z) = \int_0^\lambda \phi(u) \cos (z + \tfrac{1}{2})u \, du + \int_\lambda^\pi \phi(u) \cos (z + \tfrac{1}{2})u \, du;$$

the first integral is $O(e^{\lambda r})$. If the second integral is integrated by parts twice, the integrated terms contribute $O(e^{\lambda r})$ and the rest is

$$-(z + \tfrac{1}{2})^{-2} \int_{\lambda}^{\pi} \phi''(u) \cos (z + \tfrac{1}{2})u \, du.$$

If $\epsilon > 0$, choose $\delta > 0$ so that $\int_{\pi-\delta}^{\pi} | \phi''(u) | \, du < \epsilon$; then

$$\left| \int_{\lambda}^{\pi} \phi''(u) \cos (z + \tfrac{1}{2})u \, du \right| \le \left| \int_{\lambda}^{\pi-\delta} \right| + \epsilon e^{\pi(r+1/2)};$$

the first term is $O(e^{(\pi-\delta)r})$ and so $F(z) = o(r^{-2}e^{\pi r})$, as required.

12.6. Power series on the circle of convergence. Functions of exponential type furnish a powerful tool for discussing a power series on its circle of convergence or a Dirichlet series on the boundary of its half plane of convergence. The connection is most easily seen for power series. Let

$$f(z) = \sum_{0}^{\infty} c_n z^n$$

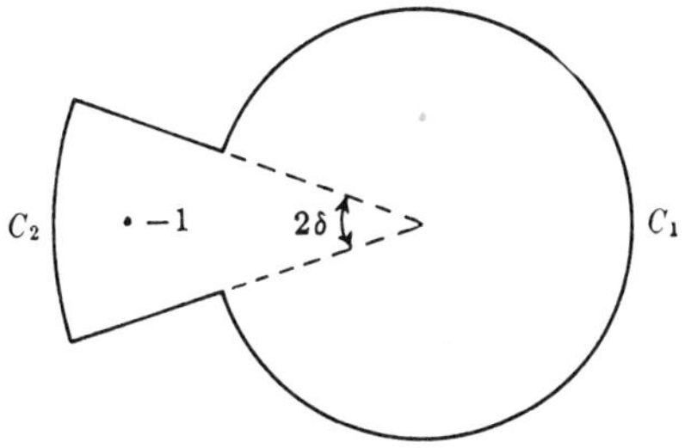

have the unit circle as its circle of convergence and suppose that $f(z)$ is regular on an arc of length 2δ centered at -1. If we write the integral expressing c_n in terms of $f(z)$, using the contour indicated in the figure, where C_1 is an arc of $| z | = r < 1$, and C_2 is an arc of $| z | = R > 1$, we have

$$\begin{aligned} 2\pi c_n = {} & \int_{-\pi+\delta}^{\pi-\delta} f(re^{i\theta}) r^{-n} e^{-in\theta} \, d\theta - i \int_{r}^{R} f(te^{i(\pi-\delta)}) t^{-n-1} e^{-in(\pi-\delta)} \, dt \\ & + i \int_{r}^{R} f(te^{i(-\pi+\delta)}) t^{-n-1} e^{-in(\pi-\delta)} \, dt \\ & + \int_{\pi-\delta}^{\pi+\delta} f(Re^{i\theta}) R^{-n} e^{-in\theta} \, d\theta. \end{aligned} \tag{12.6.1}$$

If we replace n on the right by a complex variable z, the first integral is an entire function of exponential type $\pi - \delta + \log (1/r)$, the second and third are entire functions of exponential type with the type not exceeding $\pi - \delta + \log (1/r)$ in the right-hand half plane, and the fourth is $O(R^{-x})$ as $x \to +\infty$. In other words, we have represented c_n in the form

$$c_n = F(n) + g_n, \qquad n = 0, 1, 2, \cdots, \tag{12.6.2}$$

where $F(z)$ is of exponential type $\pi - \delta + \log (1/r)$ in the right-hand half plane and $g_n = O(R^{-n})$. This holds also for $n < 0$ if we interpret c_n as 0 for $n < 0$.

Now suppose that $c_n = 0$ for certain values of n; then $F(n) = O(R^{-n})$ for these values of n. According to V. Bernstein's theorem (10.3.1), if these values of n have a density D such that πD exceeds the type of $F(z)$ for the right-hand half plane, it follows that $F(n) = O(R^{-n})$ for every n, hence $c_n = O(R^{-n})$ for every n, and this contradicts the assumption that $f(z)$ has $|z| = 1$ as its circle of convergence. Since we can make the type of $F(z)$ for the right-hand half plane arbitrarily close to $\pi - \delta$ by taking r close to 1, and since we can rotate any arc of regularity so that it is centered at -1, without changing $|c_n|$, we have proved the following theorem.

12.6.3. *Theorem.*[a] *If* $f(z) = \sum_{n=0}^{\infty} c_n z^n$ *has* $c_n = 0$ *for a sequence of density D, every arc of the circle of convergence whose central angle exceeds* $2\pi(1 - D)$ *contains a singular point of* $f(z)$.

If $D = 1$ this says that the circle of convergence is a natural boundary. We may also express the hypothesis about the density of the coefficients by saying that the nonvanishing c_n have maximum density $1-D$, which amounts to disregarding the vanishing of some of the c_n.

If $f(z)$ is not regular on an arc of the unit circle, but is continuous, or even dominated by an integrable function, in a sector of the unit circle, we still can write (12.6.1) with $R = 1$. In this case, g_n in (12.6.2) has the form

$$g_n = \int_{\pi-\delta}^{\pi+\delta} f(e^{i\theta}) e^{in\theta} \, d\theta, \tag{12.6.4}$$

and the size of g_n is determined by the boundary behavior required of $f(z)$. In particular, if $f(e^{i\theta})$ is merely integrable, (12.6.4) are the Fourier coefficients of an integrable function and so $g_n \to 0$. Hence $F(n_k) \to 0$ for any sequence $\{n_k\}$ such that $c_{n_k} = 0$. By the Duffin-Schaeffer theorem 10.5.1, if $|n_k - k/D| < L$, and $\delta > \pi(1 - D)$, $F(x) \to 0$ as $x \to \infty$, so $F(n) \to 0$ for every n, and so $c_n \to 0$. We have proved the following theorem.

12.6.5. *Theorem.*[b] *If* $f(z) = \sum_{n=0}^{\infty} c_n z^n$ *has* $c_n = 0$ *for a sequence* $\{n_k\}$ *(of density D) such that* $|n_k - k/D| < L$, *and* c_n *does not tend to* 0, *then* $f(z)$ *cannot be dominated by an integrable function in any sector, of central angle exceeding* $2\pi(1 - D)$, *of its circle of convergence; in particular,* $f(z)$ *cannot be bounded in such a sector.*

Nothing of interest results from putting $D = 1$ in 12.6.5, but the following theorem may be regarded as a limiting case, and is perhaps more interesting than 12.6.5 itself.

12.6.6. *Theorem.*[c] *If* $f(z) = \sum_{n=0}^{\infty} c_n z^n$ *has the unit circle as its circle of* *convergence, if $f(z)$ is dominated by an integrable function in some sector, and if there are only a finite number of different c_n, then $f(z)$ is a rational function.*

Here $\{c_n\}$ is certainly bounded, $g_n \to 0$, and so $F(z)$ is an entire function of exponential type less than π which is bounded for all n (positive or negative), hence (10.2.1) is bounded for real x, and hence (6.2.4) satisfies $|F(x + iy)| \leq Ae^{b|y|}$, $b < \pi$. Therefore, for any sequence of real numbers μ_n, the functions $F(z - \mu_n)$ are bounded in any bounded region and so form a normal family.

Let $a_1, \cdots, a_m$ be the values of c_n, $|a_k - a_l| \geq a > 0$, $k \neq l$. If r is a positive integer there are integers $p = p(r)$, $q = q(r)$, $p > q \geq 0$, such that $a_{p+k} = a_{q+k}$, $0 \leq k \leq r$; in other words, there is some sequence of length r that occurs twice. Either $a_{p+k} = a_{q+k}$ for all positive k, in which case the sequence $\{a_n\}$ is periodic from some point on, and $f(z)$ is rational; or there is an integer $R > r$ such that $a_{p+k} = a_{q+k}$, $0 \leq k \leq R - 1$, $a_{p+R} \neq a_{q+R}$. Assume that the second case holds for arbitrarily large r, and let

$$g_r(z) = F(z + p + R) - F(z + q + R)$$

(where p, q, R are functions of r). Then $\{g_r(z)\}$ is a normal family and contains a subsequence which tends to an entire function $g(z)$ satisfying

$$|g(x + iy)| \leq 2Ae^{b|y|}, \qquad b < \pi. \tag{12.6.7}$$

Now $R \to \infty$ when $r \to \infty$, so by (12.6.2)

$$g_r(n) = a_{n+p+R} - a_{n+q+R} + o(1),$$

for each fixed n; it follows that

$$|g(0)| = |a_{p+R} - a_{q+R}| \geq a > 0, \tag{12.6.8}$$

$$g(n) = 0, \qquad n = -1, -2, \cdots. \tag{12.6.9}$$

This is impossible since by Carlson's theorem (9.2.1), (12.6.7) and (12.6.9) imply $g(z) \equiv 0$. This excludes the second case and establishes the theorem.

Another type of theorem on singular points is obtained by supposing, not that $c_n = 0$ for a sequence of values of n, but that the sequence $\{c_n\}$ is real and changes sign at enough values of n. Taking the simplest case, suppose that $f(z)$ is regular on an arc of $|z| = 1$ around the point -1, and that $(-1)^n c_n \geq 0$, $n = 0, 1, 2, \cdots$. Then from (12.6.2) we have $(-1)^n F(n) \geq O(R^{-n})$, and so between n and $n + 1$ there is a point λ_n where $F(\lambda_n) = O(R^{-\lambda_n})$. Since $F(z)$ is of exponential type less than π in the right-hand half plane, this implies by V. Bernstein's theorem (10.3.1) that $F(x) = O(R^{-x})$, so $c_n = O(R^{-n})$, and $f(z)$ is regular in a circle larger than $|z| = 1$. Finally, $f(-z)$ has a power series with nonnegative coefficients, and so we have given a rather complicated proof of the following elementary theorem.

12.6.10. *Theorem.*[d] *If* $f(z) = \sum_{n=0}^{\infty} c_n z^n$ *has the unit circle as its circle of convergence and all* $c_n \geq 0$ *then* $f(z)$ *has a singular point at* $z = 1$.

There are much simpler proofs of 12.6.10, but our proof actually establishes much more. In the first place, we need $F(\lambda_n) = O(R^{-\lambda_n})$ only for a sequence of density 1, which means that in 12.6.10 we need only assume that c_n and c_{n+1} have the same sign for a sequence of unit density, or in other words that the integers n at which a change in sign occurs in $\{c_n\}$ have zero density.

Next, we may extend 12.6.10 by requiring fewer changes of sign and inferring merely that there is a singular point near $z = 1$.

12.6.11. *Theorem.*[e] *If* $f(z) = \sum_{n=0}^{\infty} c_n z^n$ *has the unit circle as its circle of convergence, the* c_n *are real and the integers at which no change of sign occurs in* $\{c_n\}$ *have density* D, *then* $f(z)$ *has a singular point in the arc* $|\theta| \leq \pi(1 - D)$ *of the unit circle.*

Theorem 12.6.11 illustrates the principle that a power series with sufficiently few changes of sign in its sequence of coefficients has its behavior on the circle of convergence determined by its behavior on a sufficiently large arc around the positive real point of the circle: the special case 12.6.10, for example, says that if there are no changes of sign and $f(z)$ is regular at the positive real point on a circle, it is regular on the whole circle (which consequently is not actually the circle of convergence). There are corresponding theorems for other types of behavior, for example

12.6.12. *Theorem.*[f] *If* $f(z) = \sum_{n=0}^{\infty} c_n z^n$ *has the unit circle as its circle of convergence, all* $c_n \geq 0$, *and* $f(z)$ *has boundary values belonging to* L^2 *on an arc of* $|z| = 1$ *around* $z = 1$, *then* $f(z)$ *has boundary values belonging to* L^2 *on the whole circle.*

In the same way as in our previous theorems, we could show that the same conclusion follows if the integers n_k at which no change of sign occurs satisfy $|n_k - k/D| < L$ and the boundary values belong to L^2 on an arc $|\theta| < \alpha$, $\alpha > \pi(1 - D)$.

To prove 12.6.12, we again use (12.6.2) with $R = 1$, so that $c_n = F(n) + g_n$, with g_n given by (12.6.4). Then $\sum |g_n|^2 < \infty$. As in proving 12.6.10, we find that whenever c_n and c_{n+1} have the same sign, there is a λ_n in $(n, n + 1)$ such that $F(\lambda_n) = O(g_n)$, so that $\sum |F(\lambda_n)|^2 < \infty$. Hence by 10.6.8, $\int_{-\infty}^{\infty} |F(x)|^2 \, dx < \infty$, and by 6.7.15, $\sum |F(n)|^2 < \infty$. Hence $\sum |c_n|^2 < \infty$, so $\sum_{n=0}^{\infty} c_n e^{in\theta}$ is an L^2 Fourier series and provides the required L^2 boundary values for $f(z)$.

12.7. Dirichlet series.[a] The theorems of §12.6 have analogues for the class of Dirichlet series with separated exponents,

$$(12.7.1) \qquad f(z) = \sum_{n=1}^{\infty} c_n e^{-\lambda_n z}, \qquad \lambda_n \uparrow \infty, \lambda_{n+1} - \lambda_n > \delta > 0.$$

The problems are now harder since c_n is not so simply expressible in terms of $f(z)$ as it was for power series.

Suppose that (12.7.1) converges for $x > 0$ but not for $x < 0$, and that $f(z)$ is regular in the interval $-B \leq y \leq B$ of $x = 0$, and hence for $x \geq -a$, $|y| \leq B$. Let

$$H(w) = \int_{-a}^{\infty} f(z)e^{wz}\, dz, \qquad w = u + iv.$$

This integral converges at least for $u < 0$, and we can deform the line of integration into the segment $(-a, -a + iB)$ and the line

$$(-a + iB, \infty + iB).$$

This gives, for $b > 0$,

$$(12.7.2) \qquad H(w) = \int_{-a}^{-a+iB} f(z)e^{wz}\, dz + \int_{-a+iB}^{b+iB} f(z)e^{wz}\, dz + \int_{b+iB}^{\infty+iB} e^{wz} \left\{ \sum_{n=1}^{\infty} c_n e^{-\lambda_n z} \right\} dz;$$

and since the series (12.7.1) converges absolutely for $x > 0$, because of the condition $\lambda_{n+1} - \lambda_n > \delta > 0$, we have

$$(12.7.3) \qquad H(w) = \int_{-a}^{-a+ib} f(z)e^{wz}\, dz + \int_{-a+iB}^{b+iB} f(z)e^{wz}\, dz - \sum_{n=1}^{\infty} \frac{c_n e^{(w-\lambda_n)(b+iB)}}{w - \lambda_n}.$$

Here the series converges uniformly in any bounded region which is at a positive distance from all the λ_n, and so $H(w)$ is regular in the whole plane except for poles at $w = \lambda_n$. If $F(z) = \prod (1 - z^2/\lambda_n^2)$, $G(w) = H(w)F(w)$ is an entire function such that

$$(12.7.4) \qquad G(\lambda_n) = -c_n F'(\lambda_n).$$

We now have $c_n F'(\lambda_n)$ expressed by means of an entire function. Our first requirement is an estimate for the growth of $G(z)$.

Suppose that $\{\lambda_n\}$ has density D and $B > \pi D$. Then if $w = re^{i\gamma}$, $0 < \gamma \leq \pi/2$,

$$(12.7.5) \qquad |H(w)| \leq O(e^{-ar\cos\gamma}) + O(e^{br\cos\gamma - Br\sin\gamma}),$$

since

$$\left| \sum_{n=1}^{\infty} \frac{c_n e^{(w-\lambda_n)(b+iB)}}{w - \lambda_n} \right| \leq \sum_{n=1}^{\infty} \frac{|c_n| \exp(-b\lambda_n)}{b \sin \gamma} e^{br \cos \gamma - Br \sin \gamma}.$$

If the c_n are real, $H(w)$ is real for real w and (12.7.5) holds also for $w = re^{-i\gamma}$; and in any case we can obtain (12.7.5) by reflecting the path of integration in (12.7.2) in the real axis, since this does not change $H(w)$. Hence

(12.7.6) $$|G(w)| = O\{e^{r(-a \cos \gamma + \pi D \sin \gamma + \epsilon)} + e^{r(b \cos \gamma - B \sin \gamma + \pi D \sin \gamma + \epsilon)}\}.$$

We have $B > \pi D$ and so if we take ϵ small, and choose γ so small that $\tan \gamma < a/(\pi D)$ and then b so small that $b/(B - \pi D) < \tan \gamma$, we have both exponents negative in (12.7.6), and hence

(12.7.7) $$|G(re^{\pm i\gamma})| = O(e^{-\delta r \cos \gamma}).$$

Again, by (12.7.3) $H(w)$ is of exponential type in the plane with neighborhoods of the λ_n excluded, and so $G(w)$ is of exponential type. By 5.1.2, (12.7.7) implies that $G(x) = O(e^{-\delta x})$, $x \to \infty$. Therefore, by the estimate (10.3.9) for $F'(\lambda_n)$, $c_n = O\{\exp(-\delta\lambda_n/2)\}$, which makes (12.7.1) converge for $x > -\delta/2$, contradicting the assumption that $x = 0$ is the abscissa of convergence. Hence we have proved:

12.7.8. *Theorem. If*

$$f(z) = \sum_{n=1}^{\infty} c_n e^{-\lambda_n z}, \qquad \lambda_n \uparrow \infty, \lambda_{n+1} - \lambda_n > \delta > 0,$$

with the series converging for $x > 0$ but not for $x < 0$, and $\{\lambda_n\}$ has density D, then the interval $(-iB, iB)$ of the imaginary axis contains at least one singular point of $f(z)$ if $B > \pi D$.

12.8. Gap theorems for entire functions.[a] The theorems of §12.6 show that a power series with certain special properties and a finite radius of convergence cannot have too many vanishing coefficients. There are similar theorems for power series which represent entire functions; many are analogues of theorems for power series with a finite radius of convergence, but we select as an illustration one which is not.

12.8.1. *Theorem. If $f(z)$ is an entire function of exponential type which is bounded on the real axis, and if $f^{(\lambda_n)}(0) = 0$, then $f(z)$ is a constant if both the series*

$$\sum_{\lambda_n \text{ even}} 1/\lambda_n, \qquad \sum_{\lambda_n \text{ odd}} 1/\lambda_n$$

diverge.

In particular, unless $f(z)$ is a constant its power series has nonvanishing coefficients for a set of indices of maximum density at least $\frac{1}{2}$.

Without loss of generality we may suppose that $f(x)$ belongs to L^2 (consider $z^{-1}\{f(z) - f(0)\}$). Then

$$f(z) = \int_{-\tau}^{\tau} e^{izt}\phi(t)\,dt, \qquad \phi\epsilon L^2,$$

$$f^{(\lambda_n)}(0) = \int_{-\tau}^{\tau} (it)^{\lambda_n}\phi(t)\,dt.$$

The functions

$$g_1(z) = \int_0^{\tau} t^z\{\phi(t) + \phi(-t)\}\,dt,$$

$$g_2(z) = \int_0^{\tau} t^z\{\phi(t) - \phi(-t)\}\,dt$$

vanish respectively when z is an even λ_n and when z is an odd λ_n. If both series diverge, by 12.4.4 the functions $\phi(t) \pm \phi(-t)$ vanish almost everywhere, so $\phi(t) = 0$ almost everywhere and $f(z) \equiv 0$.

12.9. Expansions of analytic functions in series of polynomials. The theorems to be discussed here depend on the expression of a set of polynomials by means of a generating function. For example, we may have the formal expansion

$$A(t)e^{zt} = \sum_{n=0}^{\infty} p_n(z)t^n, \tag{12.9.1}$$

where $A(t)$ is a formal power series

$$A(t) = \sum_{n=0}^{\infty} a_n t^n, \qquad a_0 = 1. \tag{12.9.2}$$

The $p_n(z)$ are called Appell polynomials. A more general variety of polynomial is defined by

$$A(t)e^{zH(t)} = \sum_{n=0}^{\infty} p_n(z)t^n, \qquad H(0) = 0, H'(0) = 1 \tag{12.9.3}$$

(Sheffer's "sets of type zero,"[a] Steffensen's "poweroids"[b]).

Here we shall consider some simple aspects of expansions in terms of Appell polynomials. The theory of entire functions can enter in two ways: either we may try to expand an entire function, or we may suppose that $A(t)$ is entire and try to expand more general functions. Suppose first that $f(z)$ is an entire function of exponential type τ, and that $A(t)$ and $1/A(t)$ are regular for $|z| \le \tau$. Then (12.9.1), in the form

$$e^{zt} = \sum_{n=0}^{\infty} p_n(z)t^n/A(t) \tag{12.9.4}$$

is valid with uniform convergence for $|t| \leq \tau + \epsilon$ with some positive ϵ. On the other hand, $f(z)$ has the Pólya representation

$$f(z) = \int_C e^{zt}\phi(t)\,dt, \tag{12.9.5}$$

where C is the circle $|t| = \tau + \epsilon$, and if we substitute (12.9.4) into (12.9.5) and integrate term by term, we obtain[c]

$$f(z) = \sum_{n=0}^{\infty} p_n(z) \int_C t^n \phi(t) A(t)^{-1}\,dt, \tag{12.9.6}$$

which gives us a convergent expansion for $f(z)$ as a series of the polynomials $p_n(z)$, with coefficients which can be written formally as $D^n A^{-1}(D) f(z)|_{z=0}$, $D = d/dz$. Moreover, the series (12.9.4) can still be used even when $A(t)$ has zeros inside $|z| = \tau$, but in this case the coefficients in (12.9.6) will be different when different circles C are used, so that an entire function $f(z)$ of small exponential type will have more than one convergent expansion (12.9.6). Still more generally, we can take C in (12.9.5) as a contour surrounding the conjugate indicator diagram of $f(z)$, supposing only that $1/A(t)$ can be continued analytically from its circle of regularity about 0 to a region containing C; then (12.9.4) holds on C with summability of an appropriate kind and we obtain (12.9.6) with summability instead of convergence. For example, (12.9.6) is Mittag-Leffler summable if the conjugate indicator diagram of $f(z)$ is in the Mittag-Leffler star of $1/A(t)$.

Next suppose that $f(z)$ is not necessarily entire, but is represented by a Laplace transform,

$$f(z) = \int_0^{\infty} e^{zt}\phi(t)\,dt, \tag{12.9.7}$$

for $x > -a$. If $A(t)$ is an entire function, (12.9.1) is valid for all t. Suppose that $A(t)$ is of zero exponential type. Then $1/|A(t)| \leq e^{o(t)}$ outside a set of circles whose total length out to $|t| = r$ is $o(r)$ (3.7.3). If the path of integration in (12.9.7) is modified to avoid the exceptional circles, we can substitute (12.9.4) into (12.9.7) and obtain the convergent expansion

$$f(z) = \sum_{n=0}^{\infty} p_n(z) \int t^n \phi(t) A(t)^{-1}\,dt, \qquad |z| < a, \tag{12.9.8}$$

where the integral extends over the modified contour. If $A(t)$ is of exponential type in general we can apply 3.7.1 to show that (12.9.8) holds in some circle about 0. Moreover, any function which is regular at 0 can be represented by a finite sum of functions of the form

$$f(z) = \int_0^{\infty} e^{z\alpha t}\phi(t)\,dt, \qquad |\alpha| = 1, \tag{12.9.9}$$

and so (12.9.8) can be replaced by a finite sum of series of the same form. In this way one can prove the following theorem.[d]

12.9.10.* *Theorem. Let $A(t)$ be an entire function of exponential type. If $f(z)$ is regular at 0, then if $A(t)$ is of zero type $f(z)$ is represented in its largest circle of regularity about 0 by a series of the Appell polynomials generated by $A(t)$; otherwise $f(z)$ is represented in some neighborhood of 0 provided that it is regular in a sufficiently large neighborhood of 0 (depending on $A(t)$).*

12.10. Differential equations of infinite order.[a] We consider the equation

(12.10.1) $$\sum_{k=0}^{\infty} a_k F^{(k)}(z) = G(z), \qquad a_0 \neq 0,$$

where the a_k are constants and $G(z)$ is regular at $z = 0$. The existence and properties of a solution depend on properties of the function

(12.10.2) $$A(t) = \sum_{k=0}^{\infty} a_k t^k;$$

we can write (12.10.1) formally as

(12.10.3) $$A(D)F = G, \qquad D = d/dz,$$

and so we should expect a solution in the form

(12.10.4) $$F = A(D)^{-1}G,$$

with some interpretation of the operator. In the special case where $G(z)$ is an entire function of exponential type τ, and $A(t)$ and $1/A(t)$ are regular in $|t| \leq \tau$, we can represent $G(z)$ by its Pólya representation

$$G(z) = \int_C e^{zt}\phi(t)\,dt, \qquad C = \{|t| = \tau + \epsilon\},$$

and interpret (12.10.4) as

(12.10.5) $$F(z) = \int_C e^{zt}A(t)^{-1}\phi(t)\,dt.$$

Then $F(z)$ is independent of ϵ, and is an entire function of exponential type τ. Furthermore, (2.2.12) shows that the left-hand side of (12.10.1) converges, and substitution in (12.10.5) shows that $F(z)$ is actually a solution of (12.10.1). If $A(t)^{-1} = \sum \alpha_n t^n$, we can also interpret (12.10.5) as

(12.10.6) $$\sum \alpha_n G^{(n)}(z).$$

Similarly, by representing $f(z)$ as a sum of integrals (12.9.9), we can deal with the case when $G(z)$ is regular in a finite circle, $A(t)$ is an entire function of exponential type, and $F(z)$ is to be regular at $z = 0$. Here again the theorems of §3.7 on the behavior of $1/A(t)$ are relevant.

A particular case of a differential equation (12.10.1) is the difference equation

$$F(z + \delta) - F(z) = G(z), \tag{12.10.7}$$

since

$$F(z + \delta) = \sum_{n=0}^{\infty} \delta^k F^{(k)}(z)/k! = e^{\delta D} F(z),$$

thus (12.10.7) has $A(t) = e^{\delta t} - 1$. The theory outlined above shows that for every G of exponential type less than $2\pi/\delta$ we have a solution F of the same exponential type. However, $1/A(t)$ is still regular on any contour C which does not pass through a point $2k\pi i/\delta$, so that actually we have the same conclusion for entire functions G of any exponential type; (12.10.4) is then to be interpreted as (12.10.5) and not as (12.10.6), except in terms of summability of the series.

More generally, the equation[b]

$$\sum_{k=0}^{n} \lambda_k F(z + \delta_k) = G(z)$$

is of our form with

$$A(t) = \sum_{k=0}^{n} \lambda_k \exp(t\delta_k),$$

and this leads us to consider the convolution transform

$$\int_{-\infty}^{\infty} F(z + u)\, d\Lambda(u) = G(z) \tag{12.10.8}$$

formally as an equation (12.10.3) with

$$A(t) = \int_{-\infty}^{\infty} e^{tu}\, d\Lambda(u). \tag{12.10.9}$$

Thus we should expect (12.10.8) to be inverted by $1/A(D)$, with some interpretation of the operator; this expectation can, in quite general circumstances, be justified.[c]

12.11. Approximation by entire functions.[a] Let $f(x)$ be continuous on the whole real axis. A natural generalization of the fact that a continuous function can be approximated arbitrarily closely in a finite interval by polynomials is the fact that a continuous function can be approximated arbitrarily closely on $(-\infty, \infty)$ by entire functions; in fact,

$$\sup | f(x) - g(x) | \phi(| x |)$$

can be made arbitrarily small, with $g(z)$ entire, no matter how fast the given function $\phi(x)$ grows.[b] However, $g(x)$ is generally of infinite order. A more interesting theory is obtained if we ask what bounded continuous functions can be uniformly approximated by entire functions of a given exponential type. Let us define $A_\tau[f]$ to be the minimum of $\sup_x | f(x) - g_\tau(x) |$ for all entire $g_\tau(x)$ of exponential type τ. We then have the following theorem.

12.11.1. *Theorem. If $f(x)$ is bounded on $(-\infty, \infty)$, $A_\tau[f] \to 0$ as $\tau \to \infty$ if and only if $f(x)$ is uniformly continuous in $(-\infty, \infty)$.*

Suppose first that $A_\tau[f] \to 0$. We have

$$| f(x) - g_\tau(x) | \leq A_\tau[f],$$

and since $f(x)$ is bounded, $| g_\tau(x) | \leq L$, say. Then

$$\begin{aligned} | f(x+h) - f(x) | &\leq 2A_\tau[f] + | g_\tau(x+h) - g_\tau(x) | \\ &\leq 2A_\tau[f] + h \sup | g_\tau'(x) | \\ &\leq 2A_\tau[f] + h\tau L, \end{aligned}$$

by S. Bernstein's theorem (11.1.2). Since $A_\tau[f] \to 0$ as $\tau \to \infty$, we can therefore make $\sup_x | f(x+h) - f(x) |$ arbitrarily small by taking $h = o(1/\tau)$, and so $f(x)$ is uniformly continuous.

Now suppose that $f(x)$ is uniformly continuous, and put

$$K(t) = 2\pi^{-1}(t^{-1} \sin t/2)^2,$$

so that $\int_{-\infty}^{\infty} K(t)\, dt = 1$. Consider the function

$$G(z) = \tau \int_{-\infty}^{\infty} K(\tau(t - z)) f(t)\, dt = \int_{-\infty}^{\infty} K(t) f(z + t/\tau)\, dt.$$

Since $K(t)$ is an entire function of exponential type 1, the first form of $G(z)$ shows that $G(z)$ is an entire function of exponential type τ. The second form of $G(z)$ shows that

$$\begin{aligned} | G(x) - f(x) | &= \left| \int_{-\infty}^{\infty} K(t) \{ f(x + t/\tau) - f(x) \}\, dt \right| \\ &\leq \int_{-\infty}^{\infty} K(t)\, dt \cdot \sup_x | f(x + t/\tau) - f(x) | \\ &\to 0 \end{aligned}$$

as $\tau \to \infty$, since $f(x)$ is uniformly continuous.

We also have the following result.

12.11.2.* *Theorem.*[c] *If $f(x)$ is continuous and bounded on $(-\infty, \infty)$, $A_\tau[f]$ is finite and the minimum is attained for at least one $g_\tau(x)$.*

There is an elaborate theory of approximation by entire functions of exponential type; it has many analogies to and connections with the theory of approximation by polynomials and trigonometric polynomials.

Notes for Chapter 12

12.2a. Mikusiński [1].

12.3a. Titchmarsh [2], p. 327, Crum [1], Dufresnoy [1]. A real-variable proof has recently been obtained by Mikusiński [2]. The simplicity of the present proof depends on the use of the Ahlfors-Heins theorem.

12.4a. The literature of completeness theorems, even those which can be treated by the methods of this section, is too vast to summarize here; I give references only for the theorems which I quote.

12.4b. The terminology of the subject is not uniform; some authors interchange the meanings of "closed" and "complete."

12.4c. This is Müntz's theorem. See, for example, Carleman [1], where "Carleman's theorem" is introduced for the purpose of proving Müntz's theorem; Paley and Wiener [1]. The converse is also true: if $\Sigma 1/\lambda_n$ converges the set $\{t^{\lambda_n}\}$ is not complete.

12.4d. For this and the rest of the theorems of this section see Levinson [4].

12.5a. For this and similar theorems see Levinson [4]; for almost periodic functions see Levin [4].

12.5b. Cf. Dugué [1], Boas [19].

12.5c. See Boas [22], [26] for this and related theorems.

12.6a. Theorem 12.6.3 is essentially due to Fabry, but was put in its present form by Pólya: see Pólya [2], p. 626. The case $D = 1$ is often called "Fabry's theorem."

12.6b. The theorem was suggested by Duffin and Schaeffer's theorem 12.6.6.

12.6c. Duffin and Schaeffer [3], generalizing a theorem of Szegö in which the conclusion was that the circle of convergence is a natural boundary.

12.6d. Theorem 12.6.10 is a well-known theorem of Pringsheim; the simplest proof is that of Landau [3], which establishes it in a more general form. See also V. Bernstein [1], p. 85.

12.6e. Theorem 12.6.11 is also due to Fabry and Pólya: Pólya [2], p. 626.

12.6f. Theorem 12.6.12 was proved by N. Wiener (in a more general version for Dirichlet series) in a lecture some years ago. Wiener asked whether L^2 can be replaced by L in the theorem; the answer is not known. For L^p, $1 < p < 2$, the corresponding theorem is true at least under the rather restrictive condition that $\{c_n\}$ decreases.

12.7a. This section follows Levinson [4]. Theorem 12.7.8 is due to Pólya. See also V. Bernstein [1].

12.8a. See Pólya [2] for a general discussion. Theorem 12.8.1 was obtained with generalizations in various directions, by Dzhrbashyan [2]. It should also be compared with Mandelbrojt's [2] theory of generalized quasi-analyticity.

12.9a. Sheffer [2], [3].

12.9b. Steffensen [1].

12.9c. Martin [1].

12.9d. The theorem is due to Sheffer [1]; the proof outlined here is given in more detail by Boas [18], [20].

12.10a. There is an extensive literature on differential equations of infinite order. For the first part of the discussion given here, and for further results and references, see Muggli [1], [2], Sikkema [1]; the discussion of the case where $G(z)$ is regular at 0 is amplified by Boas [21]. For recent work using other approaches see Carmichael [1], Sheffer [1], [3], Vermes [1], Gelfond [4], Carleson [1].

12.10b. The general difference equation is considered in the complex domain by Cameron and Martin [1], Yagi [1], and Strodt [1]. For differential-difference equations see Leontyev [1], [2].

12.10c. For the "real" theory of convolution transforms see Hirschman and Widder [1] (and earlier papers); Pollard [1], [2], Schoenberg [2], [3]. The theory depends on some properties of entire functions which have not been discussed in this book. Pitt [1] and Yagi [1] consider equations of the form

$$\sum_{r=0}^{R} \int_{-\infty}^{\infty} f^{(r)}(x-t)\, d\Lambda_r(t) = g(x).$$

12.11a. Theorem 12.9.1 was proved by Kober [2], who has also discussed L^p approximation [1]. The theory of best approximation by entire functions of exponential type is worked out in a series of papers by S. Bernstein [5]–[10], [15]; cf. also Ahiezer [6]. The theory is not restricted to bounded functions.

12.11b. Carleman [2], Roth [1].

12.11c. S. Bernstein [3].

BIBLIOGRAPHY

The abbreviations of names of journals are, in general, those used in *Mathematical Reviews*.

AGMON, S.

1. Functions of exponential type in an angle and singularities of Taylor series. *Trans. Amer. Math. Soc.* **70,** 492–508 (1951).

AHIEZER, N. I. (AKHYESER, AKHIEZER, ACHYESER, ETC.)

1. Sur les fonctions entières d'ordre entier. *Rend. Circ. Mat. Palermo* (1) **51,** 390–393 (1927).
2. On some properties of integral transcendental functions of exponential type. *Bull. Acad. Sci. URSS. Sér. Math.* [*Izvestiya Akad. Nauk SSSR*] **10,** 411–428 (1946). (Russian; English summary.)
3. Lectures on the theory of approximation. OGIZ, Moscow and Leningrad, 1947. (Russian.)
4. On the theory of entire functions of finite degree. *Doklady Akad. Nauk SSSR* (N.S.) **63,** 475–478 (1948). (Russian.)
5. On the interpolation of entire transcendental functions of finite degree. *Doklady Akad. Nauk SSSR* (N.S.) **65,** 781–784 (1949). (Russian.)
6. The work of Academician S. N. Bernstein on the constructive theory of functions. *Uspehi Matem. Nauk* (N.S.) **6,** No. 1(41), 3–67 (1951). (Russian.)
7. On entire transcendental functions of finite degree having a majorant on a sequence of real points. *Izvestiya Akad. Nauk SSSR. Ser. Mat.* **16,** 353–364 (1952). (Russian.)

AHLFORS, L., AND HEINS, M.

1. Questions of regularity connected with the Phragmén-Lindelöf principle. *Ann. of Math.* (2) **50,** 341–346 (1949).

AMIRÀ, B.

1. Sur un théorème de M. Wiman dans la théorie des fonctions entières. *Math. Z.* **22,** 206–221 (1925).

ARIMA, K.

1. On maximum modulus of integral functions. *J. Math. Soc. Japan* **4,** 62–66 (1952).

BELLMAN, R.

1. A generalization of a Zygmund-Bernstein theorem. *Duke Math. J.* **10,** 649–651 (1943).

BERNSTEIN, S. (BERNŠTEĬN, S. N.)

1. Sur une propriété des fonctions entières. *C. R. Acad. Sci. Paris* **176,** 1603–1605 (1923).
2. Leçons sur les propriétés extrémales et la meilleure approximation des fonctions analytiques d'une variable réelle. Gauthier-Villars, Paris, 1926.
3. Sur la meilleure approximation de $|x|^p$ par des polynômes de degrés très élevés. *Bull Acad. Sci. URSS. Sér. Math.* [*Izvestiya Akad. Nauk SSSR*] **1938,** 169–190 (Russian and French.)
4. Sur la borne supérieure du module de la dérivée d'une fonction de degré fini. *C.R. (Doklady) Acad. Sci. URSS* (N.S.) **51,** 567–568 (1946).
5. Sur la meilleure approximation sur tout l'axe réel des fonctions continues par des fonctions entières de degré fini. I, II, III, IV, V. *C.R. (Doklady) Acad. Sci. URSS* (N.S.) **51,** 331–334, 487–490; **52,** 563–566; **54,** 103–108, 475–478 (1946).

6. Démonstration nouvelle et généralisation de quelques formules de la meilleure approximation. *C.R. (Doklady) Acad. Sci. URSS* (N.S.) **54,** 663–664 (1946).
7. On limiting relations among constants of the theory of best approximation. *Doklady Akad. Nauk SSSR* (N.S.) **57,** 3–5 (1947). (Russian.)
8. On properties of homogeneous functional classes. *Doklady Akad. Nauk SSSR* (N.S.) **57,** 111–114 (1947). (Russian.)
9. Limit laws of the theory of best approximation. *Doklady Akad. Nauk SSSR* (N.S.) **58,** 525–528 (1947). (Russian.)
10. A second note on homogeneous functional classes. *Doklady Akad. Nauk SSSR* (N.S.) **59,** 1379–1384 (1948). (Russian.)
11. A generalization of an inequality of S. B. Stechkin to entire functions of finite degree. *Doklady Akad. Nauk SSSR* (N.S.) **60,** 1487–1490 (1948). (Russian.)
12. The extension of properties of trigonometric polynomials to entire functions of finite degree. *Izvestiya Akad. Nauk SSSR. Ser. Mat.* **12,** 421–444 (1948). (Russian.)
13. Remarks on my paper "The extension of properties of trigonometric polynomials to entire functions of finite degree." *Izvestiya Akad. Nauk SSSR. Ser. Mat.* **12,** 571–573 (1948). (Russian.)
14. On additive majorants of finite growth. *Doklady Akad. Nauk SSSR* (N.S.) **66,** 545–548(1949). (Russian.)
15. On some new results in the theory of approximation of functions of a real variable. *Acta Sci. Math. Szeged* **12,** Leopoldo Fejér et Frederico Riesz LXX annos natis dedicatus, Pars A, 161–169 (1950). (Russian.)
16. On weight functions. *Doklady Akad. Nauk SSSR* (N.S.) **77,** 549–552 (1951). (Russian.)

Bernstein, V.
1. Leçons sur les progrès récents de la théorie des séries de Dirichlet. Gauthier-Villars, Paris, 1933.
2. Sopra una proposizione relativa alla crescenza delle funzioni olomorfe. *Ann. Scuola Norm. Super. Pisa* (2) **2,** 381–399 (1933).

Besicovitch, A. S.
1. On integral functions of order < 1. *Math. Ann.* **97,** 677–695 (1927).

Beurling, A.
1. Some theorems on boundedness of analytic functions. *Duke Math. J.* **16,** 355–359 (1949).

Boas, R. P., Jr.
1. Some theorems on Fourier transforms and conjugate trigonometric integrals. *Trans. Amer. Math. Soc.* **40,** 287–308 (1936).
2. The derivative of a trigonometric integral. *J. London Math. Soc.* **12,** 164–165 (1937).
3. Asymptotic relations for derivatives. *Duke Math. J.* **3,** 637–646 (1937).
4. Representations for entire functions of exponential type. *Ann. of Math.* (2) **39,** 269–286 (1938); correction, *ibid.* **40,** 948 (1939).
5. Remarks on a theorem of B. Lewitan. *Rec. Math.* [*Mat. Sbornik*] (N.S.) **5(47),** 185–187 (1939).
6. Entire functions bounded on a line. *Duke Math. J.* **6,** 148–169 (1940); correction, *ibid.* **13,** 483–484 (1946).
7. Some uniqueness theorems for entire functions. *Amer. J. Math.* **62,** 319–324 (1940).
8. Univalent derivatives of entire functions. *Duke Math. J.* **6,** 719–721 (1940).

9. Expansions of analytic functions. *Trans. Amer. Math. Soc.* **48,** 467–487 (1940).
10. A note on functions of exponential type. *Bull. Amer. Math. Soc.* **47,** 750–754 (1941).
11. Entire functions of exponential type. *Bull. Amer. Math. Soc.* **48,** 839–849 (1942).
12. Representation of functions by Lidstone series. *Duke Math. J.* **10,** 239–245 (1943).
13. Functions of exponential type. I. *Duke Math. J.* **11,** 9–15 (1944).
14. Functions of exponential type. III. *Duke Math. J.* **11,** 507–511 (1944).
15. Fundamental sets of entire functions. *Ann. of Math.* (2) **47,** 21–32 (1946); correction, *ibid.* **48,** 1095 (1947).
16. The growth of analytic functions. *Duke Math. J.* **13,** 471–481 (1946).
17. Quelques généralisations d'un théorème de S. Bernstein sur la dérivée d'un polynome trigonométrique. *C. R. Acad. Sci. Paris* **227,** 618–619 (1948).
18. Exponential transforms and Appell polynomials. *Proc. Nat. Acad. Sci. U.S.A.* **34,** 481–483 (1948).
19. Sur les séries et intégrales de Fourier à coefficients positifs. *C. R. Acad. Sci. Paris* **228,** 1837–1838 (1949).
20. Polynomial expansions of analytic functions. *J. Indian Math. Soc.* (N.S.) **14,** 1–14 (1950).
21. Differential equations of infinite order. *J. Indian Math. Soc.* (N.S.) **14,** 15–19 (1950).
22. Partial sums of Fourier series. *Proc. Nat. Acad. Sci. U.S.A.* **37,** 414–417 (1951).
23. Growth of analytic functions along a line. *Proc. Nat. Acad. Sci. U.S.A.* **38,** 503–504 (1952).
24. Integrability along a line for a class of entire functions. *Trans. Amer. Math. Soc.* **73,** 191–197 (1952).
25. Inequalities between series and integrals involving entire functions. *J. Indian Math. Soc.* (N.S.) **16,** 127–135 (1952).
26. Oscillation of partial sums of Fourier series. *J. Analyse Math.* **2,** 110–126 (1952).
27. Integral functions with negative zeros. *Canadian J. Math.* **5,** 179–184 (1953).
28. Two theorems on integral functions. *J. London Math. Soc.* **28,** 194–196 (1953).
29. A Tauberian theorem for integral functions. *Proc. Cambridge Philos. Soc.* **49,** 728–730 (1953).
30. Asymptotic properties of functions of exponential type. *Duke Math. J.* **20,** 433–448 (1953).
31. Growth of analytic functions along a line. *J. Analyse Math.* **4** (to appear).

Boas, R. P., Jr., and Pollard, H.

1. Complete sets of Bessel and Legendre functions. *Ann. of Math.* (2) **48,** 366–383 (1947).

Boas, R. P., Jr., and Schaeffer, A. C.

1. A theorem of Cartwright. *Duke Math. J.* **9,** 879–883 (1942).

Boas, R. P., Jr., Buck, R. C., and Erdős, P.

1. The set on which an entire function is small. *Amer. J. Math.* **70,** 400–402 (1948).

Bonnesen, T., and Fenchel, W.

1. Theorie der konvexen Körper. Ergebnisse der Mathematik und ihrer Grenzgebiete, Vol. 3, No. 1. Springer, Berlin, 1934.

Borel, E.

1. Leçons sur les fonctions entières. Deuxième édition revue et augmentée d'une note de G. Valiron. Gauthier-Villars, Paris, 1921.

Bowen, N. A.

1. A function-theory proof of Tauberian theorems on integral functions. *Quart. J. Math., Oxford Ser.* (1) **19,** 90–100 (1948).

Bowen, N. A., and Macintyre, A. J.

1. An oscillation theorem of Tauberian type. *Quart. J. Math., Oxford Ser.* (2) **1,** 243–247 (1950).
2. Some theorems on integral functions with negative zeros. *Trans. Amer. Math. Soc.* **70,** 114–126 (1951).

Brunk, H. D.

1. On the growth of functions having poles or zeros on the positive real axis. *Pacific J. Math.* **4,** 1–19 (1954).

Buck, R. C.

1. An extension of Carlson's theorem. *Duke Math. J.* **13,** 345–349 (1946).
2. A class of entire functions. *Duke Math. J.* **13,** 541–559 (1946).
3. Interpolation and uniqueness of entire functions. *Proc. Nat. Acad. Sci. U.S.A.* **33,** 288–292 (1947).
4. Interpolation series. *Trans. Amer. Math. Soc.* **64,** 283–298 (1948).
5. Integral valued entire functions. *Duke Math. J.* **15,** 879–891 (1948).
6. On the distribution of the zeros of an entire function. *J. Indian Math. Soc.* (N.S.) **16,** 147–149 (1952).
7. On admissibility of sequences and a theorem of Pólya. *Comment. Math. Helv.* **27,** 75–80 (1953).

Cameron, R. H., and Martin, W. T.

1. Infinite linear differential equations with arbitrary real spans and first degree coefficients. *Trans. Amer. Math. Soc.* **54,** 1–22 (1943).

Carleman, T.

1. Über die Approximation analytischer Funktionen durch lineare Aggregate von vorgegebenen Potenzen. *Ark. Mat. Astr. Fys.* **17,** No. 9 (1922).
2. Sur un théorème de Weierstrass. *Ark. Mat. Astr. Fys.* **20B,** No. 4 (1927).

Carleson, L.

1. On infinite differential equations with constant coefficients. I. *Math. Scandinavica* **1,** 31–38 (1953).

Carmichael, R. D.

1. Linear differential equations of infinite order. *Bull. Amer. Math. Soc.* **42,** 193–218 (1936).

Cartwright, M. L.

1. The zeros of certain integral functions. *Quart. J. Math., Oxford Ser.* (1) **1,** 38–59 (1930).
2. The zeros of certain integral functions. II. *Quart. J. Math., Oxford Ser.* (1) **2,** 113–129 (1931).
3. On integral functions of integral order. *Proc. London Math. Soc.* (2) **33,** 209-224 (1931).
4. On functions which are regular and of finite order in an angle. *Proc. London Math. Soc.* (2) **38,** 158–179 (1934).
5. On the minimum modulus of integral functions. *Proc. Cambridge Philos. Soc.* **30,** 412–420 (1934).
6. On certain integral functions of order 1 and mean type. *Proc. Cambridge Philos. Soc.* **31,** 347–350 (1935).
7. On the directions of Borel of functions which are regular and of finite order in an angle. *Proc. London Math. Soc.* (2) **38,** 503–541 (1935).

8. Some uniqueness theorems. *Proc. London Math. Soc.* (2) **41,** 33–47 (1936).
9. On certain integral functions of order 1. *Quart. J. Math., Oxford Ser.* (1) **7,** 46–55 (1936).
10. On functions bounded at the lattice points in an angle. *Proc. London Math. Soc.* (2) **43,** 26–32 (1937).

Chandrasekharan, K.
1. On Hadamard's factorization theorem. *J. Indian Math. Soc.* (N.S.) **5,** 128–132 (1941).

Chang, Shih-Hsun.
1. On a theorem of S. Bernstein. *Proc. Cambridge Philos. Soc.* **48,** 87–92 (1952).

Chebotarëv, N. G., and Meiman, N. N. (Tschebotaröw, Čebotarëv, etc.; Meymann, Meĭman)
1. The Routh-Hurwitz problem for polynomials and entire functions. Trudy Matematicheskogo Instituta im. V.A. Steklova, No. 26. Akademiya Nauk SSSR, Moscow and Leningrad, 1949. (Russian.)

Civin, P.
1. Inequalities for trigonometric integrals. *Duke Math. J.* **8,** 656–665 (1941).

van der Corput, J. G., and Schaake, G.
1. Ungleichungen für Polynome und trigonometrische Polynome. *Compositio Math.* **2,** 321–361 (1935).

Crum, M. M.
1. On the resultant of two functions. *Quart. J. Math., Oxford Ser.* (1) **12,** 108–111 (1941).

Davis, P.
1. Completeness theorems for sets of differential operators. *Duke Math. J.* **20,** 345–357 (1953).

Delange, H.
1. Sur les suites de polynomes ou de fonctions entières à zéros réels. *Ann. Sci. École Norm. Sup.* (3) **62,** 115–183 (1945).
2. Un théorème sur les fonctions entières à zéros réels et négatifs. *J. Math. Pures Appl.* (9) **31,** 55–78 (1952).

Denjoy, A.
1. Sur un théorème de Wiman. *C. R. Acad. Sci. Paris* **193,** 828–830 (1931); correction in Sur quelques points de la théorie des fonctions, *ibid.* **194,** 44–46 (1932).

Dienes, P.
1. The Taylor series. An introduction to the theory of functions of a complex variable. Oxford University Press, 1931.

Dinghas, A.
1. Über einen Satz von Phragmén und Lindelöf. *Math. Z.* **39,** 455–461 (1934).

Duffin, R. J., and Schaeffer, A. C.
1. Some inequalities concerning functions of exponential type. *Bull. Amer. Math. Soc.* **43,** 554–556 (1937).
2. Some properties of functions of exponential type. *Bull. Amer. Math. Soc.* **44,** 236–240 (1938).
3. Power series with bounded coefficients. *Amer. J. Math.* **67,** 141–154 (1945).
4. A class of nonharmonic Fourier series. *Trans. Amer. Math. Soc.* **72,** 341–366 (1952).

Dufresnoy, J.
1. Sur le produit de composition de deux fonctions. *C. R. Acad. Sci. Paris* **225,** 857–859 (1947).

Dufresnoy, J., and Pisot, C.
1. Prolongement analytique de la série de Taylor. *Ann. Sci. École Norm. Sup.* (3) **68,** 105–124 (1951).
Dugué, D.
1. Sur certaines conséquences qu'entraîne pour une série de Fourier le fait d'avoir tous ses coefficients positifs. Complément au théorème de Weierstrass. *C. R. Acad. Sci. Paris* **228,** 1469–1470 (1949).
Dzhrbashyan, M. M. (Džrbašyan)
1. Uniqueness and representation theorems for entire functions. *Izvestiya Akad. Nauk SSSR. Ser. Mat.* **16,** 225–252 (1952). (Russian.)
2. On the integral representation and uniqueness of some classes of entire functions. *Mat. Sbornik* (N.S.) **33(75),** 485–530 (1953). (Russian.)
Eweida, M. T.
1. A note on the generalization of Taylor's expansion. *Proc. Math. Phys. Soc. Egypt* **3** (1946), No. 2, 1–7 (1947).
2. On the representation of integral functions by generalized Taylor's series. *Proc. Math. Phys. Soc. Egypt* **3,** No. 4, 39–46 (1948).
Fejér, L., and Riesz, F.
1. Über einige funktionentheoretische Ungleichungen. *Math. Z.* **11,** 305–314 (1921).
Franck, A.
1. Analytic functions of bounded type. *Amer. J. Math.* **74,** 410–422 (1952).
Fuchs, W. H. J.
1. A generalization of Carlson's theorem. *J. London Math. Soc.* **21,** 106–110 (1946).
Gabriel, R. M.
1. Some results concerning the integrals of moduli of regular functions along certain curves. *J. London Math. Soc.* **2,** 112–117 (1927).
Ganapathy Iyer, V.
1. On the Lebesgue class of integral functions along straight lines issuing from the origin. *Quart. J. Math., Oxford Ser.* (1) **7,** 294–299 (1936).
2. On the order and type of integral functions bounded at a sequence of points. *Ann of. Math.* (2) **38,** 311–320 (1937).
3. On effective sets of points in relation to integral functions. *Trans. Amer. Math. Soc.* **42,** 358–365 (1937); correction, *ibid.* **43,** 494 (1938).
4. A note on integral functions of order one. *Quart. J. Math., Oxford Ser.* (1) **8,** 103–106 (1937).
5. A note on integral functions of order 2 bounded at the lattice points. *J. London Math. Soc.* **11,** 247–249 (1936).
6. Some properties of integral functions of finite order. *Quart. J. Math., Oxford Ser.* (1) **8,** 131–141 (1937).
7. On integral functions of finite order and minimal type. *J. Indian Math. Soc.* (N.S.) **2,** 131–140 (1937).
8. A property of the zeros of the successive derivatives of integral functions. *J. Indian Math. Soc.* (N.S.) **2,** 289–294 (1937).
9. On the average radial increase of a certain class of integral functions of order one and finite type. *J. Indian Math. Soc.* (N.S.) **3,** 87–95 (1938).
10. Some theorems on functions regular in an angle. *Quart. J. Math., Oxford Ser.* (1) **9,** 206–215 (1938).
11. The behaviour of integral functions at the lattice-points. *J. London Math. Soc.* **13,** 91–94 (1938).

12. The Phragmén-Lindelöf theorem in the critical angle. *J. London Math. Soc.* **14,** 286–292 (1939).
13. The influence of zeros on the magnitude of functions regular in an angle. *J. Indian Math. Soc.* (N.S.) **7,** 1–16 (1943).

GELFOND, A. O. (GEL'FOND)

1. Sur une application du calcul des différences finies à l'étude des fonctions entières. *Rec. Math.* [*Mat. Sbornik*] **36,** 173–183 (1929).
2. Interpolation et unicité des fonctions entières. *Rec. Math.* [*Mat. Sbornik*] (N.S.) **4(46),** 115–147 (1938).
3. On the Taylor series associated with an integral function. *C. R.* (*Doklady*) *Acad. Sci. URSS* (N.S.) **23,** 756–758 (1939).
4. Linear differential equations of infinite order with constant coefficients and asymptotic periods of entire functions. Trudy Matematicheskogo Instituta im. V.A. Steklova, No. 38, pp. 42–67 (1951) (Russian); translated as *Amer. Math. Soc. Translation* No. 84 (1953).

GELFOND, A. O., AND IBRAGIMOV, I. I.

1. On functions whose derivatives are zero at two points. *Izvestiya Akad. Nauk SSSR. Ser. Mat.* **11,** 547–560 (1947). (Russian.)

GERMAY, R. H. J.

1. Sur une application des théorèmes de Weierstrass et de Mittag-Leffler de la théorie générale des fonctions. *Ann. Soc. Sci. Bruxelles* Sér. I. **60,** 190–195 (1946).

GIACCARDI, F.

1. Su di una condizioni perchè una funzione analitica periodica si riduca ad un polinomio trigonometrico. *Atti. Accad. Lincei. Rend. Cl. Sci. Fis. Mat. Nat.* (6) **25,** 555–559 (1937).

GONTCHAROFF, W. (GONČAROV, V.)

1. Détermination des fonctions entières par interpolation. Actualités Scientifiques et Industrielles, No. 465. Hermann, Paris, 1937.

GRANDJOT, K.

1. Über Polynome, die in Einheitswurzeln beschränkt sind. *Jber. Deutsch. Math. Verein.* **34,** 80–86 (1925); correction, *ibid.* **35,** 112 (1926).

GRONWALL, T. H.

1. A sequence of polynomials connected with the nth roots of unity. *Bull. Amer. Math. Soc.* **27,** 275–279 (1921).

GURIN, L. S.

1. On an interpolation problem. *Mat. Sbornik* (N.S.) **22(64),** 425–438 (1948). (Russian.)

HARVEY, A. R.

1. The mean of a function of exponential type. *Amer. J. Math.* **70,** 181–202 (1948).

HAYMAN, W. K.

1. The minimum modulus of large integral functions. *Proc. London Math. Soc.* (3) **2,** 469–512 (1952).

HEINS, M.

1. On the Phragmén-Lindelöf principle. *Trans. Amer. Math. Soc.* **60,** 238–244 (1946).
2. Entire functions with bounded minimum modulus; subharmonic function analogues. *Ann. of Math.* (2) **49,** 200–213 (1948).

HERVÉ, M.

1. Sur quelques applications de la notion d'ordre précisé. *Bull. Sci. Math.* (2) **66,** 17–24, 31–48 (1942).

HILLE, E., AND TAMARKIN, J. D.

1. On the absolute integrability of Fourier transforms. *Fund. Math.* **25,** 329–352 (1935).

HIRSCHMAN, I. I., JR., AND WIDDER, D. V.

1. Convolution transforms with complex kernels. *Pacific J. Math.* **1,** 211–225 (1951).

HOHEISEL, G.

1. Über das Verhalten einer analytischen Funktion in einer Teilumgebung eines singulären Punktes. *S.-B. Preuss. Akad. Wiss. Berlin. Phys.-Math. Kl.* **1923,** 177–180.

HUBER, A.

1. Über Wachstumseigenschaften gewisser Klassen von subharmonischen Funktionen. *Comment. Math. Helv.* **26,** 81–116 (1952).

IBRAGIMOV, I. I. (IBRAGUIMOFF)

1. Sur quelques systèmes complets de fonctions analytiques. *Bull. Acad. Sci. URSS. Sér. Math.* [*Izvestiya Akad. Nauk SSSR*] **3,** 553–568 (1939). (Russian; French summary.)

IKEHARA, S.

1. On integral functions with real negative zeros. *J. Math. Phys.* **10,** 84–91 (1931).

INOUE, M.

1. Sur le module minimum des fonctions sousharmoniques et des fonctions entières d'ordre < ½. *Mem. Fac. Sci. Kyūsyū Univ.* A. **4,** 183–193 (1949).
2. On the growth of subharmonic functions and its applications to a study of the minimum modulus of integral functions. *J. Inst. Polytech. Osaka City Univ.* **1,** No. 2, Ser. A, 71–82 (1950).
3. A note on minimum modulus of integral functions of lower order < ½. *Math. Japonicae* **2,** 41–47 (1950).

JAIN, S. P.

1. An analogue of a theorem of Phragmén-Lindelöf. *J. Indian Math. Soc.* **19,** 241–245 (1932).

JUNNILA, A.

1. Über das Anwachsen einer analytischen Funktion in einer gegebenen Punktfolge. *Ann. Acad. Sci. Fennicae* A. **48,** No. 2 (1936).

KAHANE, J.-P.

1. Extension du théorème de Carlson et applications. *C.R. Acad. Sci. Paris* **234,** 2038–2040 (1952).

KAWATA, T.

1. Remarks on the representation of entire functions of exponential type. *Proc. Imp. Acad. Japan* **14,** 266–269 (1938).

KJELLBERG, B.

1. On certain integral and harmonic functions. A study in minimum modulus. Thesis, University of Uppsala, 1948.
2. On integral functions bounded on a given set. *Mat. Tidsskrift* B. **1952,** 92–99.

KNOPP, K.

1. Theory and application of infinite series. Blackie, London and Glasgow, 1928.

KOBER, H.

1. On the approximation to integrable functions by integral functions. *Trans. Amer. Math. Soc.* **54,** 70–82 (1943).
2. Approximation of continuous functions by integral functions of finite order. *Trans. Amer. Math. Soc.* **61,** 293–306 (1947).

KOREVAAR, J.

1. An inequality for entire functions of exponential type. *Nieuw Arch. Wiskunde* (2) **23,** 55–62 (1949).
2. A simple proof of a theorem of Pólya. *Simon Stevin* **26,** 81–89 (1949).
3. Approximation and interpolation applied to entire functions. Thesis, University of Leiden, 1949. The second part appeared also as Functions of exponential type bounded on sequences of points, *Ann. Soc. Polon. Math.* **22** (1949), 207–234 (1950).
4. The zeros of approximating polynomials and the canonical representation of an entire function. *Duke Math. J.* **18,** 573–592 (1951).

KREIN, M. G. (KREĬN)

1. On the representation of functions by Fourier-Stieltjes integrals. *Uchenye Zapiski Kuĭbyshevskogo Pedagogicheskogo Instituta imeni V. V. Kuĭbysheva* **7,** 123ff. (1943). (Russian.)
2. A contribution to the theory of entire functions of exponential type. *Bull. Acad. Sci. URSS. Sér. Math.* [*Izvestiya Akad. Nauk SSSR*] **11,** 309–326 (1947). (Russian; English summary.)

LAKSHMINARASIMHAN, T. V.

1. A Tauberian theorem for the type of an entire function. *J. Indian Math. Soc.* (N.S.) **17,** 55–58 (1953).

LAMMEL, E.

1. Zum Interpolationsproblem von Funktionen, welche in einfach zusammenhängenden Bereichen regulär und von endlicher Ordnung sind. *Math. Ann.* **115,** 68–74 (1937).

LANDAU, E.

1. Handbuch der Lehre von der Verteilung der Primzahlen. Teubner, Leipzig and Berlin, 1909.
2. Vorlesungen über Zahlentheorie. Hirzel, Leipzig, 1927.
3. Darstellung und Begründung einiger neuerer Ergebnisse der Funktionentheorie. 2d ed. Springer, Berlin, 1929.

LANGER, R. E.

1. On the zeros of exponential sums and integrals. *Bull. Amer. Math. Soc.* **37,** 213–239 (1931).

LELONG-FERRAND, J.

1. Étude au voisinage de la frontière des fonctions surharmoniques positives dans un demi-espace. *Ann. Sci. École Norm. Sup.* (3) **66,** 125–159 (1949).

LEONTYEV, A. F. (LEONT'EV)

1. Differential-difference equations. *Mat. Sbornik* (N.S.) **24(66),** 347–374 (1949) (Russian); translated as *Amer. Math. Soc. Translation* No. 78 (1952).
2. Series of Dirichlet polynomials and their generalizations. Trudy Matematicheskogo Instituta im. V. A. Steklova, No. 39. Akademiya Nauk SSSR, Moscow, 1951. (Russian.)

LEVIN, B. YA. (LÉVINE, B. J.)

1. Sur la croissance d'une fonction entière suivant un rayon et la distribution de ses zéros suivant leurs arguments. *Rec. Math.* [*Mat. Sbornik*] (N.S.) **2(44),** 1097–1142 (1937). (Russian; French summary.)
2. Sur certaines applications de la série d'interpolation de Lagrange dans la théorie des fonctions entières. *Rec. Math.* [*Mat. Sbornik*] (N.S.) **8(50),** 437–454 (1940). (Russian; French summary.)
3. On functions of finite degree, bounded on a sequence of points. *Doklady Akad. Nauk SSSR* (N.S.) **65,** 265–268 (1949). (Russian.)

4. On functions determined by their values on an interval. *Doklady Akad. Nauk SSSR* (N.S.) **70,** 757-760 (1950). (Russian.)
5. On entire functions of finite degree which are of regular growth. *Doklady Akad. Nauk SSSR* (N.S.) **71,** 601–604 (1950). (Russian.)
6. On a special class of entire functions and on related extremal properties of entire functions of finite degree. *Izvestiya Akad. Nauk SSSR. Ser. Mat.* **14,** 45–84 (1950). (Russian.)
7. On a class of entire functions. *Doklady Akad. Nauk SSSR* (N.S.) **78,** 1085–1088 (1951). (Russian.)
8. The general form of special operators on entire functions of finite degree. *Doklady Akad. Nauk SSSR* (N.S.) **79,** 397–400 (1951). (Russian.)

LEVINSON, N.
1. On a problem of Pólya. *Amer. J. Math.* **58,** 791–798 (1936).
2. On certain theorems of Pólya and Bernstein. *Bull. Amer. Math. Soc.* **42,** 702–706 (1936).
3. On the growth of analytic functions. *Trans. Amer. Math. Soc.* **43,** 240–257 (1938).
4. Gap and density theorems. American Mathematical Society, New York, 1940.
5. A theorem of Boas. *Duke Math. J.* **8,** 181–182 (1941).
6. An integral inequality of the Phragmén Lindelöf type. *J. Math. Phys.* **20,** 89–98 (1941).

LINDELÖF, E.
1. Mémoire sur la théorie des fonctions entières de genre fini. *Acta Soc. Sci. Fennicae* **31,** No. 1 (1902).
2. Sur les fonctions entières d'ordre entier. *Ann. Sci. École Norm. Sup.* (3) **22,** 369–395 (1905).

LOHIN, I. F.
1. Concerning the representation of entire analytic functions. *Doklady Akad. Nauk SSSR* (N.S.) **66,** 157–160 (1949). (Russian.)
2. On a representation of an entire analytic function of the first order of normal type. *Doklady Akad. Nauk SSSR* (N.S.) **72,** 629–632 (1950). (Russian.)

MACINTYRE, A. J.
1. Laplace's transformation and integral functions. *Proc. London Math. Soc.* (2) **45,** 1–20 (1938).
2. Wiman's method and the "flat regions" of integral functions. *Quart. J. Math., Oxford Ser.* (1) **9,** 81–88 (1938).
3. The minimum modulus of integral functions of finite order. *Quart. J. Math., Oxford Ser.* (1) **9,** 182–184 (1938).
4. Interpolation series for integral functions of exponential type. *Trans. Amer. Math. Soc.* **76,** 1–13 (1954).

MACINTYRE, A. J., AND MACINTYRE, S. S.
1. Theorems on the convergence and asymptotic validity of Abel's series. *Proc. Roy. Soc. Edinburgh* Sect. A. **63,** 222–231 (1952).

MACINTYRE, S. S. (SCOTT, S.)
1. On the asymptotic periods of integral functions. *Proc. Cambridge Philos. Soc.* **31,** 543–554 (1935).
2. An upper bound for the Whittaker constant *W*. *J. London Math. Soc.* **22** (1947), 305–311 (1948).
3. On the zeros of successive derivatives of integral functions. *Trans. Amer. Math. Soc.* **67,** 241–251 (1949).

4. Overconvergence properties of some interpolation series. *Quart. J. Math., Oxford Ser.* (2) **2,** 109–120 (1951).
5. Some generalizations of two-point expansions. *Proc. Cambridge Philos. Soc.* **48,** 583–586 (1952).
6. An interpolation series for integral functions. *Proc. Edinburgh Math. Soc.* (2) **9,** 1–6 (1950).

MACPHAIL, M. S.

1. Entire functions bounded on a set. *Trans. Roy. Soc. Canada.* Sect. III. **37,** 31–38 (1943).

MAITLAND, B. J.

1. The flat regions of integral functions of finite order. *Quart. J. Math., Oxford Ser.* (1) **15,** 84–96 (1944).

MANDELBROJT, S.

1. Séries de Fourier et classes quasi-analytiques de fonctions. Gauthier-Villars, Paris, 1935.
2. Séries adhérentes. Régularisation des suites. Applications. Gauthier-Villars, Paris, 1952.

MANDELBROJT, S., AND ULRICH, F. E.

1. Regions of flatness for analytic functions and their derivatives. *Duke Math. J.* **18,** 549–556 (1951).

MARTIN, W. T.

1. On expansions in terms of a certain general class of functions. *Amer. J. Math.* **58,** 407–420 (1936).

MAZURKIEWICZ, S.

1. Sur le terme maximum d'une fonction entière. *C.R. Soc. Sci. Varsovie.* Cl. III. **29,** 1–6 (1936).

MEIMAN, N. N. (MEYMANN, MEĬMAN)

1. Differential inequalities and some questions of the distribution of zeros of entire and single-valued analytic functions. *Uspehi Matem. Nauk* (N.S.) **7,** No. 3(49), 3–62 (1952). (Russian.)

MIKUSIŃSKI, J. G.-

1. Remarks on the moment problem and on a theorem of Picone. *Colloquium Math.* **2,** 138–141 (1951).
2. A new proof of Titchmarsh's theorem on convolution. *Studia Math.* **13,** 56–58 (1953).
3. On the Paley-Wiener theorem. *Studia Math.* **13,** 287–295 (1953).

MUGGLI, H.

1. Differentialgleichungen unendlich hoher Ordnung mit konstanten Koeffizienten. *Comment. Math. Helv.* **11,** 151–179 (1938).
2. Differentialgleichungen unendlich hoher Ordnung. *Comment. Math. Helv.* **14,** 381–393 (1942).

NEVANLINNA, F., AND NEVANLINNA, R.

1. Über die Eigenschaften analytischer Funktionen in der Umgebung einer singulären Stelle oder Linie. *Acta Soc. Sci. Fennicae* **50,** No. 5 (1922).

NEVANLINNA, R.

1. Über die Eigenschaften meromorpher Funktionen in einem Winkelraum. *Acta Soc. Sci. Fennicae* **50,** No. 12 (1925).
2. Eindeutige analytische Funktionen. Springer, Berlin, 1936.

NIKOLSKII, S. M. (NIKOLSKY, NIKOL'SKIĬ)

1. Generalization of an inequality of S. N. Bernstein. *Doklady Akad. Nauk SSSR* (N.S.) **60,** 1507–1510 (1948). (Russian.)

2. Inequalities for entire functions of finite degree and their application in the theory of differentiable functions of several variables. Trudy Mathematicheskogo Instituta im. V. A. Steklova, No. **38,** 244–278 (1951). (Russian.)

NOBLE, M. E.

1. Extensions and applications of a Tauberian theorem due to Valiron. *Proc. Cambridge Philos. Soc.* **47,** 22–37 (1951).
2. Non-measurable interpolation sets. I. Integral functions. *Proc. Cambridge Philos. Soc.* **47,** 713–732 (1951).
3. Non-measurable interpolation sets. II. Functions regular in an angle. *Proc. Cambridge Philos. Soc.* **47,** 733–740 (1951).
4. Non-measurable interpolation sets. III. A theorem of B. J. Maitland. *Quart. J. Math., Oxford Ser.* (2) **4,** 11–18 (1953).

OGURA, K.

1. Sur la théorie de l'interpolation de Stirling et les zéros des fonctions entières. *Bull. Sci. Math.* (2) **45,** 31–40 (1921).

OKAMURA, H.

1. Sur la croissance des séries entières. *Mem. Coll. Sci. Kyoto* A. **19,** 253–269 (1936).

PALEY, R. E. A. C., AND WIENER, N.

1. Fourier transforms in the complex domain. American Mathematical Society, New York, 1934.

PENNYCUICK, K.

1. On a theorem of Besicovitch. *J. London Math. Soc.* **10,** 210–212 (1935).
2. Extension of a theorem of Faber-Pólya. *J. London Math. Soc.* **12,** 267–272 (1937).

PFLUGER, A.

1. On analytic functions bounded at the lattice points. *Proc. London Math. Soc.* (2) **42,** 305–315 (1936).
2. Über das Anwachsen von Funktionen, die in einem Winkelraum regulär und vom Exponentialtypus sind. *Compositio Math.* **4,** 367–372 (1937).
3. Die Wertverteilung und das Verhalten von Betrag und Argument einer speziellen Klasse analytischer Funktionen. *Comment. Math. Helv.* **11,** 180–213 (1938); **12,** 25–65 (1939).
4. Über Interpolation ganzer Funktionen. *Comment. Math. Helv.* **14,** 314–349 (1942).
5. Über gewisse ganze Funktionen vom Exponentialtypus. *Comment. Math. Helv.* **16,** 1–18 (1944).
6. Über ganze Funktionen ganzer Ordnung. *Comment. Math. Helv.* **18,** 177–203 (1946).

PHRAGMÉN, E., AND LINDELÖF, E.

1. Sur une extension d'un principe classique de l'analyse et sur quelques propriétés des fonctions monogènes dans le voisinage d'un point singulier. *Acta Math.* **31,** 381–406 (1908).

PISOT, C.

1. Über ganzwertige ganze Funktionen. *Jber. Deutsch. Math. Verein.* **52,** 95–102 (1942).
2. Sur les fonctions arithmétiques analytiques à croissance exponentielle. *C.R. Acad. Sci. Paris* **222,** 988–990 (1946).
3. Sur les fonctions analytiques arithmétiques et presque arithmétiques. *C.R. Acad. Sci. Paris* **222,** 1027–1028 (1946).

PITT, H. R.

1. On a class of linear integro-differential equations. *Proc. Cambridge Philos. Soc.* **43,** 153–163 (1947).

Plancherel, M., and Pólya, G.

1. Fonctions entières et intégrales de Fourier multiples. *Comment. Math. Helv.* **9,** 224–248 (1937); **10,** 110–163 (1938).

Pollard, H.

1. Integral transforms. *Duke Math. J.* **13,** 307–330 (1946).
2. Integral transforms. II. *Ann. of Math.* (2) **49,** 956–965 (1948).

Pólya, G.

1. Bemerkungen über unendliche Folgen und ganze Funktionen. *Math. Ann.* **88,** 169–183 (1923).
2. Untersuchungen über Lücken und Singularitäten von Potenzreihen. *Math. Z.* **29,** 549–640 (1929).
3. On the zeros of the derivatives of a function and its analytic character. *Bull. Amer. Math. Soc.* **49,** 178–191 (1943).
4. Remarks on characteristic functions. Proceedings of the Berkeley Symposium on Mathematical Statistics and Probability, 1945, 1946, pp. 115–123. University of California Press, 1949.

Pólya, G., and Szegö, G.

1. Aufgaben und Lehrsätze aus der Analysis. Springer, Berlin, 1925.

Rademacher, H.

1. Über die asymptotische Verteilung gewisser konvergenzerzeugender Faktoren. *Math. Z.* **11,** 276–288 (1921).

Radó, T.

1. Subharmonic functions. Ergebnisse der Mathematik und ihrer Grenzgebiete, Vol. 5, No. 1. Springer, Berlin, 1937.

Rajagopal, C. T.

1. On inequalities for analytic functions. *Amer. Math. Monthly* **60,** 693–695 (1953).

Redheffer, R. M.

1. Remarks on the incompleteness of $\{e^{i\lambda_n x}\}$, nonaveraging sets, and entire functions. *Proc. Amer. Math. Soc.* **2,** 365–369 (1951).
2. On even entire functions with zeros having a density. *Trans. Amer. Math. Soc.* **74,** (to appear).
3. On a theorem of Plancherel and Pólya. *Pacific J. Math.* **3,** 823–835 (1953).

Riesz, M.

1. Sur le principe de Phragmén-Lindelöf. *Proc. Cambridge Philos. Soc.* **20,** 205–207 (1920); correction, *ibid.* **21,** 6 (1921).

Roth, A.

1. Approximationseigenschaften und Strahlengrenzwerte meromorpher und ganzer Funktionen. *Comment. Math. Helv.* **11,** 77–125 (1938).

Schaeffer, A. C.

1. Inequalities of A. Markoff and S. Bernstein for polynomials and related functions. *Bull. Amer. Math. Soc.* **47,** 565–579 (1941).
2. Entire functions and trigonometric polynomials. *Duke Math. J.* **20,** 77–88 (1953).

Schaeffer, A. C., and Szegö, G.

1. Inequalities for harmonic polynomials in two and three dimensions. *Trans. Amer. Math. Soc.* **50,** 187–225 (1941).

Schmidli, S.

1. Über gewisse Interpolationsreihen. Thesis, Eidgenössische Technische Hochschule in Zürich, 1942.

Schoenberg, I. J.

1. On certain two-point expansions of integral functions of exponential type. *Bull. Amer. Math. Soc.* **42,** 284–288 (1936).

2. On Pólya frequency functions. I. The totally positive functions and their Laplace transforms. *J. Analyse Math.* **1,** 331–374 (1951).
3. On Pólya frequency functions. II. Variation-diminishing operators of the convolution type. *Acta Sci. Math. Szeged* **12,** Leopoldo Fejér et Frederico Riesz LXX annos natis dedicatus, Pars B, 97–106 (1950).

SELBERG, A.
1. Über ganzwertige ganze transzendente Funktionen. I, II. *Arch. Math. Naturvid.* **44,** 45–52, 171–181 (1941).
2. Über einen Satz von A. Gelfond. *Arch. Math. Naturvid.* **44,** 159–170 (1941).

SELBERG, H. L.
1. Bemerkungen zum Wigertschen Satz. *Avh. Norske Vid. Akad. Oslo* **1932,** No. 9.

SHAH, S. M.
1. On integral functions of perfectly regular growth. *J. London Math. Soc.* **14,** 293–302 (1939).
2. A theorem on integral functions of integral order. *J. London Math. Soc.* **15,** 23–31 (1940).
3. A theorem on integral functions of integral order. II. *J. Indian Math. Soc.* (N.S.) **5,** 179–188 (1941).
4. A note on the classification of integral functions. *Math. Student* **9,** 63–67 (1941).
5. Note on a theorem of Pólya. *J. Indian Math. Soc.* (N.S.) **5,** 189–191 (1941).
6. On integral functions of integral or zero order. *Bull. Amer. Math. Soc.* **48,** 329–334 (1942).
7. The lower order of the zeros of an integral function. *J. Indian Math. Soc.* (N.S.) **6,** 63–68 (1942).
8. The lower order of the zeros of an integral function. II. *Proc. Indian Acad. Sci.* Sect. A. **21,** 162–174 (1945).
9. On the relations between the lower order and the exponent of convergence of zeros of an integral function. *J. Univ. Bombay* **11,** Part III, 10–13 (1942).
10. The maximum term of an entire series. *Math. Student* **10,** 80–82 (1942).
11. The maximum term of an entire series. II. *J. Indian Math. Soc.* (N.S.) **9,** 54–55 (1944).
12. The maximum term of an entire series. III. *Quart. J. Math., Oxford Ser.* (1) **19,** 220–223 (1948).
13. The maximum term of an entire series. IV. *Quart. J. Math., Oxford Ser.* (2) **1,** 112–116 (1950).
14. The maximum term of an entire series. V. *J. Indian Math. Soc.* (N.S.) **13,** 60–64 (1949).
15. The maximum term of an entire series. VI. *J. Indian Math. Soc.* (N.S.) **14,** 21–28 (1950).
16. The maximum term of an entire series. VII. *Gaṇita* **1,** 82–85 (1950).
17. A note on the maximum modulus of the derivative of an integral function. *J. Univ. Bombay* **13,** Part III, 1–3 (1944).
18. On proximate orders of integral functions. *Bull. Amer. Math. Soc.* **52,** 326–328 (1946).
19. On the lower order of integral functions. *Bull. Amer. Math. Soc.* **52,** 1046–1052 (1946).
20. A note on the minimum modulus of a class of integral functions. *Bull. Amer. Math. Soc.* **53,** 524–529 (1947).
21. A note on the derivatives of integral functions. *Bull. Amer. Math. Soc.* **53,** 1156–1163 (1947).
22. A note on lower proximate orders. *J. Indian Math. Soc.* (N.S.) **12,** 31–32 (1948).

23. A note on uniqueness sets for entire functions. *Proc. Indian Acad. Sci.* Sect. A. **28,** 519–526 (1948).
24. On the coefficients of an entire series of finite order. *J. London Math. Soc.* **26,** 45–46 (1951).
25. A note on entire functions of perfectly regular growth. *Math. Z.* **56,** 254–257 (1952).

SHEFFER, I. M.
1. Concerning Appell sets and associated linear functional equations. *Duke Math. J.* **3,** 593–609 (1937).
2. Some properties of polynomial sets of type zero. *Duke Math. J.* **5,** 590–622 (1939).
3. Some applications of certain polynomial classes. *Bull. Amer. Math. Soc.* **47,** 885–898 (1941).

SHTEINBERG, N. S. (ŠTEĬNBERG)
1. On the interpolation of entire functions. *Mat. Sbornik* (N.S.) **30(72),** 559–574 (1952). (Russian.)

SIDDIQI, J. A.
1. Quelques théorèmes d'unicité. *C.R. Acad. Sci. Paris* **236,** 1727–1729 (1953).

SIKKEMA, P. C.
1. Differential operators and differential equations of infinite order with constant coefficients. Researches in connection with integral functions of finite order. Noordhoff, Groningen-Djakarta, 1953.

SINGH, S. K.
1. A note on entire functions. *J. Univ. Bombay* (N.S.) **20,** Part 5, Sect. A, 1–7 (1952).
2. A note on a paper of R. C. Buck. *Proc. Indian Acad. Sci.* Sect. A. **38,** 120–121 (1953).

SRIVASTAVA, P. L.
1. On two theorems of Akhyeser and a theorem of Cramér. *Rend. Circ. Mat. Palermo* (1) **55,** 116–120 (1931).
2. A theorem on integral functions. *Bull. Acad. Sci. Allahabad* **1,** 43–44 (1932); review by E. Ullrich, *Zentralblatt für Math.* **6,** 409 (1933).
3. On the Phragmén-Lindelöf principle. *Proc. Nat. Acad. Sci. India* (Allahabad) **6,** 241–243 (1936).

STECHKIN, S. B. (STEČKIN)
1. A generalization of some inequalities of S. N. Bernstein. *Doklady Akad. Nauk SSSR* (N.S.) **60,** 1511–1514 (1948). (Russian.)

STEFFENSEN, J. F.
1. The poweroid, an extension of the mathematical notion of power. *Acta Math.* **73,** 333–366 (1941).

STRODT, W.
1. Linear difference equations and exponential polynomials. *Trans. Amer. Math. Soc.* **64,** 439–466 (1948).

SZEGÖ, G.
1. Zur Theorie der fastperiodischen Funktionen. *Math. Ann.* **96,** 378–382 (1926).
2. Über einen Satz des Herrn Serge Bernstein. *Schr. Königsberg. Gel. Ges. Nat.-Wiss. Kl.* **5,** 59–70 (1928).
3. On conjugate trigonometric polynomials. *Amer. J. Math.* **65,** 532–536 (1943).

SZ.-NAGY, B. (VON SZ. NAGY)
1. Über gewisse Extremalfragen bei transformierten trigonometrischen Entwick-

lungen. I. Periodischer Fall. II. Nichtperiodischer Fall. *Ber. Math.-Phys. Kl. Sächs. Akad. Wiss. Leipzig* **90**, 103–134 (1938); **91**, 3–24 (1939).
2. Über die Ungleichung von H. Bohr. *Math. Nachr.* **9**, 255–259 (1953).

SZ.-NAGY, B., AND STRAUSZ, A.
1. Egy Bohr-féle tételről (Über einen Satz von H. Bohr). *Math. Nat. Anzeiger Ungar. Akad. Wiss.* **57**, 121–135 (1938). (Hungarian; German summary.)

TIMAN, A. F.
1. On interference phenomena in the behavior of entire functions of finite degree. *Doklady Akad. Nauk SSSR* (N. S.) **89**, 17–20 (1953). (Russian.)

TITCHMARSH, E. C.
1. On integral functions with real negative zeros. *Proc. London Math. Soc.* (2) **26**, 185–200 (1927).
2. Introduction to the theory of Fourier integrals. Oxford University Press, 1937.
3. The theory of functions. 2d ed. Oxford University Press, 1939.

TSUJI, M.
1. Wiman's theorem on integral functions of order $<\frac{1}{2}$. *Proc. Japan Acad.* **26**, No. 2–5, 117–130 (1950).

TURETSKII, A. H. (TOURETSKY, TURECKIĬ)
1. Sur quelques propriétés extrémales des polynomes. *Comm. Inst. Sci. Math. Méc. Univ. Kharkoff* [*Zapiski Inst. Mat. Mech.*] (4) **17**, 45–63 (1940). (Russian; French summary.)
2. On an extremal property of trigonometric polynomials which satisfy a differential relation at isolated points of an interval. *Doklady Akad. Nauk SSSR* (N.S.) **50**, 65–68 (1945). (Russian.)

UHL, W.
1. Über die Darstellung ganzer Funktionen mittels der Stirling'schen Reihe bei Hermite'scher Interpolation. *Mitt. Math. Sem. Univ. Giessen* **33** (1944).

VALIRON, G.
1. Sur les fonctions entières d'ordre fini et d'ordre nul, et en particulier les fonctions à correspondance régulière. *Ann. Fac. Sci. Univ. Toulouse* (3) **5**, 117–257 (1914).
2. Sur les fonctions entières d'ordre fini. *Bull. Sci. Math.* (2) **45**, 258–270 (1921).
3. Lectures on the general theory of integral functions. Privat, Toulouse, 1923.
4. A propos d'un mémoire de M. Pólya. *Bull. Sci. Math.* (2) **48**, 9–12 (1924).
5. Fonctions entières et fonctions méromorphes d'une variable. Mémorial des Sciences Mathématiques, No. 2. Gauthier-Villars, Paris, 1925.
6. Sur la formule d'interpolation de Lagrange. *Bull. Sci. Math.* (2) **49**, 181–192, 203–224 (1925).
7. Sur un théorème de M. Wiman. Opuscula Mathematica A. Wiman Dedicata, Lund, 1930, 1–12.
8. Sur une classe de fonctions entières admettant deux directions de Borel d'ordre divergent. *Compositio Math.* **1**, 193–206 (1934).
9. Sur le minimum du module des fonctions entières d'ordre inférieur à un. *Mathematica, Cluj* **11**, 264–269 (1935).
10. Directions de Borel des fonctions méromorphes. Mémorial des Sciences Mathématiques, No. 89. Gauthier-Villars, Paris, 1938.

VAN DER CORPUT, J. G. *See* CORPUT.

VERMES, P.
1. Note on certain differential equations of infinite order. *Nederl. Akad. Wetensch. Proc.* Ser. A. **55** = *Indagationes Math.* **14**, 28–31 (1952).

VIDENSKII, V. S. (VIDENSKIĬ)
1. Consequences of a proposition of S. N. Bernstein on entire functions of genus zero. *Doklady Akad. Nauk SSSR* (N.S.) **84,** 421–424 (1952). (Russian.)

WHITTAKER, J. M.
1. The lower order of integral functions. *J. London Math. Soc.* **8,** 20–27 (1933).
2. Interpolatory function theory. Cambridge Tracts in Mathematics and Mathematical Physics, No. 33. Cambridge University Press, 1935.

WILSON, R.
1. Directions of strongest growth of the product of integral functions of finite order and mean type. *J. London Math. Soc.* **28,** 185–193 (1953).

WIMAN, A.
1. Über eine Eigenschaft der ganzen Funktionen von der Höhe Null. *Math. Ann.* **76,** 197–211 (1915).

WISHARD, A.
1. Functions of bounded type. *Duke Math. J.* **9,** 663–676 (1942).

YAGI, F.
1. On certain Stieltjes integral equations. *Univ. Washington Publ. Math.* **3,** No. 1, 21–30 (1948).

ZYGMUND, A.
1. A remark on conjugate series. *Proc. London Math. Soc.* (2) **34,** 392–400 (1932).

ADDENDA

BARI, N. K.
1. Generalization of inequalities of S. N. Bernstein and A. A. Markov. *Izvestiya Akad. Nauk SSSR. Ser. Mat.* **18,** 159–176 (1954). [Pp. 210 ff.]

BIEBERBACH, L.
1. Über einen Satz Pólyascher Art. *Arch. Math.* **4,** 23–27 (1953). [Note 9.12b, p. 177.]

GAIER, D.
1. Zur Frage der Indexverschiebung beim Borel-Verfahren. *Math. Z.* **58,** 453–455 (1953). [11.3.4, p. 212.]

SINGH, U. N.
1. Fonctions entières et transformée de Fourier généralisée. *C. R. Acad. Sci. Paris* **237,** 14–16 (1953). [P. 107.]

INDEX

INDEX OF NOTATIONS